MULTIVARIATE STATISTICS

T0214590

Wolfgang Härdle
Zdeněk Hlávka

Multivariate Statistics:

Exercises and Solutions

 Springer

Wolfgang Härdle
Wirtschaftswissenschaftliche Fakultätlnst
Statistik und Ökonometrie
Humboldt-Universität zu Berlin, Berlin
1 Spandauer Str.
Berlin 10178
Germany
stat@wiwi.hu-berlin.de

Zdeněk Hlávka
Dept. Mathematics
Charles University in Prague
Prague
Czech Republic
83 Sokolovska
Praha 8 186 75

Library of Congress Control Number: 2007929450

ISBN 978-0-387-70784-6   e-ISBN 978-0-387-73508-5

Printed on acid-free paper.

9 8 7 6 5 4 3 2 1

springer.com

Für meine Familie

Mé rodině

To our families

Preface

> There can be no question, my dear Watson, of the value of exercise before breakfast.
> Sherlock Holmes in "The Adventure of Black Peter"

The statistical analysis of multivariate data requires a variety of techniques that are entirely different from the analysis of one-dimensional data. The study of the joint distribution of many variables in high dimensions involves matrix techniques that are not part of standard curricula. The same is true for transformations and computer-intensive techniques, such as projection pursuit.

The purpose of this book is to provide a set of exercises and solutions to help the student become familiar with the techniques necessary to analyze high-dimensional data. It is our belief that learning to apply multivariate statistics is like studying the elements of a criminological case. To become proficient, students must not simply follow a standardized procedure, they must compose with creativity the parts of the puzzle in order to see the big picture. We therefore refer to Sherlock Holmes and Dr. Watson citations as typical descriptors of the analysis.

> Puerile as such an exercise may seem, it sharpens the faculties of observation, and teaches one where to look and what to look for.
> Sherlock Holmes in "Study in Scarlet"

Analytic creativity in applied statistics is interwoven with the ability to see and change the involved software algorithms. These are provided for the student via the links in the text. We recommend doing a small number of problems from this book a few times a week. And, it does not hurt to redo an exercise, even one that was mastered long ago. We have implemented in these links software quantlets from XploRe and R. With these quantlets the student can reproduce the analysis on the spot.

This exercise book is designed for the advanced undergraduate and first-year graduate student as well as for the data analyst who would like to learn the various statistical tools in a multivariate data analysis workshop.

The chapters of exercises follow the ones in Härdle & Simar (2003). The book is divided into three main parts. The first part is devoted to graphical techniques describing the distributions of the variables involved. The second part deals with multivariate random variables and presents from a theoretical point of view distributions, estimators, and tests for various practical situations. The last part is on multivariate techniques and introduces the reader to the wide selection of tools available for multivariate data analysis. All data sets are downloadable at the authors' Web pages. The source code for generating all graphics and examples are available on the same Web site. Graphics in the printed version of the book were produced using XploRe. Both XploRe and R code of all exercises are also available on the authors' Web pages. The names of the respective programs are denoted by the symbol ⊙.

In Chapter 1 we discuss boxplots, graphics, outliers, Flury-Chernoff faces, Andrews' curves, parallel coordinate plots and density estimates. In Chapter 2 we dive into a level of abstraction to relearn the matrix algebra. Chapter 3 is concerned with covariance, dependence, and linear regression. This is followed by the presentation of the ANOVA technique and its application to the multiple linear model. In Chapter 4 multivariate distributions are introduced and thereafter are specialized to the multinormal. The theory of estimation and testing ends the discussion on multivariate random variables.

The third and last part of this book starts with a geometric decomposition of data matrices. It is influenced by the French school of data analysis. This geometric point of view is linked to principal component analysis in Chapter 9. An important discussion on factor analysis follows with a variety of examples from psychology and economics. The section on cluster analysis deals with the various cluster techniques and leads naturally to the problem of discrimination analysis. The next chapter deals with the detection of correspondence between factors. The joint structure of data sets is presented in the chapter on canonical correlation analysis, and a practical study on prices and safety features of automobiles is given. Next the important topic of multidimensional scaling is introduced, followed by the tool of conjoint measurement analysis. Conjoint measurement analysis is often used in psychology and marketing to measure preference orderings for certain goods. The applications in finance (Chapter 17) are numerous. We present here the CAPM model and discuss efficient portfolio allocations. The book closes with a presentation on highly interactive, computationally intensive, and advanced nonparametric techniques.

A book of this kind would not have been possible without the help of many friends, colleagues, and students. For many suggestions on how to formulate the exercises we would like to thank Michal Benko, Szymon Borak, Ying

Chen, Sigbert Klinke, and Marlene Müller. The following students have made outstanding proposals and provided excellent solution tricks: Jan Adamčák, David Albrecht, Lütfiye Arslan, Lipi Banerjee, Philipp Batz, Peder Egemen Baykan, Susanne Böhme, Jan Budek, Thomas Diete, Daniel Drescher, Zeno Enders, Jenny Frenzel, Thomas Giebe, LeMinh Ho, Lena Janys, Jasmin John, Fabian Kittman, Lenka Komárková, Karel Komorád, Guido Krbetschek, Yulia Maletskaya, Marco Marzetti, Dominik Michálek, Alena Myšičková, Dana Novotny, Björn Ohl, Hana Pavlovičová, Stefanie Radder, Melanie Reichelt, Lars Rohrschneider, Martin Rolle, Elina Sakovskaja, Juliane Scheffel, Denis Schneider, Burcin Sezgen, Petr Stehlík, Marius Steininger, Rong Sun, Andreas Uthemann, Aleksandrs Vatagins, Manh Cuong Vu, Anja Weiß, Claudia Wolff, Kang Xiaowei, Peng Yu, Uwe Ziegenhagen, and Volker Ziemann. The following students of the computational statistics classes at Charles University in Prague contributed to the R programming: Alena Babiaková, Blanka Hamplová, Tomáš Hovorka, Dana Chromíková, Kristýna Ivanková, Monika Jakubcová, Lucia Jarešová, Barbora Lebdušková, Tomáš Marada, Michaela Maršálková, Jaroslav Pazdera, Jakub Pečánka, Jakub Petrásek, Radka Picková, Kristýna Sionová, Ondřej Šedivý, Tereza Těšitelová, and Ivana Žohová.

We acknowledge support of MSM 0021620839 and the teacher exchange program in the framework of Erasmus/Sokrates.

We express our thanks to David Harville for providing us with the LaTeX sources of the starting section on matrix terminology (Harville 2001). We thank John Kimmel from Springer Verlag for continuous support and valuable suggestions on the style of writing and the content covered.

Berlin and Prague,   *Wolfgang K. Härdle*
April 2007   *Zdeněk Hlávka*

Contents

Part III Multivariate Techniques

Symbols and Notation

I can't make bricks without clay.
Sherlock Holmes in "The Adventure of The Copper Beeches"

Basics

X, Y	random variables or vectors
$X_1, X_2, \ldots, X_p$	random variables
$X = (X_1, \ldots, X_p)^\top$	random vector
$X \sim \cdot$	X has distribution $\cdot$
$\mathcal{A}, \mathcal{B}$	matrices
Γ, Δ	matrices
$\mathcal{X}, \mathcal{Y}$	data matrices
Σ	covariance matrix
1_n	vector of ones $\underbrace{(1, \ldots, 1)}_{n\text{-times}}^\top$
0_n	vector of zeros $\underbrace{(0, \ldots, 0)}_{n\text{-times}}^\top$
$\mathcal{I}_p$	identity matrix
$\mathbf{I}(.)$	indicator function, for a set M is $\mathbf{I} = 1$ on M, $\mathbf{I} = 0$ otherwise
i	$\sqrt{-1}$
$\Rightarrow$	implication
$\Leftrightarrow$	equivalence
$\approx$	approximately equal
$\otimes$	Kronecker product
iff	if and only if, equivalence

Characteristics of Distribution

$f(x)$	pdf or density of X
$f(x, y)$	joint density of X and Y
$f_X(x), f_Y(y)$	marginal densities of X and Y
$f_{X_1}(x_1), \ldots, f_{X_p}(x_p)$	marginal densities of $X_1, \ldots, X_p$
$\hat{f}_h(x)$	histogram or kernel estimator of $f(x)$
$F(x)$	cdf or distribution function of X
$F(x, y)$	joint distribution function of X and Y
$F_X(x), F_Y(y)$	marginal distribution functions of X and Y
$F_{X_1}(x_1), \ldots, F_{X_p}(x_p)$	marginal distribution functions of $X_1, \ldots, X_p$
$f_{Y|X=x}(y)$	conditional density of Y given $X = x$
$\varphi_X(t)$	characteristic function of X
m_k	kth moment of X
κ_j	cumulants or semi-invariants of X

Moments

EX, EY	mean values of random variables or vectors X and Y
$E(Y|X = x)$	conditional expectation of random variable or vector Y given $X = x$
$\mu_{Y|X}$	conditional expectation of Y given X
$\text{Var}(Y|X = x)$	conditional variance of Y given $X = x$
$\sigma^2_{Y|X}$	conditional variance of Y given X
$\sigma_{XY} = \text{Cov}(X, Y)$	covariance between random variables X and Y
$\sigma_{XX} = Var(X)$	variance of random variable X
$\rho_{XY} = \dfrac{\text{Cov}(X, Y)}{\sqrt{Var(X)\, Var(Y)}}$	correlation between random variables X and Y
$\Sigma_{XY} = \text{Cov}(X, Y)$	covariance between random vectors X and Y, i.e., $\text{Cov}(X, Y) = E(X - EX)(Y - EY)^{\top}$
$\Sigma_{XX} = \text{Var}(X)$	covariance matrix of the random vector X

Samples

x, y	observations of X and Y
$x_1, \ldots, x_n = \{x_i\}_{i=1}^n$	sample of n observations of X
$\mathcal{X} = \{x_{ij}\}_{i=1,\ldots,n;j=1,\ldots,p}$	$(n \times p)$ data matrix of observations of $X_1, \ldots, X_p$ or of $X = (X_1, \ldots, X_p)^T$
$x_{(1)}, \ldots, x_{(n)}$	the order statistic of $x_1, \ldots, x_n$
$\mathcal{H}$	centering matrix, $\mathcal{H} = \mathcal{I}_n - n^{-1} 1_n 1_n^{\top}$

Empirical Moments

$$\bar{x} = \frac{1}{n} \sum_{i=1}^{n} x_i \qquad\qquad \text{average of } X \text{ sampled by } \{x_i\}_{i=1,\ldots,n}$$

$$s_{XY} = \frac{1}{n} \sum_{i=1}^{n} (x_i - \bar{x})(y_i - \bar{y}) \quad \text{empirical covariance of random variables } X$$
$$\text{and } Y \text{ sampled by } \{x_i\}_{i=1,\ldots,n} \text{ and}$$
$$\{y_i\}_{i=1,\ldots,n}$$

$$s_{XX} = \frac{1}{n} \sum_{i=1}^{n} (x_i - \bar{x})^2 \qquad \text{empirical variance of random variable } X$$
$$\text{sampled by } \{x_i\}_{i=1,\ldots,n}$$

$$r_{XY} = \frac{s_{XY}}{\sqrt{s_{XX}s_{YY}}} \qquad\qquad \text{empirical correlation of } X \text{ and } Y$$

$$\mathcal{S} = \{s_{X_i X_j}\} \qquad\qquad \text{empirical covariance matrix of } X_1, \ldots, X_p \text{ or}$$
$$\text{of the random vector } X = (X_1, \ldots, X_p)^\top$$

$$\mathcal{R} = \{r_{X_i X_j}\} \qquad\qquad \text{empirical correlation matrix of } X_1, \ldots, X_p \text{ or}$$
$$\text{of the random vector } X = (X_1, \ldots, X_p)^\top$$

Distributions

$\varphi(x)$	density of the standard normal distribution
$\Phi(x)$	distribution function of the standard normal distribution
$N(0,1)$	standard normal or Gaussian distribution
$N(\mu, \sigma^2)$	normal distribution with mean μ and variance σ^2
$N_p(\mu, \Sigma)$	p-dimensional normal distribution with mean μ and covariance matrix Σ
$\xrightarrow{\mathcal{L}}$	convergence in distribution
$\xrightarrow{P}$	convergence in probability
CLT	Central Limit Theorem
χ_p^2	χ^2 distribution with p degrees of freedom
$\chi_{1-\alpha;p}^2$	$1 - \alpha$ quantile of the χ^2 distribution with p degrees of freedom
t_n	t-distribution with n degrees of freedom
$t_{1-\alpha/2;n}$	$1 - \alpha/2$ quantile of the t-distribution with n degrees of freedom
$F_{n,m}$	F-distribution with n and m degrees of freedom
$F_{1-\alpha;n,m}$	$1 - \alpha$ quantile of the F-distribution with n and m degrees of freedom

Mathematical Abbreviations

$\text{tr}(\mathcal{A})$	trace of matrix $\mathcal{A}$
$\text{diag}(\mathcal{A})$	diagonal of matrix $\mathcal{A}$
$\text{rank}(\mathcal{A})$	rank of matrix $\mathcal{A}$
$\det(\mathcal{A})$ or $|\mathcal{A}|$	determinant of matrix $\mathcal{A}$
$\text{hull}(x_1, \ldots, x_k)$	convex hull of points $\{x_1, \ldots, x_k\}$
$\text{span}(x_1, \ldots, x_k)$	linear space spanned by $\{x_1, \ldots, x_k\}$

Some Terminology

I consider that a man's brain originally is like a little empty attic, and you have to stock it with such furniture as you choose. A fool takes in all the lumber of every sort that he comes across, so that the knowledge which might be useful to him gets crowded out, or at best is jumbled up with a lot of other things so that he has a difficulty in laying his hands upon it. Now the skilful workman is very careful indeed as to what he takes into his brain-attic. He will have nothing but the tools which may help him in doing his work, but of these he has a large assortment, and all in the most perfect order. It is a mistake to think that that little room has elastic walls and can distend to any extent. Depend upon it there comes a time when for every addition of knowledge you forget something that you knew before. It is of the highest importance, therefore, not to have useless facts elbowing out the useful ones.
Sherlock Holmes in "Study in Scarlet"

This section contains an overview of some terminology that is used throughout the book. We thank David Harville, who kindly allowed us to use his TeX files containing the definitions of terms concerning matrices and matrix algebra; see Harville (2001). More detailed definitions and further explanations of the statistical terms can be found, e.g., in Breiman (1973), Feller (1966), Härdle & Simar (2003), Mardia, Kent & Bibby (1979), or Serfling (2002).

adjoint matrix The *adjoint matrix* of an $n \times n$ matrix $\mathcal{A} = \{a_{ij}\}$ is the transpose of the cofactor matrix of $\mathcal{A}$ (or equivalently is the $n \times n$ matrix whose ijth element is the cofactor of a_{ji}).

asymptotic normality A sequence $X_1, X_2, \ldots$ of random variables is *asymptotically normal* if there exist sequences of constants $\{\mu_i\}_{i=1}^{\infty}$ and $\{\sigma_i\}_{i=1}^{\infty}$ such that $\sigma_n^{-1}(X_n - \mu_n) \xrightarrow{\mathcal{L}} N(0,1)$. The asymptotic normality means that for sufficiently large n, the random variable X_n has approximately $N(\mu_n, \sigma_n^2)$ distribution.

bias Consider a random variable X that is parametrized by $\theta \in \Theta$. Suppose that there is an estimator $\widehat{\theta}$ of θ. The *bias* is defined as the systematic difference between $\widehat{\theta}$ and θ, $E\{\widehat{\theta} - \theta\}$. The estimator is unbiased if $E\widehat{\theta} = \theta$.

characteristic function Consider a random vector $X \in \mathbb{R}^p$ with pdf f. The *characteristic function* (cf) is defined for $t \in \mathbb{R}^p$:

$$\varphi_X(t) - E[\exp(it^\top X)] = \int \exp(it^\top X) f(x) dx.$$

The cf fulfills $\varphi_X(0) = 1$, $|\varphi_X(t)| \le 1$. The pdf (density) f may be recovered from the cf: $f(x) = (2\pi)^{-p} \int \exp(-it^\top X) \varphi_X(t) dt$.

characteristic polynomial (and equation) Corresponding to any $n \times n$ matrix $\mathcal{A}$ is its characteristic polynomial, say $p(.)$, defined (for $-\infty < \lambda < \infty$) by $p(\lambda) = |\mathcal{A} - \lambda \mathcal{I}|$, and its characteristic equation $p(\lambda) = 0$ obtained by setting its characteristic polynomial equal to 0; $p(\lambda)$ is a polynomial in λ of degree n and hence is of the form $p(\lambda) = c_0 + c_1 \lambda + \cdots + c_{n-1} \lambda^{n-1} + c_n \lambda^n$, where the coefficients $c_0, c_1, \ldots, c_{n-1}, c_n$ depend on the elements of $\mathcal{A}$.

cofactor (and minor) The *cofactor* and *minor* of the ijth element, say a_{ij}, of an $n \times n$ matrix $\mathcal{A}$ are defined in terms of the $(n-1) \times (n-1)$ submatrix, say $\mathcal{A}_{ij}$, of $\mathcal{A}$ obtained by striking out the ith row and jth column (i.e., the row and column containing a_{ij}): the minor of a_{ij} is $|\mathcal{A}_{ij}|$, and the cofactor is the "signed" minor $(-1)^{i+j} |\mathcal{A}_{ij}|$.

cofactor matrix The *cofactor matrix* (or matrix of cofactors) of an $n \times n$ matrix $\mathcal{A} = \{a_{ij}\}$ is the $n \times n$ matrix whose ijth element is the cofactor of a_{ij}.

conditional distribution Consider the joint distribution of two random vectors $X \in \mathbb{R}^p$ and $Y \in \mathbb{R}^q$ with pdf $f(x, y): \mathbb{R}^{p+1} \longrightarrow \mathbb{R}$. The marginal density of X is $f_X(x) = \int f(x, y) dy$ and similarly $f_Y(y) = \int f(x, y) dx$. The *conditional density* of X given Y is $f_{X|Y}(x|y) = f(x, y)/f_Y(y)$. Similarly, the conditional density of Y given X is $f_{Y|X}(y|x) = f(x, y)/f_X(x)$.

conditional moments Consider two random vectors $X \in \mathbb{R}^p$ and $Y \in \mathbb{R}^q$ with joint pdf $f(x, y)$. The *conditional moments* of Y given X are defined as the moments of the conditional distribution.

contingency table Suppose that two random variables X and Y are observed on discrete values. The two-entry frequency table that reports the simultaneous occurrence of X and Y is called a *contingency table*.

critical value Suppose one needs to test a hypothesis $H_0: \theta = \theta_0$. Consider a test statistic T for which the distribution under the null hypothesis is given by P_{θ_0}. For a given significance level α, the *critical value* is c_α such that $P_{\theta_0}(T > c_\alpha) = \alpha$. The critical value corresponds to the threshold that a test statistic has to exceed in order to reject the null hypothesis.

cumulative distribution function (cdf) Let X be a p-dimensional random vector. The *cumulative distribution function* (cdf) of X is defined by $F(x) = P(X \leq x) = P(X_1 \leq x_1, X_2 \leq x_2, \ldots, X_p \leq x_p)$.

derivative of a function of a matrix The *derivative of a function f of an* $m \times n$ *matrix* $\mathcal{X} = \{x_{ij}\}$ of mn "independent" variables is the $m \times n$ matrix whose ijth element is the partial derivative $\partial f / \partial x_{ij}$ of f with respect to x_{ij} when f is regarded as a function of an mn-dimensional column vector x formed from $\mathcal{X}$ by rearranging its elements; the derivative of a function f of an $n \times n$ symmetric (but otherwise unrestricted) matrix of variables is the $n \times n$ (symmetric) matrix whose ijth element is the partial derivative $\partial f / \partial x_{ij}$ or $\partial f / \partial x_{ji}$ of f with respect to x_{ij} or x_{ji} when f is regarded as a function of an $n(n+1)/2$-dimensional column vector x formed from any set of $n(n+1)/2$ nonredundant elements of $\mathcal{X}$.

determinant The *determinant* of an $n \times n$ matrix $\mathcal{A} = \{a_{ij}\}$ is (by definition) the (scalar-valued) quantity $\sum (-1)^{|\tau|} a_{1\tau(1)} \cdots a_{n\tau(n)}$, where $\tau(1), \ldots, \tau(n)$ is a permutation of the first n positive integers and the summation is over all such permutations.

eigenvalues and eigenvectors An *eigenvalue* of an $n \times n$ matrix $\mathcal{A}$ is (by definition) a scalar (real number), say λ, for which there exists an $n \times 1$ vector, say x, such that $\mathcal{A}x = \lambda x$, or equivalently such that $(\mathcal{A} - \lambda \mathcal{I})x = \mathbf{0}$; any such vector x is referred to as an *eigenvector* (of $\mathcal{A}$) and is said to belong to (or correspond to) the eigenvalue λ. Eigenvalues (and eigenvectors), as defined herein, are restricted to real numbers (and vectors of real numbers).

eigenvalues (not necessarily distinct) The characteristic polynomial, say $p(.)$, of an $n \times n$ matrix $\mathcal{A}$ is expressible as

$$p(\lambda) = (-1)^n (\lambda - d_1)(\lambda - d_2) \cdots (\lambda - d_m) q(\lambda) \qquad (-\infty < \lambda < \infty),$$

where $d_1, d_2, \ldots, d_m$ are not-necessarily-distinct scalars and $q(.)$ is a polynomial (of degree $n - m$) that has no real roots; $d_1, d_2, \ldots, d_m$ are referred to as the *not-necessarily-distinct eigenvalues* of $\mathcal{A}$ or (at the possible risk of confusion) simply as the eigenvalues of $\mathcal{A}$. If the spectrum of $\mathcal{A}$ has k members, say $\lambda_1, \ldots, \lambda_k$, with algebraic multiplicities of $\gamma_1, \ldots, \gamma_k$, respectively, then $m = \sum_{i=1}^k \gamma_i$, and (for $i = 1, \ldots, k$) γ_i of the m not-necessarily-distinct eigenvalues equal λ_i.

empirical distribution function Assume that $X_1, \ldots, X_n$ are iid observations of a p-dimensional random vector. The *empirical distribution function* (edf) is defined through $F_n(x) = n^{-1} \sum_{i=1}^n I(X_i \leq x)$.

empirical moments The moments of a random vector X are defined through $m_k = E(X^k) = \int x^k dF(x) = \int x^k f(x) dx$. Similarly, the *empirical moments* are defined through empirical distribution function $F_n(x) = n^{-1} \sum_{i=1}^n I(X_i \leq x)$. This leads to $\widehat{m}_k = n^{-1} \sum_{i=1}^n X_i^k = \int x^k dF_n(x)$.

estimate An *estimate* is a function of the observations designed to approximate an unknown parameter value.

estimator An *estimator* is the prescription (on the basis of a random sample) of how to approximate an unknown parameter.

expected (or mean) value For a random vector X with pdf f the *mean* or *expected value* is $E(X) = \int x f(x) dx$.

gradient (or gradient matrix) The *gradient* of a vector $f = (f_1, \ldots, f_p)^\top$ of functions, each of whose domain is a set in $\mathcal{R}^{m \times 1}$, is the $m \times p$ matrix $[(\mathcal{D}f_1)^\top, \ldots, (\mathcal{D}f_p)^\top]$, whose jith element is $D_j f_i$. The gradient of f is the transpose of the Jacobian matrix of f.

gradient vector The *gradient vector* of a function f, with domain in $\mathcal{R}^{m \times 1}$, is the m-dimensional column vector $(\mathcal{D}f)^\top$ whose jth element is the partial derivative $D_j f$ of f.

Hessian matrix The *Hessian matrix* of a function f, with domain in $\mathcal{R}^{m \times 1}$, is the $m \times m$ matrix whose ijth element is the ijth partial derivative $D_{ij}^2 f$ of f.

idempotent matrix A (square) matrix $\mathcal{A}$ is *idempotent* if $\mathcal{A}^2 = \mathcal{A}$.

Jacobian matrix The *Jacobian matrix* of a p-dimensional vector $f = (f_1, \ldots, f_p)^\top$ of functions, each of whose domain is a set in $\mathcal{R}^{m \times 1}$, is the $p \times m$ matrix $(D_1 f, \ldots, D_m f)$ whose ijth element is $D_j f_i$; in the special case where $p = m$, the determinant of this matrix is referred to as the Jacobian (or Jacobian determinant) of f.

kernel density estimator The *kernel density estimator* $\widehat{f}$ of a pdf f, based on a random sample $X_1, X_2, \ldots, X_n$ from f, is defined by

$$\widehat{f}(x) = \frac{1}{nh} \sum_{i=1}^n K_h \left(\frac{x - X_i}{h} \right).$$

The properties of the estimator $\widehat{f}(x)$ depend on the choice of the kernel function $K(.)$ and the bandwidth h. The kernel density estimator can be seen as a smoothed histogram; see also Härdle, Müller, Sperlich & Werwatz (2004).

likelihood function Suppose that $\{x_i\}_{i=1}^n$ is an iid sample from a population with pdf $f(x; \theta)$. The *likelihood function* is defined as the joint pdf of the observations $x_1, \ldots, x_n$ considered as a function of the parameter θ, i.e., $L(x_1, \ldots, x_n; \theta) = \prod_{i=1}^n f(x_i; \theta)$. The log-likelihood function,

$\ell(x_1, \ldots, x_n; \theta) = \log L(x_1, \ldots, x_n; \theta) = \sum_{i=1}^{n} \log f(x_i; \theta)$, is often easier to handle.

linear dependence or independence A nonempty (but finite) set of matrices (of the same dimensions $(n \times p)$), say $\mathcal{A}_1, \mathcal{A}_2, \ldots, \mathcal{A}_k$, is (by definition) *linearly dependent* if there exist scalars $x_1, x_2, \ldots, x_k$, not all 0, such that $\sum_{i=1}^{k} x_i \mathcal{A}_i = 0_n 0_p^\top$; otherwise (if no such scalars exist), the set is linearly independent. By convention, the empty set is linearly independent.

marginal distribution For two random vectors X and Y with the joint pdf $f(x, y)$, the *marginal pdfs* are defined as $f_X(x) = \int f(x, y) dy$ and $f_Y(y) = \int f(x, y) dx$.

marginal moments The *marginal moments* are the moments of the marginal distribution.

mean The *mean* is the first-order empirical moment $\overline{x} = \int x dF_n(x) = n^{-1} \sum_{i=1}^{n} x_i = \widehat{m}_1$.

mean squared error (MSE) Suppose that for a random vector C with a distribution parametrized by $\theta \in \Theta$ there exists an estimator $\widehat{\theta}$. The *mean squared error* (MSE) is defined as $E_X(\widehat{\theta} - \theta)^2$.

median Suppose that X is a continuous random variable with pdf $f(x)$. The *median* $\widetilde{x}$ lies in the center of the distribution. It is defined as $\int_{-\infty}^{\widetilde{x}} f(x) dx = \int_{\widetilde{x}}^{+\infty} f(x) dx - 0.5$.

moments The *moments* of a random vector X with the distribution function $F(x)$ are defined through $m_k = E(X^k) = \int x^k dF(x)$. For continuous random vectors with pdf $f(x)$, we have $m_k = E(X^k) = \int x^k f(x) dx$.

normal (or Gaussian) distribution A random vector X with the *multinormal distribution* $N(\mu, \Sigma)$ with the mean vector μ and the variance matrix Σ is given by the pdf

$$f_X(x) = |2\pi\Sigma|^{-1/2} \exp\left\{-\frac{1}{2}(x - \mu)^\top \Sigma^{-1}(x - \mu)\right\}.$$

orthogonal complement The *orthogonal complement* of a subspace $\mathcal{U}$ of a linear space $\mathcal{V}$ is the set comprising all matrices in $\mathcal{V}$ that are orthogonal to $\mathcal{U}$. Note that the orthogonal complement of $\mathcal{U}$ depends on $\mathcal{V}$ as well as $\mathcal{U}$ (and also on the choice of inner product).

orthogonal matrix An $(n \times n)$ matrix $\mathcal{A}$ is *orthogonal* if $\mathcal{A}^\top \mathcal{A} = \mathcal{A} \mathcal{A}^\top = \mathcal{I}_n$.

partitioned matrix A *partitioned matrix*, say $\begin{pmatrix} \mathcal{A}_{11} & \mathcal{A}_{12} & \ldots & \mathcal{A}_{1c} \\ \mathcal{A}_{21} & \mathcal{A}_{22} & \ldots & \mathcal{A}_{2c} \\ \vdots & \vdots & & \vdots \\ \mathcal{A}_{r1} & \mathcal{A}_{r2} & \ldots & \mathcal{A}_{rc} \end{pmatrix}$, is a

matrix that has (for some positive integers r and c) been subdivided into rc submatrices $\mathcal{A}_{ij}$ ($i = 1, 2, \ldots, r$; $j = 1, 2, \ldots, c$), called *blocks*, by implicitly superimposing on the matrix $r - 1$ horizontal lines and $c - 1$ vertical lines (so that all of the blocks in the same "row" of blocks have the same number of rows and all of those in the same "column" of blocks have the same number of columns). In the special case where $c = r$, the blocks $\mathcal{A}_{11}, \mathcal{A}_{22}, \ldots, \mathcal{A}_{rr}$ are referred to as the diagonal blocks (and the other blocks are referred to as the off-diagonal blocks).

probability density function (pdf) For a continuous random vector X with cdf F, the *probability density function* (pdf) is defined as $f(x) = \partial F(x)/\partial x$.

quantile For a random variable X with pdf f the α *quantile* q_α is defined through: $\int_{-\infty}^{q_\alpha} f(x)dx = \alpha$.

p-value The critical value c_α gives the critical threshold of a test statistic T for rejection of a null hypothesis $H_0 : \theta = \theta_0$. The probability $P_{\theta_0}(T > c_\alpha) = p$ defines that *p-value*. If the p-value is smaller than the significance level α, the null hypothesis is rejected.

random variable and vector Random events occur in a probability space with a certain even structure. A *random variable* is a function from this probability space to $\mathbb{R}$ (or $\mathbb{R}^p$ for random vectors) also known as the state space. The concept of a random variable (vector) allows one to elegantly describe events that are happening in an abstract space.

scatterplot A *scatterplot* is a graphical presentation of the joint empirical distribution of two random variables.

Schur complement In connection with a partitioned matrix $\mathcal{A}$ of the form $\mathcal{A} = \begin{pmatrix} T & \mathcal{U} \\ \mathcal{V} & \mathcal{W} \end{pmatrix}$ or $\mathcal{A} = \begin{pmatrix} \mathcal{W} & \mathcal{V} \\ \mathcal{U} & T \end{pmatrix}$, the matrix $\mathcal{Q} = \mathcal{W} - \mathcal{V}T^-\mathcal{U}$ is referred to as the *Schur complement* of T in $\mathcal{A}$ relative to T^- or (especially in a case where $\mathcal{Q}$ is invariant to the choice of the generalized inverse T^-) simply as the Schur complement of T in $\mathcal{A}$ or (in the absence of any ambiguity) even more simply as the Schur complement of T.

singular value decomposition (SVD) An $m \times n$ matrix $\mathcal{A}$ of rank r is expressible as

$$\mathcal{A} = \mathcal{P}\begin{pmatrix} \mathcal{D}_1 & 0 \\ 0 & 0 \end{pmatrix}\mathcal{Q}^\top = \mathcal{P}_1\mathcal{D}_1\mathcal{Q}_1^\top = \sum_{i=1}^{r} s_i p_i q_i^\top = \sum_{j=1}^{k} \alpha_j \mathcal{U}_j,$$

where $\mathcal{Q} = (q_1, \ldots, q_n)$ is an $n \times n$ orthogonal matrix and $\mathcal{D}_1 = \text{diag}(s_1, \ldots, s_r)$ an $r \times r$ diagonal matrix such that $\mathcal{Q}^\top\mathcal{A}^\top\mathcal{A}\mathcal{Q} = \begin{pmatrix} \mathcal{D}_1^2 & 0 \\ 0 & 0 \end{pmatrix}$, where $s_1, \ldots, s_r$ are (strictly) positive, where $\mathcal{Q}_1 = (q_1, \ldots, q_r)$, $\mathcal{P}_1 = (p_1, \ldots, p_r) = \mathcal{A}\mathcal{Q}_1\mathcal{D}_1^{-1}$, and, for any $m \times (m-r)$ matrix $\mathcal{P}_2$ such that $\mathcal{P}_1^\top\mathcal{P}_2 = \mathbf{0}$, $\mathcal{P} = (\mathcal{P}_1, \mathcal{P}_2)$, where $\alpha_1, \ldots, \alpha_k$ are the distinct values represented among

$s_1, \ldots, s_r$, and where (for $j = 1, \ldots, k$) $\mathcal{U}_j = \sum_{\{i \,:\, s_i = \alpha_j\}} p_i q_i^\top$; any of these four representations may be referred to as the *singular value decomposition* of $\mathcal{A}$, and $s_1, \ldots, s_r$ are referred to as the singular values of $\mathcal{A}$. In fact, $s_1, \ldots, s_r$ are the positive square roots of the nonzero eigenvalues of $\mathcal{A}^\top \mathcal{A}$ (or equivalently $\mathcal{A}\mathcal{A}^\top$), $q_1, \ldots, q_n$ are eigenvectors of $\mathcal{A}^\top \mathcal{A}$, and the columns of $\mathcal{P}$ are eigenvectors of $\mathcal{A}\mathcal{A}^\top$.

spectral decomposition A $p \times p$ symmetric matrix $\mathcal{A}$ is expressible as

$$\mathcal{A} = \Gamma \Lambda \Gamma^\top = \sum_{i=1}^{p} \lambda_i \gamma_i \gamma_i^\top$$

where $\lambda_1, \ldots, \lambda_p$ are the not-necessarily-distinct eigenvalues of $\mathcal{A}$, $\gamma_1, \ldots, \gamma_p$ are orthonormal eigenvectors corresponding to $\lambda_1, \ldots, \lambda_p$, respectively, $\Gamma = (\gamma_1, \ldots, \gamma_p)$, $\mathcal{D} = \mathrm{diag}(\lambda_1, \ldots, \lambda_p)$.

subspace A *subspace* of a linear space $\mathcal{V}$ is a subset of $\mathcal{V}$ that is itself a linear space.

Taylor expansion The *Taylor series* of a function $f(x)$ in a point a is the power series $\sum_{n=0}^{\infty} \frac{f^{(n)}(a)}{n!}(x - a)^n$. A truncated Taylor series is often used to approximate the function $f(x)$.

Part I

Descriptive Techniques

1

Comparison of Batches

> Like all other arts, the Science of Deduction and Analysis is one which can only be acquired by long and patient study nor is life long enough to allow any mortal to attain the highest possible perfection in it. Before turning to those moral and mental aspects of the matter which present the greatest difficulties, let the enquirer begin by mastering more elementary problems.
> Sherlock Holmes in "Study in Scarlet"

The aim of this chapter is to describe and discuss the basic graphical techniques for a representation of a multidimensional data set. These descriptive techniques are explained in detail in Härdle & Simar (2003).

The graphical representation of the data is very important for both the correct analysis of the data and full understanding of the obtained results. The following answers to some frequently asked questions provide a gentle introduction to the topic.

We discuss the role and influence of outliers when displaying data in boxplots, histograms, and kernel density estimates. Flury-Chernoff faces—a tool for displaying up to 32 dimensional data—are presented together with parallel coordinate plots. Finally, Andrews' curves and draftman plots are applied to data sets from various disciplines.

EXERCISE 1.1. *Is the upper extreme always an outlier?*

An outlier is defined as an observation which lies beyond the outside bars of the boxplot, the outside bars being defined as:

$$F_U + 1.5d_F$$
$$F_L - 1.5d_F,$$

where F_L and F_U are the lower and upper fourths, respectively, and d_F is the interquartile range. The upper extreme is the maximum of the data set. These two terms could be sometimes mixed up! As the minimum or maximum do not have to lie outside the bars, they are not always the outliers.

Plotting the boxplot for the car data given in Table A.4 provides a nice example ◘ SMSboxcar.

EXERCISE 1.2. *Is it possible for the mean or the median to lie outside of the fourths or even outside of the outside bars?*

The median lies between the fourths per definition. The mean, on the contrary, can lie even outside the bars because it is very sensitive with respect to the presence of extreme outliers.

Thus, the answer is: NO for the median, but YES for the mean. It suffices to have only one extremely high outlier as in the following sample: 1, 2, 2, 3, 4, 99. The corresponding depth values are $1, 2, 3, 3, 2, 1$. The median depth is $(6 + 1)/2 = 3.5$. The depth of F is (depth of median+1)$/2 = 2.25$. Here, the median and the mean are:

$$x_{0.5} = \frac{2 + 3}{2} = 2.5,$$
$$\overline{x} = 18.5.$$

The fourths are $F_L = 2$, $F_U = 4$. The outside bars therefore are $2 - 2 \times 1.5 = -1$ and $4 + 2 \times 1.5 = 7$. The mean clearly falls outside the boxplot's outside bars.

EXERCISE 1.3. *Assume that the data are normally distributed $N(0, 1)$. What percentage of the data do you expect to lie outside the outside bars?*

In order to solve this exercise, we have to make a simple calculation.

For sufficiently large sample size, we can expect that the characteristics of the boxplots will be close to the theoretical values. Thus the mean and the median are expected to lie very close to 0, the fourths F_L and F_U should be lying close to standard normal quartiles $z_{0.25} = -0.675$ and $z_{0.75} = 0.675$.

The expected percentage of outliers is then calculated as the probability of having an outlier. The upper bound for the outside bar is then

$$c = F_U + 1.5d_F = -(F_L - 1.5d_F) \approx 2.7,$$

where d_F is the interquartile range. With Φ denoting the cumulative distribution function (cdf) of a random variable X with standard normal distribution $N(0, 1)$, we can write

$$P(X \notin [-c, c]) = 1 - P(X \in [-c, c])$$
$$= 1 - \{\Phi(c) - \Phi(-c)\}$$
$$= 2\{1 - \Phi(c)\}$$
$$= 2\{1 - \Phi(2.7)\}$$
$$= 2\{1 - 0.9965)\}$$
$$\approx 0.007$$

Thus, on average, 0.7 percent of the data will lie outside of the outside bars.

EXERCISE 1.4. *What percentage of the data do you expect to lie outside the outside bars if we assume that the data are normally distributed $N(0, \sigma^2)$ with unknown variance σ^2?*

From the theory we know that σ changes the scale, i.e., for large sample sizes the fourths F_L and F_U are now close to -0.675σ and 0.675σ. One could therefore guess that the percentage of outliers stays the same as in Exercise 1.3 since the change of scale affects the outside bars and the observations in the same way.

Our guess can be verified mathematically. Let X denote random variable with distribution $N(0, \sigma^2)$. The expected percentage of outliers can now be calculated for $c = F_U + 1.5d_F = -(F_L - 1.5d_F) \approx 2.7\sigma$ as follows:

$$P(X \notin [-c, c]) = 1 - P(X \in [-c, c])$$
$$= 1 - P\left(\frac{X}{\sigma} \in \left[-\frac{c}{\sigma}, \frac{c}{\sigma}\right]\right)$$
$$= 1 - \left\{\Phi\left(\frac{c}{\sigma}\right) - \Phi\left(-\frac{c}{\sigma}\right)\right\}$$
$$= 2\left\{1 - \Phi\left(\frac{c}{\sigma}\right)\right\}$$
$$= 2\{1 - \Phi(2.7)\}$$
$$\approx 0.007.$$

Again, 0.7 percent of the data lie outside of the bars.

EXERCISE 1.5. *How would the Five Number Summary of the 15 largest U.S. cities differ from that of the 50 largest U.S. cities? How would the five-number summary of 15 observations of $N(0, 1)$-distributed data differ from that of 50 observations from the same distribution?*

In the Five Number Summary, we calculate the upper fourth or upper quartile F_U, the lower fourth (quartile) F_L, the median and the extremes. The Five Number Summary can be graphically represented by a boxplot.

15 largest cities	
Minimum	77355
25% Quartile	84650
Median	104091
75% Quartile	134319
Maximum	591004

All 50 cities	
Minimum	1212
25% Quartile	36185
Median	56214
75% Quartile	83564
Maximum	591004

Taking 50 instead of 15 largest cities results in a decrease of all characteristics in the five-number summary except for the upper extreme, which stays the same (we assume that there are not too many cities of an equal size).

15 observations	
Minimum	−2.503
25% Quartile	−1.265
Median	−0.493
75% Quartile	−0.239
Maximum	1.950

50 observations	
Minimum	−2.757
25% Quartile	−1.001
Median	−0.231
75% Quartile	0.209
Maximum	2.444

In the case of the normally distributed data, the obtained result depends on the randomly generated samples. The median and the fourths should be, on average, of the same magnitude in both samples and they should lie a bit closer to the theoretical values $\Phi^{-1}(0.25) = -0.6745$ and $\Phi^{-1}(0.75) = 0.6745$ in the bigger sample.

We can expect that the extremes will lie further from the center of the distribution in the bigger sample.

EXERCISE 1.6. *Is it possible that all five numbers of the five-number summary could be equal? If so, under what conditions?*

Yes, it is possible. This can happen only if the maximum is equal to the minimum, i.e., if **all** observations are equal. Such a situation is in practice rather unusual.

EXERCISE 1.7. *Suppose we have 50 observations of $X \sim N(0,1)$ and another 50 observations of $Y \sim N(2,1)$. What would the 100 Flury-Chernoff faces (Chernoff 1973, Flury & Riedwyl 1981) look like if X and Y define the face line and the darkness of hair? Do you expect any similar faces? How many faces look like observations of Y even though they are X observations?*

One would expect many similar faces, because for each of these random variables 47.7% of the data lie between 0 and 2.

You can see the resulting Flury-Chernoff faces plotted on Figures 1.1 and 1.2. The "population" in Figure 1.1 looks thinner and the faces in Figure 1.2 have

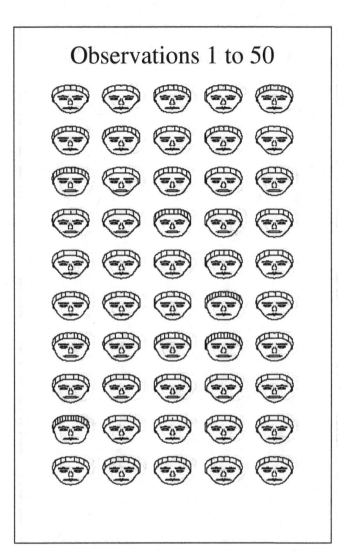

Fig. 1.1. Flury-Chernoff faces of the 50 $N(0,1)$ distributed data. ◘ SMSfacenorm

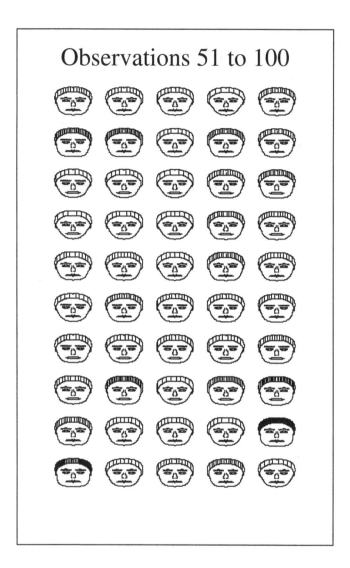

Fig. 1.2. Flury-Chernoff faces of the 50 $N(2,1)$ distributed data. ◘ SMSfacenorm

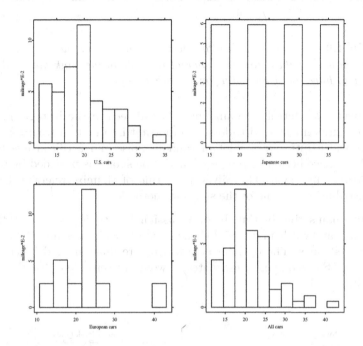

Fig. 1.3. Histograms for the mileage of the U.S. (top left), Japanese (top right), European (bottom left) and all (bottom right) cars. Q SMShiscar

darker hair. However, many faces could claim that they are coming from the other sample without arousing any suspicion.

EXERCISE 1.8. *Draw a histogram for the mileage variable of the car data (Table A.4). Do the same for the three groups (U.S., Japan, Europe). Do you obtain a similar conclusion as in the boxplots on Figure 1.3 in Härdle & Simar (2003)?*

The histogram is a density estimate which gives us a good impression of the shape distribution of the data.

The interpretation of the histograms in Figure 1.3 doesn't differ too much from the interpretation of the boxplots as far as only the European and the U.S. cars are concerned.

The distribution of mileage of Japanese cars appears to be multimodal—the amount of cars which achieve a high fuel economy is considerable as well as the amount of cars which achieve a very low fuel economy. In this case, the median and the mean of the mileage of Japanese cars don't represent the data properly since the mileage of most cars lies relatively far away from these values.

EXERCISE 1.9. *Use some bandwidth selection criterion to calculate the optimally chosen bandwidth h for the diagonal variable of the bank notes. Would it be better to have one bandwidth for the two groups?*

The bandwidth h controls the amount of detail seen in the histogram. Too large bandwidths might lead to loss of important information whereas a too small bandwidth introduces a lot of random noise and artificial effects. A reasonable balance between "too large" and "too small" is provided by bandwidth selection methods. The Silverman's rule of thumb—referring to the normal distribution—is one of the simplest methods.

Using Silverman's rule of thumb for Gaussian kernel, $h_{opt} = \hat{\sigma} n^{-1/5} 1.06$, the optimal bandwidth is 0.1885 for the genuine banknotes and 0.2352 for the counterfeit ones. The optimal bandwidths are different and indeed, for comparison of the two density estimates, it would be sensible to use the same bandwidth.

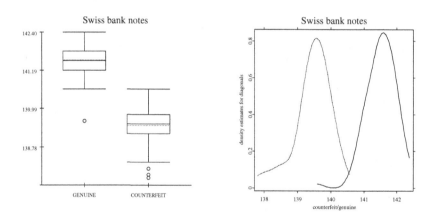

Fig. 1.4. Boxplots and kernel densities estimates of the diagonals of genuine and counterfeit bank notes.
Q SMSboxbank6 Q SMSdenbank

EXERCISE 1.10. *In Figure 1.4, the densities overlap in the region of diagonal ≈ 140.4. We partially observe this also in the boxplots. Our aim is to separate the two groups. Will we be able to do this effectively on the basis of this diagonal variable alone?*

No, using the variable diagonal alone, the two groups cannot be effectively separated since the densities overlap too much. However, the length of the diagonal is a very good predictor of the genuineness of the banknote.

EXERCISE 1.11. *Draw a parallel coordinates plot for the car data.*

Parallel coordinates plots (PCP) are a handy graphical method for displaying multidimensional data. The coordinates of the observations are drawn in a system of parallel axes. Index j of the coordinate is mapped onto the horizontal axis, and the $(0, 1)$ normalized value x_j is mapped onto the vertical axis. The PCP of the car data set is drawn in Figure 1.5. Different line styles allow to visualize the differences between groups and/or to find suspicious or outlying observations. The styles scheme in Figure 1.5 shows that the European and Japanese cars are quite similar. American cars, on the other hand, show much larger values of the 7th up to 11th variable. The parallelism of the lines in this region shows that there is a positive relationship between these variables. Checking the variable names in Table A.4 reveals that these variables describe the size of the car. Indeed, U.S. cars tend to be larger than European or Japanese cars.

The large amount of intersecting lines between the first and the second axis proposes a negative relationship between the first and the second variable, price and mileage.

The disadvantage of PCP is that the type of relationship between two variables can be seen clearly only on neighboring axes. Thus, we recommend that also some other type of graphics, e.g. scatterplot matrix, complements the analysis.

EXERCISE 1.12. *How would you identify discrete variables (variables with only a limited number of possible outcomes) on a parallel coordinates plot?*

Discrete variables on a parallel coordinates plot can be identified very easily since for discrete variable all the lines join in a small number of knots.

Look for example at the last variable, X_{13} = company headquarters, on the PCP for the car data in Figure 1.5.

EXERCISE 1.13. *Is the height of the bars of a histogram equal to the relative frequency with which observations fall into the respective bin?*

The histogram is constructed by counting the number of observations in each bin and then standardizing it to integrate to 1. The statement is therefore true.

EXERCISE 1.14. *Must the kernel density estimate always take on values only between 0 and 1?*

Cars data

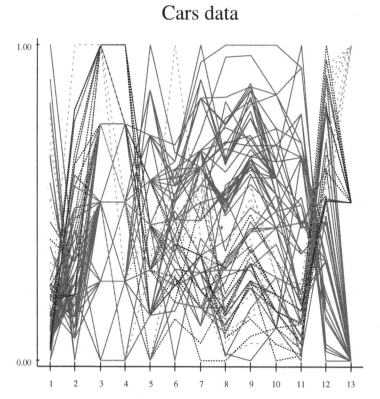

Fig. 1.5. Parallel coordinates plot for the car data. The full line marks U.S. cars, the dotted line marks Japanese cars and the dashed line marks European cars.
Q SMSpcpcar

False. The values of the density itself can lie anywhere between 0 and $+\infty$. Only the integral of the density has to be equal to one.

EXERCISE 1.15. *Let the following data set represent the heights (in m) of 13 students taking a multivariate statistics course:*

$$1.72, 1.83, 1.74, 1.79, 1.94, 1.81, 1.66, 1.60, 1.78, 1.77, 1.85, 1.70, 1.76.$$

1. *Find the corresponding five-number summary.*

2. *Construct the boxplot.*

3. Draw a histogram for this data set.

Let us first sort the data set in ascending order:

$$1.60, 1.66, 1.70, 1.72, 1.74, 1.76, 1.77, 1.78, 1.79, 1.81, 1.83, 1.85, 1.94.$$

As the number of observations is $n = 13$, the depth of the median is $(13 + 1)/2 = 7$ and the median is equal to the 7th observation $x_{(7)} = 1.77$. Next, the depth of fourths is defined as $\frac{[\text{depth of median}+1]}{2} = \frac{7+1}{2} = 4$ and the fourths are $F_U = x_{(4)} = 1.72$ and $F_L = x_{(10)} = 1.81$. This leads the following Five Number Summary:

Height	
Minimum	1.60
25% Quartile	1.72
Median	1.77
75% Quartile	1.81
Maximum	1.94

In order to construct the boxplot, we have to compute the outside bars. The F-spread is $d_F = F_U - F_L = 1.81 - 1.72 = 0.09$ and the outside bars are equal to $F_L - 1.5d_F = 1.585$ and $F_U + 1.5d_F = 1.945$. Apparently, there are no outliers, so the boxplot consists only of the box itself, the mean and median lines, and from the whiskers.

The histogram is plotted on Figure 1.6. The binwidth $h = 5\text{cm} = 0.05\text{m}$ seems to provide a nice picture here.

EXERCISE 1.16. *Analyze data that contain unemployment rates of all German federal states (Table A.16) using various descriptive techniques.*

A good way to describe one-dimensional data is to construct a boxplot. In the same way as in Exercise 1.15, we sort the data in ascending order,

$$5.8, 6.2, 7.7, 7.9, 8.7, 9.8, 9.8, 9.8, 10.4, 13.9, 15.1, 15.8, 16.8, 17.1, 17.3, 19.9,$$

and construct the boxplot. There are $n = 16$ federal states, the depth of the median is therefore $(16 + 1).2 = 8.5$ and the depth of fourths is 4.5.

The median is equal to the average of the 8th and 9th smallest observation, i.e., $M = \frac{1}{2}\left(x_{\left(\frac{n}{2}\right)} + x_{\left(\frac{n}{2}+1\right)}\right) = 10.1$ and the lower and upper fourths (quartiles) are $F_L = \frac{1}{2}(x_{(4)} + x_{(5)}) = 8.3$, $F_U = \frac{1}{2}(x_{(12)} + x_{(13)}) = 16.3$.

The outside bars are $F_U + 1.5d_F = 28.3$ and $F_L - 1.5d_F = -3.7$ and hence we can conclude that there are no outliers. The whiskers end at 5.8 and 19.9, the most extreme points that are not outliers.

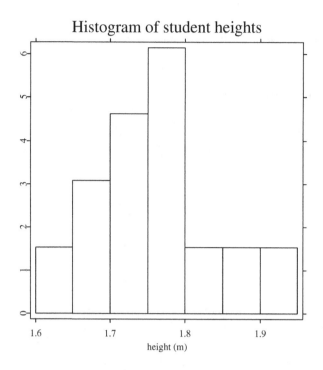

Fig. 1.6. Histogram of student heights. Q SMShisheights

The resulting boxplot for the complete data set is shown on the left hand side of Figure 1.7. The mean is greater than the median, which implies that the distribution of the data is not symmetric. Although 50% of the data are smaller than 10.1, the mean is 12. This indicates that there are a few observations that are much bigger than the median. Hence, it might be a good idea to explore the structure of the data in more detail. The boxplots calculated only for West and East Germany show a large discrepancy in unemployment rate between these two regions. Moreover, some outliers appear when these two subsets are plotted separately.

EXERCISE 1.17. *Using the yearly population data in Table A.11, generate*

1. *a boxplot (choose one of variables),*

2. *an Andrews' Curve (choose ten data points),*

3. *a scatterplot,*

4. *a histogram (choose one of the variables).*

Fig. 1.7. Boxplots for the unemployment data. ⌑ SMSboxunemp

What do these graphs tell you about the data and their structure?

A boxplot can be generated in the same way as in the previous examples. However, plotting a boxplot for time series data might mislead us since the distribution changes every year and the upward trend observed in this data makes the interpretation of the boxplot very difficult.

A histogram gives us a picture about how the distribution of the variable looks like, including its characteristics such as skewness, heavy tails, etc. In contrast to the boxplot it can also show multimodality. Similarly as the boxplot, a histogram would not be a reasonable graphical display for this time series data.

In general, for time series data in which we expect serial dependence, any plot omitting the time information may be misleading.

Andrews' curves are calculated as a linear combination of sine and cosine curves with different frequencies, where the coefficients of the linear combination are the multivariate observations from our data set (Andrews 1972). Each

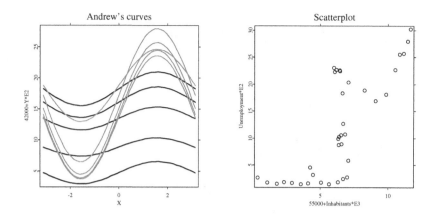

Fig. 1.8. Andrews' curves ⚏ SMSandcurpopu and scatterplot of unemployment against number of inhabitants ⚏ SMSscapopu for population data.

multivariate observation is represented by one curve. Differences between various observations lead to curves with different shapes. In this way, Andrews' curves allow to discover homogeneous subgroups of the multivariate data set and to identify outliers.

Andrews' curves for observations from years 1970–1979 are presented in Figure 1.8. Apparently, there are two periods. One period with higher (years 1975–1979) and the other period with lower (years 1970–1974) values.

A scatterplot is a two-dimensional graph in which each of two variables is put on one axis and data points are drawn as single points (or other symbols). The result for the analyzed data can be seen on Figure 1.8. From a scatterplot you can see whether there is a relationship between the two investigated variables or not. For this data set, the scatterplot in Figure 1.8 provides a very informative graphic. Plotted against the population (that increased over time) one sees the sharp oil price shock recession.

EXERCISE 1.18. *Make a draftman plot for the car data with the variables*

$$X_1 = price,$$
$$X_2 = mileage,$$
$$X_8 = weight,$$
$$X_9 = length.$$

Move the brush into the region of heavy cars. What can you say about price, mileage and length? Move the brush onto high fuel economy. What are the differences among the Japanese, European and U.S. American cars?

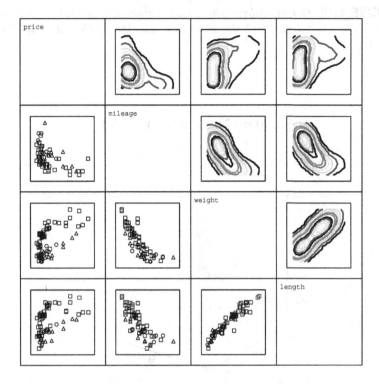

Fig. 1.9. Draftman plot and density contour plots for the car data. In scatterplots, the squares mark U.S. cars, the triangles mark Japanese cars and the circles mark European cars. ⊙ SMSdrafcar

The so-called draftman plot is a matrix consisting of all pairwise scatterplots. Clearly, the matrix is symmetric and hence we display also estimated density contour plots in the upper right part of the scatterplot matrix in Figure 1.9.

The heaviest cars in Figure 1.9 are all American, and any of these cars is characterized by high values of price, mileage, and length. Europeans and Japanese prefer smaller, more economical cars.

EXERCISE 1.19. *What is the form of a scatterplot of two independent normal random variables X_1 and X_2?*

The point cloud has circular shape and the density of observations is highest in the center of the circle. This corresponds to the density of two-dimensional normal distribution which is discussed in Härdle & Simar (2003, chapter 5).

EXERCISE 1.20. *Rotate a three-dimensional standard normal point cloud in 3D space. Does it "almost look the same from all sides"? Can you explain why or why not?*

Standard Normal point cloud

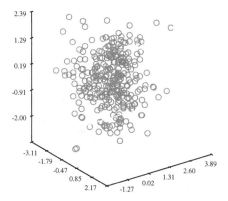

Fig. 1.10. A 3D scatterplot of the standard normal distributed data (300 observations). ⌕ SMSscanorm3.

The standard normal point cloud in 3D space, see Figure 1.10, looks almost the same from all sides, because it is a realization of random variables whose variances are equal and whose covariances are zero.

The density of points corresponds to the density of a three-dimensional normal distribution which has spherical shape. Looking at the sphere from any point of view, the cloud of points always has a circular (spherical) shape.

Part II

Multivariate Random Variables

A Short Excursion into Matrix Algebra

Dirty-looking rascals, but I suppose every one has some little immortal spark concealed about him.
Sherlock Holmes in "The Sign of Four"

In statistics, data sets come in matrix form and the characteristics of the data can be written in terms of matrix operations. Understanding matrix algebra is crucial for investigating the properties of the observed data.

The importance of matrix algebra lies in a great simplification of many mathematical formulas and derivations. The spectral decomposition is one of the most commonly used tools in multivariate statistics because it allows a nice representation of large dimensional matrices in terms of their eigenvalues and eigenvectors.

Calculation of the determinant of partitioned matrices helps us in calculating constrained maximum likelihood estimators and testing of hypothesis. Properties of projection matrices are useful in least squares regression analysis, and iso-distance ellipsoids help us to understand covariance structures.

EXERCISE 2.1. *Compute the determinant for a* (3×3) *matrix.*

For a square matrix $\mathcal{A}$, the determinant is defined as:

$$\det(\mathcal{A}) = |\mathcal{A}| = \sum (-1)^{|\tau|} a_{1\tau(1)} \cdots a_{p\tau(p)},$$

the summation is over all permutations τ of $\{1, 2, \ldots, p\}$, and $(-1)^{|\tau|}$ denotes the sign of the permutation τ. For a three-dimensional matrix $\mathcal{A}_{3\times 3} = \{a_{ij}\}$, the determinant of $\mathcal{A}$ becomes

$$|\mathcal{A}| = a_{11}a_{22}a_{33} + a_{12}a_{23}a_{31} + a_{13}a_{21}a_{32} - a_{31}a_{22}a_{13} - a_{32}a_{23}a_{11} - a_{33}a_{21}a_{12}.$$

In terms of the eigenvalues λ_1, λ_2, and λ_3, the determinant can be written as $|\mathcal{A}| = \lambda_1\lambda_2\lambda_3$.

EXERCISE 2.2. *Suppose that* $|\mathcal{A}| = 0$. *Is it possible that all eigenvalues of* $\mathcal{A}$ *are positive?*

Given $\mathcal{A}_{n \times n}$, the eigenvalues λ_i, for $i = 1, ..., n$ are the roots of the polynomial

$$|\mathcal{A} - \lambda \mathcal{I}| = 0. \tag{2.1}$$

If $|\mathcal{A}| = 0$ then one of the solutions of (2.1) is $\lambda = 0$. Hence, if $|\mathcal{A}| = 0$ then there exists at least one eigenvalue such that $\lambda_i = 0$.

EXERCISE 2.3. *Suppose that all eigenvalues of some (square) matrix* $\mathcal{A}$ *are different from zero. Does the inverse* $\mathcal{A}^{-1}$ *of* $\mathcal{A}$ *exist?*

The fact that all eigenvalues are different from zero implies that also the determinant $|\mathcal{A}| = \prod_i \lambda_i \neq 0$ and the inverse matrix $\mathcal{A}^{-1}$ can be calculated as $\mathcal{A}^{-1} = |\mathcal{A}|^{-1}\mathcal{C}$, where $\mathcal{C}$ is the so-called adjoint matrix of $\mathcal{A}$, see the introductory section on terminology for more details.

EXERCISE 2.4. *Write a program that calculates the spectral decomposition of the matrix*

$$\mathcal{A} = \begin{pmatrix} 1\,2\,3 \\ 2\,1\,2 \\ 3\,2\,1 \end{pmatrix}.$$

Check the properties of the spectral decomposition numerically, i.e., calculate $|\mathcal{A}|$ *as in Exercise 2.1 and check that it is equal to* $\lambda_1\lambda_2\lambda_3$.

We obtain the following matrix of eigenvectors

$$\Gamma = (\gamma_1, \gamma_3, \gamma_3) = \begin{pmatrix} 0.3645\ 0.6059\ \text{-}0.7071 \\ \text{-}0.8569\ 0.5155\ 0.0000 \\ 0.3645\ 0.6059\ 0.7071 \end{pmatrix}$$

and the following eigenvalues

$$\Lambda = \begin{pmatrix} \text{-}0.7016\ 0.0000\ \ \ 0.0000 \\ 0.0000\ 5.7016\ \ \ 0.0000 \\ 0.0000\ 0.0000\ \text{-}2.0000 \end{pmatrix}.$$

Now it can be easily verified that $\Gamma \Lambda \Gamma^\top = \mathcal{A}$, $\Gamma^\top \Gamma = \mathcal{I}$, $\operatorname{tr}\mathcal{A} = \lambda_1 + \lambda_2 + \lambda_3$, $|\mathcal{A}| = \lambda_1\lambda_2\lambda_3$, etc. **Q** SMSjordandec

EXERCISE 2.5. *Prove that* $\frac{\partial a^\top x}{\partial x} = a$, $\frac{\partial x^\top \mathcal{A} x}{\partial x} = 2\mathcal{A}x$, *and* $\frac{\partial^2 x^\top \mathcal{A} x}{\partial x \partial x^\top} = \frac{\partial 2\mathcal{A}x}{\partial x} = 2\mathcal{A}$.

Recall the gradient vector definition from the introductory section on terminology. The kth element of the vector of partial derivatives $\frac{\partial a^\top x}{\partial x}$ is equal to $\frac{\partial a^\top x}{\partial x_k} = a_k$. It follows immediately that

$$\frac{\partial a^\top x}{\partial x} = a.$$

Similarly, differentiating

$$\frac{\partial x^\top \mathcal{A} x}{\partial x} = \frac{\partial(\sum_{i=1}^p \sum_{j=1}^p a_{ij} x_i x_j)}{\partial x}$$

with respect to x_k gives

$$\frac{\partial(.)}{\partial x_k} = \frac{\partial a_{kk} x_k^2}{\partial x_k} + \frac{\partial \sum_{i \neq k} a_{ik} x_i x_k}{\partial x_k} + \frac{\partial \sum_{j \neq k} a_{kj} x_k x_j}{\partial x_k} = 2 \sum_{j=1}^p a_{kj} x_j,$$

which is just the kth element of vector $2\mathcal{A}x$.

Using the above two properties, we have the following for the last formula

$$\frac{\partial^2 x^\top \mathcal{A} x}{\partial x \partial x^\top} = \frac{\partial 2\mathcal{A} x}{\partial x} = 2\mathcal{A}.$$

EXERCISE 2.6. *Show that a projection (idempotent) matrix has eigenvalues only in the set $\{0, 1\}$.*

$\mathcal{A}$ is a projection matrix if $\mathcal{A} = \mathcal{A}^2 = \mathcal{A}^\top$. Let λ_i be an eigenvalue of $\mathcal{A}$ and γ_i its corresponding eigenvector:

$$\mathcal{A}\gamma_i = \lambda_i \gamma_i$$
$$\mathcal{A}^2 \gamma_i = \lambda_i \mathcal{A}\gamma_i$$
$$\mathcal{A}\gamma_i = \lambda_i \mathcal{A}\gamma_i$$
$$\mathcal{A}\gamma_i = \lambda_i^2 \gamma_i$$
$$\lambda_i \gamma_i = \lambda_i^2 \gamma_i$$
$$\lambda_i = \lambda_i^2.$$

It is obvious that $\lambda_i = \lambda_i^2$ only if λ_i is equal to 1 or 0.

EXERCISE 2.7. *Draw some iso-distance ellipsoids $\{x \in \mathbb{R}^p | (x - x_0)^\top \mathcal{A}(x - x_0) = d^2\}$ for the metric $\mathcal{A} = \Sigma^{-1}$, where $\Sigma = \begin{pmatrix} 1 & \rho \\ \rho & 1 \end{pmatrix}$.*

The eigenvalues of Σ are solutions to:

$$\begin{vmatrix} 1-\lambda & \rho \\ \rho & 1-\lambda \end{vmatrix} = 0.$$

Hence, $\lambda_1 = 1 + \rho$ and $\lambda_2 = 1 - \rho$. Notice, that the eigenvalues of matrix $\mathcal{A}$ are equal to λ_1^{-1} and λ_2^{-1}. The eigenvector corresponding to $\lambda_1 = 1 + \rho$ can be computed from the system of linear equations:

$$\begin{pmatrix} 1 & \rho \\ \rho & 1 \end{pmatrix} \begin{pmatrix} x_1 \\ x_2 \end{pmatrix} = (1 + \rho) \begin{pmatrix} x_1 \\ x_2 \end{pmatrix}$$

or

$$x_1 + \rho x_2 = x_1 + \rho x_1$$
$$\rho x_1 + x_2 = x_2 + \rho x_2$$

and thus $x_1 = x_2$. The first (standardized) eigenvector is

$$\gamma_1 = \begin{pmatrix} 1/\sqrt{2} \\ 1/\sqrt{2} \end{pmatrix}.$$

The second eigenvector (orthogonal to γ_1) is

$$\gamma_2 = \begin{pmatrix} 1/\sqrt{2} \\ -1/\sqrt{2} \end{pmatrix}.$$

The axes of the ellipsoid point in the directions provided by the eigenvectors. The length of each axis is equal to $d\sqrt{\lambda_i}$.

Four ellipses for varying values of d and ρ are plotted in Figure 2.1.

EXERCISE 2.8. *Find a formula for* $|\mathcal{A} + aa^\top|$ *and for* $(\mathcal{A} + aa^\top)^{-1}$.

We define matrix $\mathcal{B} = \begin{pmatrix} 1 & -a^\top \\ a & \mathcal{A} \end{pmatrix}$ and apply the formulas for determinant and inverse of a partitioned matrix. The determinant of $\mathcal{B}$ can be written in two ways as

$$|\mathcal{B}| = |1||\mathcal{A} + aa^\top| \tag{2.2}$$
$$|\mathcal{B}| = |\mathcal{A}||1 + a^\top \mathcal{A}^{-1} a|. \tag{2.3}$$

Comparing (2.2) and (2.3) implies that

$$|\mathcal{A} + aa^\top| = |\mathcal{A}||1 + a^\top \mathcal{A}^{-1} a|.$$

Next, using the formula for inverse of the partitioned matrix $\mathcal{B}$, we obtain

$$(\mathcal{A} + aa^\top)^{-1} = \mathcal{A}^{-1} - \frac{\mathcal{A}^{-1} aa^\top \mathcal{A}^{-1}}{1 + a^\top \mathcal{A}^{-1} a}.$$

This result will prove to be useful in the derivation of the variance efficient portfolios discussed in Exercises 17.1 and 17.3.

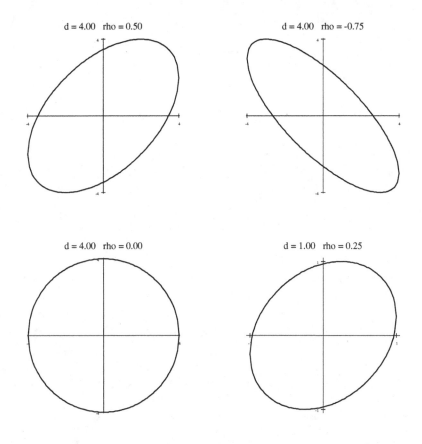

Fig. 2.1. Ellipses for varying ρ and d. ◘ SMSellipse

EXERCISE 2.9. *Prove the binomial inverse theorem for two non-singular matrices $\mathcal{A}(p \times p)$ and $\mathcal{B}(p \times p)$: $(\mathcal{A}+\mathcal{B})^{-1} = \mathcal{A}^{-1} - \mathcal{A}^{-1}(\mathcal{A}^{-1}+\mathcal{B}^{-1})^{-1}\mathcal{A}^{-1}$.*

Let us define $\mathcal{C} = \begin{pmatrix} \mathcal{A} & \mathcal{I}_p \\ -\mathcal{I}_p & \mathcal{B}^{-1} \end{pmatrix}$. Considering the formula for the inverse of a partitioned matrix, the submatrix $\mathcal{C}^{11}$ of $\mathcal{C}^{-1}$ can be obtained in two ways:

$$\mathcal{C}^{11} = (\mathcal{A} + \mathcal{I}\mathcal{B}\mathcal{I})^{-1}$$
$$= (\mathcal{A} + \mathcal{B})^{-1} \tag{2.4}$$
$$\mathcal{C}^{11} = \mathcal{A}^{-1} + \mathcal{A}^{-1}\mathcal{I}(\mathcal{B}^{-1} - \mathcal{I}\mathcal{A}^{-1}\mathcal{I})^{-1}\mathcal{I}\mathcal{A}^{-1}$$
$$= \mathcal{A}^{-1} - \mathcal{A}^{-1}(\mathcal{A}^{-1} + \mathcal{B}^{-1})^{-1}\mathcal{A}^{-1}. \tag{2.5}$$

Comparing expressions (2.4) and (2.5) proves the binomial inverse theorem.

3

Moving to Higher Dimensions

At first it was but a lurid spark upon the stone pavement. Then it
lengthened out until it became a yellow line, and then without any
warning or sound, a gash seemed to open and a hand appeared, ...
"The Red-Headed League"

The basic tool used for investigating dependencies between the ith and jth
components of a random vector X is the covariance

$$\sigma_{X_i X_j} = Cov(X_i, X_j) = E(X_i X_j) - (EX_i)(EX_j).$$

From a data set, the covariance between the ith and jth columns can be
estimated as

$$s_{X_i X_j} = \frac{1}{n} \sum_{k=1}^{n} (x_{ik} - \overline{x}_i)(x_{jk} - \overline{x}_j).$$

The covariance tells us how one variable depends linearly on another variable.
The concept of covariance and correlation is therefore strongly tied to linear
statistical modeling. The significance of correlation is measured via Fisher's
Z-transform, and the fit of regression lines is judged by the coefficient of determination. The ANOVA variance decomposition helps us understand nested
linear models.

We discuss here linear models for a marketing example (the sales of classic
blue pullovers) and study theoretical properties of covariance and correlation.
The least squares method is revisited and analyzed with analytical tools.

Analysis of variance (ANOVA) can be seen as a special case of the linear
model with an appropriately selected design matrix. Similarly, the test of the
ANOVA hypothesis of the equality of mean effects in more treatment groups
can be seen as a special case of an F-test in the linear model formulation.

Fig. 3.1.   Scatterplot of variables X_4 vs. X_5 of the entire bank data set.
Q SMSscabank45

EXERCISE 3.1. *The covariance $s_{X_4 X_5}$ between X_4 and X_5 for the entire bank data set is positive. Given the definitions of X_4 and X_5, we would expect a negative covariance. Using Figure 3.1 can you explain why $s_{X_4 X_5}$ is positive?*

Variables X_4 and X_5 are defined as the distance of the inner frame to the lower or upper border, respectively. In general, small deviations in the position of the center picture would lead to negative dependencies between variables X_4 and X_5.

Surprisingly, the empirical covariance is equal to 0.16.

An explanation is shown in Figure 3.1. We observe in fact two clouds of points, the counterfeit and the genuine banknotes. The relationship between X_4 and X_5 is negative inside these groups. The calculation of the empirical covariance ignores this information and it is confused by the relative position of these two groups of observations.

EXERCISE 3.2. *Consider the two sub-clouds of counterfeit and genuine bank notes in Figure 3.1 separately. Do you still expect $s_{X_4 X_5}$ (now calculated separately for each cloud) to be positive?*

Considering the covariance of X_4 and X_5 for the full bank data set gives a result which does not have any meaningful interpretation. As expected, the covariances for the first hundred observations (-0.26) and for the second hundred observations (-0.49) are negative.

EXERCISE 3.3. *It is well known that for two normal random variables, zero covariance implies independence. Why does this not apply to the following situation: $X \sim N(0,1)$, $Cov(X, X^2) = EX^3 - EXEX^2 = 0 - 0 = 0$ but obviously X^2 is totally dependent on X?*

It is easy to show that independence of two random variables implies zero covariance:

$$Cov(X, Y) = E(XY) - EXEY \stackrel{indep.}{=} EXEY - EXEY = 0.$$

The opposite is true only if X and Y are jointly normally distributed which can be checked by calculating the joint density and the product of the marginals.

From above we see that, for standard normally distributed random variable X, we have $Cov(X, X^2) = 0$. In this example, zero covariance does not imply independence since the random variable X^2 is not normally distributed.

EXERCISE 3.4. *Compute the covariance between the variables*

$$X_2 = \text{ miles per gallon,}$$
$$X_8 = \text{ weight}$$

from the car data set (Table A.4). What sign do you expect the covariance to have?

The empirical covariance is -3732. It is negative as expected since heavier cars tend to consume more gasoline and this leads to lower mileage. The negative covariance corresponds to a negative slope that could be observed in a scatterplot.

It is very difficult to judge the strength of the dependency between weight and mileage on the basis of the covariance. A more appropriate measure is the correlation which is a scale independent version of the covariance.

Correlation lies always between -1 and 1. Values close to 1 or -1 indicate strong positive or negative relationship, respectively. Correlation $r_{X_2 X_8} = -0.823$ between weight and mileage suggests rather strong negative relationship. Using Fisher's Z-transformation, we can prove the statistical significance of $r_{X_2 X_8}$, see Härdle & Simar (2003, example 3.5).

EXERCISE 3.5. *Compute the correlation matrix of the variables in "classic blue" pullover data set (Table A.6). Comment on the sign of the correlations and test the hypothesis*

$$\rho_{X_1 X_2} = 0.$$

The correlation matrix is

$$\mathcal{R} = \begin{pmatrix} 1.000 & -0.168 & 0.867 & 0.633 \\ -0.168 & 1.000 & 0.121 & -0.464 \\ 0.867 & 0.121 & 1.000 & 0.308 \\ 0.633 & -0.464 & 0.308 & 1.000 \end{pmatrix}.$$

The correlation $r_{X_1 X_2} = -0.168$ says that the relationship between sales and prices is negative as predicted by the economic theory. On the other hand, we observe positive correlation of sales with advertisement and presence of a sale assistant which suggests that investments in advertisement and sale assistants increase the sales.

Using the Fisher Z-transformation and standardizing the transformed value, we obtain the value $z = -0.4477$ and hence we cannot reject the null hypothesis $H_0 : \rho = 0$ since this is a nonsignificant value.

Considering the small sample size, $n = 10$, we can improve the test using Hotelling's transformation

$$w^* = w - \frac{3w + \tanh(w)}{4(n-1)} = -0.1504$$

which is also nonsignificant since $-0.1504 \sqrt{n-1} = -0.4513 < -1.96$.

EXERCISE 3.6. *Suppose you have observed a set of observations $\{x_i\}_{i=1}^n$ with $\bar{x} = 0$, $s_{XX} = 1$ and $n^{-1} \sum_{i=1}^n (x_i - \bar{x})^3 = 0$. Define the variable $y_i = x_i^2$. Can you immediately tell whether $r_{XY} \neq 0$?*

Plugging $y_i = x_i^2$ into the following formula for calculating the empirical covariance

$$s_{XY} = \frac{1}{n} \sum_{i=1}^n x_i y_i - \overline{xy}$$

we obtain

$$s_{XY} = s_{XX^2} = \frac{1}{n} \sum_{i=1}^n x_i^2 x_i - \overline{yx} = \frac{1}{n} \sum_{i=1}^n x_i^3 = \frac{1}{n} \sum_{i=1}^n (x_i^3 - \bar{x}) = 0.$$

We remark that this calculation holds for any finite value of s_{XX}.

EXERCISE 3.7. *Find the values $\widehat{\alpha}$ and $\widehat{\beta}$ that minimize the sum of squares*

$$\sum_{i=1}^{n}(y_i - \alpha - \beta x_i)^2 \tag{3.1}$$

The values $\widehat{\alpha}$ and $\widehat{\beta}$ are actually estimates of an intercept and a slope, respectively, of a regression line fitted to data $\{(x_i, y_i)\}_{i=1}^{n}$ by the least squares method. More formally, the estimators can be expressed as

$$(\widehat{\alpha}, \widehat{\beta}) = \arg\min_{(\alpha, \beta)} \sum_{i=1}^{n}(y_i - \alpha - \beta x_i)^2.$$

One has to understand that $\widehat{\alpha}$ and $\widehat{\beta}$ are random variables since they can be expressed as functions of random observations x_i and y_i. Random variables $\widehat{\alpha}$ and $\widehat{\beta}$ are called estimators of the true unknown (fixed) parameters α and β.

The estimators can be obtained by differentiating the sum of squares (3.1) with respect to α and β and by looking for a zero point of the derivative. We obtain

$$\frac{\partial \sum_{i=1}^{n}(y_i - \alpha - \beta x_i)^2}{\partial \alpha} = -2\sum_{i=1}^{n}(y_i - \alpha - \beta x_i) = 0, \tag{3.2}$$

$$\alpha = n^{-1}\sum_{i=1}^{n}y_i - n^{-1}\beta\sum_{i=1}^{n}x_i, \tag{3.3}$$

and

$$\frac{\partial \sum_{i=1}^{n}(y_i - \alpha - \beta x_i)^2}{\partial \beta} = -2\sum_{i=1}^{n}(y_i - \alpha - \beta x_i)x_i = 0. \tag{3.4}$$

Substituting for α leads to

$$0 = \sum_{i=1}^{n}y_i x_i - n^{-1}\sum_{i=1}^{n}y_i\sum_{i=1}^{n}x_i + n^{-1}\beta\left(\sum_{i=1}^{n}x_i\right)^2 - \beta\sum_{i=1}^{n}x_i^2.$$

Solving the above equation in β gives the following estimate

$$\begin{aligned}
\beta &= \frac{n^{-1}\sum_{i=1}^{n}y_i\sum_{i=1}^{n}x_i - \sum_{i=1}^{n}y_i x_i}{n^{-1}\left(\sum_{i=1}^{n}x_i\right)^2 - \sum_{i=1}^{n}x_i^2} \\
&= \frac{\sum_{i=1}^{n}y_i x_i - n^{-1}\sum_{i=1}^{n}y_i\sum_{i=1}^{n}x_i}{\sum_{i=1}^{n}x_i^2 - n^{-1}\left(\sum_{i=1}^{n}x_i\right)^2} \\
&= \frac{\sum_{i=1}^{n}y_i x_i - n\overline{y}\overline{x}}{\sum_{i=1}^{n}x_i^2 - n\overline{x}^2} \\
&= \frac{s_{XY}}{s_{XX}}.
\end{aligned}$$

Hence, the sum of squares is minimized for $\alpha = \widehat{\alpha} = \overline{y} - \widehat{\beta}\overline{x}$ and $\beta = \widehat{\beta} = \frac{s_{XY}}{s_{XX}}$.

EXERCISE 3.8. *How many sales does the textile manager expect with a "classic blue" pullover price of $x = 105$?*

The least squares estimates of the intercept and slope are

$$\widehat{\alpha} = 210.774 \qquad \text{and} \qquad \widehat{\beta} = -0.364$$

and the estimated linear regression model can be written as

$$\text{Sales} = 210.774 - 0.364 \times \text{Price} + \varepsilon.$$

Plugging in the pullover price 120 leads to expected sales equal to $210.774 - 0.364 \times 120 = 167.094$. This value can be interpreted also as the conditional expected value of the random variable "sales" conditioned on the event {price $= 120$}.

EXERCISE 3.9. *What does a scatterplot of two random variables look like for $r^2 = 1$ and $r^2 = 0$?*

The coefficient of determination, r^2 is defined as

$$r^2 = \frac{\sum_{i=1}^{n}(\widehat{y}_i - \overline{y})^2}{\sum_{i=1}^{n}(y_i - \overline{y})^2},$$

i.e., it is a ratio of the explained sum of squares and the total sum of squares. The coefficient r^2 is equal to one only if the numerator and denominator are equal. Now, the decomposition of the total sum of squares

$$\sum_{i=1}^{n}(y_i - \overline{y})^2 = \sum_{i=1}^{n}(y_i - \widehat{y}_i)^2 + \sum_{i=1}^{n}(\widehat{y}_i - \overline{y})^2 \qquad (3.5)$$

implies that this can happen only if the first term on the right hand side of (3.5) is equal to zero, i.e., if $y_i = \widehat{y}_i$ for all $1 \leq i \leq n$. Hence, $r^2 = 1$ if and only if all y_i's plotted as a function of the corresponding x_i's are lying on a straight line.

Similarly, we can see that $r^2 = 0$ only if $\sum_{i=1}^{n}(\widehat{y}_i - \overline{y})^2 = 0$. This can happen only if all $\widehat{y}_i$'s are equal to each other. In other words, this happens if we do not observe any trend in the scatterplot of y_i's plotted against the x_i's.

Interestingly, observations lying on a straight horizontal line satisfy both of the above conditions. Closer look at the definition of the coefficient of determination reveals that in this case, it is not defined.

EXERCISE 3.10. *Prove the variance decomposition (3.5) and show that the coefficient of determination is the square of the simple correlation between X and Y.*

First, as

$$\sum_{i=1}^{n}(y_i - \overline{y})^2 = \sum_{i=1}^{n}(y_i - \widehat{y}_i + \widehat{y}_i - \overline{y})^2$$

$$= \sum_{i=1}^{n}(\widehat{y}_i - \overline{y})^2 + \sum_{i=1}^{n}(y_i - \widehat{y}_i)^2 + 2\sum_{i=1}^{n}(\widehat{y}_i - \overline{y})(y_i - \widehat{y}_i),$$

it is enough to show that the last term on the right hand side is equal to zero. This follows immediately from the first order conditions (3.2) and (3.4) once we rewrite the expression a bit:

$$\sum_{i=1}^{n}(\widehat{y}_i - \overline{y})(y_i - \widehat{y}_i) = \sum_{i=1}^{n}(\widehat{a} - \overline{y})(y_i - \widehat{a} - \widehat{b}x_i) + \sum_{i=1}^{n}\widehat{b}x_i(y_i - \widehat{a} - \widehat{b}x_i).$$

Note that it implies $\overline{\widehat{y}} = \overline{y}$ and $\sum_{i=1}^{n}x_i(y_i - \widehat{y}_i) = 0$.

Next, we shall prove that $r^2 = r_{XY}^2$, i.e.,

$$\frac{\sum_{i=1}^{n}(\widehat{y}_i - \overline{y})^2}{\sum_{i=1}^{n}(y_i - \overline{y})^2} = \frac{\left(\sum_{i=1}^{n}(y_i - \overline{y})(x_i - \overline{x})\right)^2}{\sum_{i=1}^{n}(y_i - \overline{y})^2 \sum_{i=1}^{n}(x_i - \overline{x})^2}.$$

Using the conclusions reached above, this reduces to

$$1 = \frac{\{\sum_{i=1}^{n}(\widehat{y}_i - \overline{\widehat{y}})(x_i - \overline{x})\}^2}{\sum_{i=1}^{n}(\widehat{y}_i - \overline{\widehat{y}})^2 \sum_{i=1}^{n}(x_i - \overline{x})^2} = r_{\widehat{Y}X}^2.$$

This holds by definition since $\widehat{y}_i = \widehat{\alpha} + \widehat{\beta}x_i$, $i = 1\ldots,n$, is a linear function of x_i.

EXERCISE 3.11. *Make a boxplot for the residuals $\varepsilon_i = y_i - \widehat{\alpha} - \widehat{\beta}x_i$ for the "classic blue" pullover data (Table A.6). If there are outliers, identify them and run the linear regression again without them. Do you obtain a stronger influence of price on sales?*

The boxplot of the residuals ε_i is plotted in the right graphics in Figure 3.2. The left graphics in Figure 3.2 shows the dependency of pullover sales on the price, the regression with the outliers (dashed line) and the regression without the outliers (full line). The two outliers are marked by red triangles. Performing the regression without the outliers shows evidence for stronger influence of price on sales.

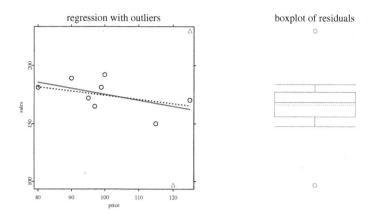

Fig. 3.2. Linear regression (dashed black line) and the corrected linear regression without outliers (full blue line). The second graph shows a boxplot of the residuals.
Q SMSlinregpull

In this case, the influence of the outliers on the regression line does not seem to be too large. Deleting only one of the outliers would lead to much different results. Moreover, such an outlier might influence the regression line so that it is not detectable using only the residuals.

EXERCISE 3.12. *Under what circumstances would you obtain the same coefficients from the linear regression lines of Y on X and of X on Y?*

Let us recall the formulas derived in Exercise 3.7:

$$\widehat{\alpha} = \overline{y} - \widehat{\beta}\overline{x} \text{ and } \widehat{\beta} = \frac{s_{XY}}{s_{XX}}.$$

From the formula for the slope of the regression line, $\widehat{\beta}$, it follows that the slopes are identical if the variances of X and Y are equal, $s_{XX} = s_{YY}$, or if the covariance between X and Y is equal to zero, $s_{XY} = 0$.

If the slopes are equal then it is obvious from the formula for the intercept of the regression line $\widehat{\alpha}$ that the intercepts are equal if and only if the means of X and Y are the same.

EXERCISE 3.13. *Compute an approximate confidence interval for the correlation coefficient $\rho_{X_4 X_1}$ between the presence of the sales assistants (X_4) and the number of sold pullovers (X_1). Hint: start from a confidence interval for $\tanh^{-1}(\rho_{X_4 X_1})$ and then apply the inverse transformation.*

The estimate of the correlation is $r_{X_4 X_1} = 0.633$. In order to calculate the approximate confidence interval, we can apply the Fisher's Z-transformation

$$W = \tanh^{-1}(r_{X_4 X_1}) = \frac{1}{2} \log \left(\frac{1 + r_{X_4 X_1}}{1 - r_{X_4 X_1}} \right)$$

which has approximately a normal distribution with the expected value $EW = \frac{1}{2} \log\{(1 + \rho_{X_4 X_1})/(1 - \rho_{X_4 X_1})\}$ and the variance $Var\, W = 1/(n-3)$, see Härdle & Simar (2003, section 3.2).

Hence, with $\tanh(x) = (e^{2x} - 1)/(e^{2x} + 1)$:

$$1 - \alpha \approx P\left(|\sqrt{n-3}\{\tanh^{-1}(r_{X_4 X_1}) - \tanh^{-1}(\rho_{X_4 X_1})\}| \leq \Phi^{-1}(1 - \alpha/2) \right)$$

$$= P\left\{ \left(\tanh^{-1}(r_{X_4 X_1}) \pm \frac{\Phi^{-1}(1 - \alpha/2)}{\sqrt{n-3}} \right) \ni \tanh^{-1}(\rho_{X_4 X_1}) \right\}$$

$$= P\left\{ \tanh\left(\tanh^{-1}(r_{X_4 X_1}) \pm \frac{\Phi^{-1}(1 - \alpha/2)}{\sqrt{n-3}} \right) \ni \rho_{X_4 X_1} \right\}$$

and we can say that the (random) interval

$$\left(\tanh\left\{ \tanh^{-1}(r_{X_4 X_1}) \pm \frac{\Phi^{-1}(1 - \alpha/2)}{\sqrt{n-3}} \right\} \right)$$

covers the unknown value of the true correlation coefficient $\rho_{X_4 X_1}$ with probability approximately $1 - \alpha$.

For our example, we choose $\alpha = 0.05$ which implies that $\Phi^{-1}(1 - \alpha/2) = 1.96$ and with $r_{X_4 X_1} = 0.633$ and $n = 10$, we obtain the approximate 95% confidence interval $(0.0055, 0.9028)$.

EXERCISE 3.14. *Using the exchange rate of 1 EUR = 106 JPY, compute the empirical covariance between pullover sales and prices in Japanese Yen rather than in Euros. Is there a significant difference? Why?*

The covariance is $s_{X_1 X_2}^{EUR} = -80.02$ in Euro and $s_{X_1 X_2}^{JPY} = -8482.14$ in Japanese Yen. The difference is caused entirely by the change of scale. The covariance in Yen can be expressed from the covariance in Euro as

$$s_{X_1 X_2}^{JPY} = s_{X_1 X_2}^{EUR} \times 106 = -80.02 \times 106 = -8482.12.$$

The remaining small difference 0.02 is due to the rounding error.

Notice that the calculation would look differently for covariance between the price (X_2) and advertisement cost (X_3) since

$$s_{X_2 X_3}^{JPY} = s_{X_1 X_2}^{EUR} \times 106^2.$$

Here, we change the scale of both variables by factor 106 and, hence, we have to multiply the covariance by 106^2.

EXERCISE 3.15. *Why does the correlation have the same sign as the covariance?*

The correlation is defined as

$$\rho_{XY} = \frac{\mathrm{Cov}(X,Y)}{\sqrt{Var(X)\ Var(Y)}}$$

and the denominator $\sqrt{Var(X)\ Var(Y)}$ is a non-negative quantity. Hence, the correlation is equal to the covariance multiplied by a positive constant. Notice that the correlation is defined only if the variances of X and Y are greater than 0.

EXERCISE 3.16. *Show that* $\mathrm{rank}(\mathcal{H}) = \mathrm{tr}(\mathcal{H}) = n - 1$, *where* $\mathcal{H} = \mathcal{I}_n - n^{-1}1_n1_n^\top$ *is the so-called centering matrix.*

The centering matrix $\mathcal{H}$ has dimension $n \times n$. and its diagonal elements are $h_{ii} = \frac{n-1}{n}$, $i = 1, ..., n$. Hence, $\mathrm{tr}(\mathcal{H}) = \sum_{i=1}^{n} h_{ii} = n\frac{n-1}{n} = n - 1$.

Notice that $\mathcal{HH} = (\mathcal{I}_n - n^{-1}1_n1_n^\top)(\mathcal{I}_n - n^{-1}1_n1_n^\top) = \mathcal{I}_n - 2n^{-1}1_n1_n^\top + n^{-2}n1_n1_n^\top = \mathcal{H}$. This means that the matrix $\mathcal{H}$ is idempotent which implies that its eigenvalues, λ_i, $i = 1, \ldots, n$ can be only 0 or 1, see Exercise 2.6. The rank of the centering matrix $\mathcal{H}$ is equal to the number of nonzero eigenvalues, i.e., to the number of eigenvalues which are equal to 1. Now, using the fact that the trace of a matrix is equal to the sum of its eigenvalues, we can write

$$\mathrm{rank}(\mathcal{H}) = \sum_{i=1}^{n} \lambda_i = \mathrm{tr}(\mathcal{H}) = n - 1.$$

EXERCISE 3.17. *Define* $\mathcal{X}_* = \mathcal{HXD}^{-1/2}$, *where* $\mathcal{X}$ *is a* $(n \times p)$ *matrix,* $\mathcal{H}$ *is the centering matrix, and* $\mathcal{D}^{-1/2} = diag(s_{11}^{-1/2}, \ldots, s_{pp}^{-1/2})$. *Show that* $\mathcal{X}_*$ *is the standardized data matrix, i.e.,* $\overline{x}_* = 0_p$ *and* $\mathcal{S}_{\mathcal{X}_*} = \mathcal{R}_{\mathcal{X}}$, *the correlation matrix of* $\mathcal{X}$.

The vector of means, $\overline{x}_*$, can be expressed as

$$\begin{aligned}
\overline{x}_* &= 1_n^\top \mathcal{X}_*/n \\
&= 1_n^\top \mathcal{HXD}^{-1/2}/n \\
&= 1_n^\top (\mathcal{I}_n - n^{-1}1_n1_n^\top)\mathcal{XD}^{-1/2}/n \\
&= (1_n^\top - 1_n^\top n^{-1}1_n1_n^\top)\mathcal{XD}^{-1/2}/n \\
&= (1_n^\top - 1_n^\top)\mathcal{XD}^{-1/2}/n \\
&= 0_p.
\end{aligned}$$

Similarly, we have for the variance matrix, $\mathcal{S}_{\mathcal{X}_*}$, of $\mathcal{X}_*$ that

$$S_{\mathcal{X}_*} = \mathrm{Var}(\mathcal{H}\mathcal{X}\mathcal{D}^{-1/2})$$
$$= \mathrm{Var}(\mathcal{I}_n\mathcal{X}\mathcal{D}^{-1/2}) + \mathrm{Var}(n^{-1}1_n1_n^\top)\mathcal{X}\mathcal{D}^{-1/2})$$
$$= \mathcal{D}^{-1/2}\,\mathrm{Var}(\mathcal{X})\mathcal{D}^{-1/2})$$
$$= \mathcal{D}^{-1/2}S_\mathcal{X}\mathcal{D}^{-1/2})$$
$$= \mathcal{R}_\mathcal{X}.$$

Closer inspection of the above formulas reveals that multiplication from the left by the centering matrix $\mathcal{H}$ subtracts the columns means whereas the multiplication from the right by the matrix $\mathcal{D}^{-1/2}$ divides each column by the estimated standard deviation.

EXERCISE 3.18. *Compute for the pullover data (Table A.6) the regression of X_1 on X_2, X_3 and of X_1 on X_2, X_4. Which one has the better coefficient of determination?*

Performing the calculation in any statistical software leads to coefficients of determination $r^2_{X_2,X_3} = 0.8275$ and $r^2_{X_2,X_4} = 0.4207$. A better coefficient of determination is achieved by the regression of sales (X_1) on price and assistant hours $(X_2$ and $X_3)$.

From the following output for dependency on price and assistant hours, we see that the parameter corresponding to assistant hours (X_3), denoted by b[2,] in the computer output, is highly significant.

```
A  N  O  V  A              SS       df     MSS        F-test    P-value
----------------------------------------------------------------------
Regression            8583.747     2    4291.874     16.799    0.0021
Residuals             1788.353     7     255.479
Total Variation      10372.100     9    1152.456


Multiple R       = 0.90971
R^2              = 0.82758
Adjusted R^2     = 0.77832
Standard Error   = 15.98371


PARAMETERS       Beta         SE       StandB    t-test    P-value
----------------------------------------------------------------------
b[ 0,]=        176.6919     36.5078    0.0000     4.840     0.0019
b[ 1,]=         -0.6013      0.3434   -0.2768    -1.751     0.1235
b[ 2,]=          0.5663      0.0994    0.9008     5.697     0.0007
```

Q SMSdeterpull

EXERCISE 3.19. *Compare for the pullover data the coefficient of determina-tion for the regression of X_1 on X_2, of X_1 on X_2, X_3 (Exercise 3.18) and of*

X_1 on X_2, X_3, X_4. *Observe that the coefficient of determination is increasing with the number of predictor variables. Is this always the case?*

The coefficients of determination for the models are: $r^2_{X_2} = 0.02808$, $r^2_{X_2,X_3} = 0.82758$, and $r^2_{X_2,X_3,X_4} = 0.90671$.

The coefficient of determination is defined as the ratio of the explained and total variation. Including more variables in the model has not any effect on the total variation (of the dependent variable) and can not decrease the explained variation. Hence, adding more variables cannot decrease the coefficient of determination.

◻ SMSdete2pull

EXERCISE 3.20. *A company decides to compare the effect of three marketing strategies*

1. *advertisement in local newspaper,*

2. *presence of sales assistant,*

3. *special presentation in shop windows,*

on the sales of their portfolio in 30 shops. The 30 shops were divided into 3 groups of 10 shops. The sales using the strategies 1, 2, and 3 were $y_1 = (9, 11, 10, 12, 7, 11, 12, 10, 11, 13)^\top$, $y_2 = (10, 15, 11, 15, 15, 13, 7, 15, 13, 10)^\top$, and $y_3 = (18, 14, 17, 9, 14, 17, 16, 14, 17, 15)^\top$, respectively. Define x_i as the index of the shop, i.e., $x_i = i, i = 1, 2, \ldots, 30$. Using this notation, the null hypothesis corresponds to a constant regression line, $EY = \mu$. What does the alternative hypothesis involving a regression curve look like?

There are $p = 3$ factors and $n = 30$ observations in the data set. The company wants to know whether all three marketing strategies have the same effect or whether there is a difference. The null hypothesis is $H_0 : \mu_1 = \mu_2 = \mu_3$ and the alternative hypothesis is $H_1 : \mu_l \neq \mu_{l'}$ for some l and l'. The standard approach to this problem is the analysis of variance (ANOVA) technique which leads to an F-test.

In this exercise, we use an alternative and in fact equivalent approach based on the regression model. The null hypothesis can be tested in a regression model that has explanatory variables defined as $z_{2i} = \boldsymbol{I}(x_i \in (11, 20))$ and $z_{3i} = \boldsymbol{I}(x_i \in (21, 30))$. These two variables now allow to describe the difference in sales due to the marketing strategies.

The regression model can be written as

$$\begin{pmatrix} x_1 \\ x_2 \\ x_3 \end{pmatrix} = \begin{pmatrix} 1_{10} & 0_{10} & 0_{10} \\ 1_{10} & 1_{10} & 0_{10} \\ 1_{10} & 0_{10} & 1_{10} \end{pmatrix} \begin{pmatrix} \beta_1 \\ \beta_2 \\ \beta_3 \end{pmatrix} + \varepsilon.$$

Here, the regression curve corresponding to the alternative hypothesis in the ANOVA model looks like three horizontal lines, each of them corresponding to one marketing strategy.

The F-test for testing the null hypothesis $H_0 : \beta_2 = \beta_3 = 0$ corresponds to the test of the null hypothesis that the effect of the three marketing strategies is the same.

```
A  N  O  V  A            SS      df    MSS      F-test   P-value
---------------------------------------------------------------
Regression            102.600    2    51.300    8.783    0.0012
Residuals             157.700   27     5.841
Total Variation       260.300   29     8.976


Multiple R      = 0.62782
R^2             = 0.39416
Adjusted R^2    = 0.34928
Standard Error  = 2.41676


PARAMETERS       Beta        SE        StandB    t-test   P-value
----------------------------------------------------------------
b[ 0,]=        10.6000     0.7642     0.0000    13.870    0.0000
b[ 1,]=         1.8000     1.0808     0.2881     1.665    0.1074
b[ 2,]=         4.5000     1.0808     0.7202     4.164    0.0003
```

◘ SMSanovapull

The above computer output shows that the value of the F-statistic for our null hypothesis is 8.783, the corresponding p-value is smaller than 0.05. Thus, on the usual confidence level 95%, the null hypothesis is rejected.

The computer output also contains the mean sales of all three marketing strategies. The mean sales for the first marketing strategy were 10.6, for the second strategy $10.6 + 1.8 = 12.4$, and for the third strategy $10.6 + 4.5 = 15.1$.

EXERCISE 3.21. *Perform the test in Exercise 3.20 for the shop example with a 0.99 significance level. Do you still reject the hypothesis of equal marketing strategies?*

From the p-value (0.0012), we can immediately tell that the null hypothesis is rejected also on the 0.99 significance level.

EXERCISE 3.22. *Consider the ANOVA problem again. Establish the constraint matrix A for testing $H_0 : \mu_1 = \mu_2$ against $H_1 : \mu_1 \neq \mu_2$ and test the hypothesis.*

Using the constraint matrix $A = (1, -1, 0)^\top$, the null hypothesis $H_0 : \mu_1 = \mu_2$ hypothesis can be expressed in the following form: $H_0 : A^\top \mu = 0$. Formally, the test can be performed by comparing the sum of squares under the null and alternative hypothesis. Under the null hypothesis, the F-statistics

$$F = \frac{\{||y - \mathcal{X}\hat{\beta}_{H_0}||^2 - ||y - \mathcal{X}\hat{\beta}_{H_1}||^2\}/r}{||y - \mathcal{X}\hat{\beta}_{H_1}||^2/(n-r)} \tag{3.6}$$

has F-distribution with r and $n - r$ degrees of freedom, where r denotes the difference in the number of parameters of the null and alternative linear model.

In our testing problem, the F-statistics, 2.77, is smaller than the appropriate critical value $F_{0.95;1,27} = 4.21$. The null hypothesis is not rejected at a 0.95 significance level and we can say that the difference between the effect of the first and the second marketing strategy is not statistically significant.

EXERCISE 3.23. *The linear model can be written as*

$$Y = \mathcal{X}\beta + \varepsilon, \tag{3.7}$$

where $\mathcal{X}$ is of full rank and ε are the random errors. Show that the least squares solution,

$$\hat{\beta} = \arg\min_{\beta} (Y - \mathcal{X}\beta)^\top (Y - \mathcal{X}\beta) = \arg\min_{\beta} \varepsilon^\top \varepsilon, \tag{3.8}$$

can be expressed as $\hat{\beta} = (\mathcal{X}^\top \mathcal{X})^{-1} \mathcal{X}^\top Y$.

We define the function $f(\beta) = (Y - \mathcal{X}\beta)^\top (Y - \mathcal{X}\beta)$, i.e.,

$$f(\beta) = Y^\top Y - 2\beta^\top \mathcal{X}^\top Y + \beta^\top \mathcal{X}^\top \mathcal{X}\beta.$$

The minimum of $f(\beta)$ can be found by searching for the zero of its derivative

$$\frac{\partial f(\beta)}{\partial \beta} = \frac{\partial Y^\top Y - 2\beta^\top \mathcal{X}^\top Y + \beta^\top \mathcal{X}^\top \mathcal{X}\beta}{\partial \beta} = -2\mathcal{X}^\top Y + 2\mathcal{X}^\top \mathcal{X}\beta = 0.$$

It follows that the solution, $\hat{\beta}$, has to satisfy $\hat{\beta} = (\mathcal{X}^\top \mathcal{X})^{-1} \mathcal{X}^\top Y$.

Let us now verify that we have found the minimum by calculating the second derivative of the function $f(\beta)$ in the point $\hat{\beta}$:

$$\frac{\partial^2 f(\beta)}{\partial \beta \partial \beta^\top} = \frac{\partial(-2\mathcal{X}^\top Y + 2\mathcal{X}^\top \mathcal{X}\beta)}{\partial \beta} = 2\mathcal{X}^\top \mathcal{X}.$$

The matrix $\mathcal{X}$ has full rank, therefore the matrix $\mathcal{X}^\top \mathcal{X}$ is positive definite and, hence, $\hat{\beta}$ is indeed the location of the minimum of the residual square function $f(\beta)$.

EXERCISE 3.24. *Consider the linear model* $Y = \mathcal{X}\beta + \varepsilon$ *where the estimator* $\hat{\beta} = \arg\min_{\beta} \varepsilon^{\top}\varepsilon$ *is subject to the linear constraint* $A\hat{\beta} = a$, *where* $A(q \times p), (q \leq p)$ *is of rank* q *and* a *is of dimension* $(q \times 1)$.

Show that

$$\hat{\beta} = \hat{\beta}_{OLS} - (\mathcal{X}^{\top}\mathcal{X})^{-1}A^{\top}\left\{A(\mathcal{X}^{\top}\mathcal{X})^{-1}A^{\top}\right\}^{-1}\left(A\hat{\beta}_{OLS} - a\right)$$

where $\hat{\beta}_{OLS} = (\mathcal{X}^{\top}\mathcal{X})^{-1}\mathcal{X}^{\top}Y$ *is the unconstrained (ordinary) least squares estimator.*

Similarly, as in the previous exercise, we define

$$f(\beta, \lambda) = (Y - \mathcal{X}\beta)^{\top}(Y - \mathcal{X}\beta) - \lambda^{\top}(A\beta - a),$$

where $\lambda \in \mathbb{R}^q$ and solve the system of equations:

$$\frac{\partial f(\beta, \lambda)}{\partial \beta} = 0$$

$$\frac{\partial f(\beta, \lambda)}{\partial \lambda} = 0.$$

Evaluating the derivatives, we obtain the system of equations:

$$\frac{\partial f(\beta, \lambda)}{\partial \beta} = -2\mathcal{X}^{\top}Y + 2\mathcal{X}^{\top}\mathcal{X}\hat{\beta} - A^{\top}\hat{\lambda} = 0, \tag{3.9}$$

$$\frac{\partial f(\beta, \lambda)}{\partial \lambda} = -(A\hat{\beta} - a) = 0.$$

Rearranging (3.9) with respect to $\hat{\beta}$ leads to

$$\hat{\beta} = (\mathcal{X}^{\top}\mathcal{X})^{-1}\mathcal{X}^{\top}Y + \frac{1}{2}(\mathcal{X}^{\top}\mathcal{X})^{-1}A^{\top}\hat{\lambda}, \tag{3.10}$$

$$A\hat{\beta} = A\hat{\beta}_{OLS} + \frac{1}{2}A(\mathcal{X}^{\top}\mathcal{X})^{-1}A^{\top}\hat{\lambda}. \tag{3.11}$$

Next, rearranging (3.11) with respect to $\hat{\lambda}$ implies that

$$\hat{\lambda} = 2\left\{A(\mathcal{X}^{\top}\mathcal{X})^{-1}A^{\top}\right\}^{-1}\left(a - A\hat{\beta}_{OLS}\right). \tag{3.12}$$

Plugging (3.12) in (3.10) finally leads to the desired formula

$$\hat{\beta} = \hat{\beta}_{OLS} - (\mathcal{X}^{\top}\mathcal{X})^{-1}A^{\top}\left\{A(\mathcal{X}^{\top}\mathcal{X})^{-1}A^{\top}\right\}^{-1}\left(A\hat{\beta}_{OLS} - a\right).$$

EXERCISE 3.25. *Compute the covariance matrix* $S = \text{Cov}(\mathcal{X})$ *where* $\mathcal{X}$ *denotes the matrix of observations on the counterfeit bank notes. Make a spectral decomposition of* S. *Why are all of the eigenvalues positive?*

The covariance matrix of all 6 variables in the bank notes data set is

$$
V = \begin{pmatrix}
0.142 & 0.031 & 0.023 & -0.103 & -0.019 & 0.084 \\
0.031 & 0.130 & 0.108 & 0.216 & 0.105 & -0.209 \\
0.023 & 0.108 & 0.163 & 0.284 & 0.130 & -0.240 \\
-0.103 & 0.216 & 0.284 & 2.087 & 0.165 & -1.037 \\
-0.019 & 0.105 & 0.130 & 0.165 & 0.645 & -0.550 \\
0.084 & -0.209 & -0.240 & -1.037 & -0.550 & 1.328
\end{pmatrix}.
$$

The eigenvalues of V, $(0.195, 0.085, 0.036, 3.000, 0.936, 0.243)$ are, indeed, all positive.

In general, the eigenvalues of any variance matrix are always nonnegative. This property can be demonstrated by realizing that, for arbitrary vector a, we have for the linear combination $\mathcal{X}a$ that its variance $\mathrm{Var}(\mathcal{X}a) = a^\top \mathrm{Var}\,\mathcal{X}a \geq 0$. This implies that any variance matrix is positive semidefinite and, hence, it cannot have any negative eigenvalues.

⊙ SMScovbank

EXERCISE 3.26. *Compute the covariance of the counterfeit notes after they are linearly transformed by the vector* $a = (1, 1, 1, 1, 1, 1)^\top$.

The variance of the sum of all lengths for the counterfeit variables is $\mathrm{Var}(\mathcal{X}_f a) = 1.7423$.

As explained in Exercise 3.25, the relation $\mathrm{Var}(\mathcal{X}_f a) = a^\top \mathrm{Var}(\mathcal{X}_f)a$ and the nonnegativity of the variance imply the positive semidefiniteness of the variance matrix $\mathrm{Var}\,\mathcal{X}_f$.

⊙ SMScovbank

Multivariate Distributions

Individuals vary, but percentages remain constant. So says the statistician.
Sherlock Holmes in "The Sign of Four"

A random vector is a vector of random variables. A random vector $X \in \mathbb{R}^p$ has a multivariate cumulative distribution function (cdf) and a multivariate probability density function (pdf). They are defined as:

$$
\begin{aligned}
F_X(x) &= P(X \leq x) \\
&= P(X_1 \leq x_1, X_2 \leq x_2, \ldots, X_p \leq x_p) \\
&= \int_{-\infty}^{\infty} \cdots \int_{-\infty}^{\infty} f_X(x_1, x_2, \ldots, x_p) dx_1 dx_2 \ldots dx_p,
\end{aligned}
$$

and if the cdf $F_X(.)$ is differentiable, the pdf $f_X(.)$ is

$$
f_X(x) = \frac{\partial^p F(x)}{\partial x_1 \ldots \partial x_p}.
$$

Important features that can be extracted from $F_X(.)$ and $f_X(.)$ are the mutual dependencies of the elements of X, moments, and multivariate tail behavior.

In the multivariate context the first moment, the expected value, is a vector EX of the same dimension p as X. The generalization of the one-dimensional variance to the multivariate case leads to the $(p \times p)$ covariance matrix $\Sigma = \text{Var}(x)$ containing the covariances of all pairs of components of X.

Another important feature that needs to be considered is the behavior of a random vector after it is (nonlinearly) transformed and the conditional distribution given other elements of the random vector.

In this chapter, we discuss a variety of exercises on moment and dependence calculations. We also study in depth the characteristics of the cdf and pdf of the transformed random vectors. In particular, we present the CLT of transformed statistics and calculate several examples for conditional distributions.

EXERCISE 4.1. *Assume that the random vector Y has the following normal distribution: $Y \sim N_p(0, \mathcal{I})$. Transform it to create $X \sim N(\mu, \Sigma)$ with mean $\mu = (3, 2)^\top$ and $\Sigma = \begin{pmatrix} 1 & -1.5 \\ -1.5 & 4 \end{pmatrix}$. How would you implement the resulting formula on a computer?*

Let us consider the transformation

$$X = \mu + \Sigma^{1/2} Y.$$

We know that a linearly transformed normally distributed random vector is again normally distributed. From the rules for the mean and variance matrix of the linearly transformed random variable we know that $EX = \mu + \Sigma^{1/2} EY = \mu$ and $\text{Var } X = \Sigma^{1/2} \text{Var } Y (\Sigma^{1/2})^\top = \Sigma$.

On a computer, the square root matrix $\Sigma^{1/2}$ can be easily calculated from Σ using spectral decomposition:

$$\Sigma^{1/2} = \begin{pmatrix} -0.38 & 0.92 \\ 0.92 & 0.38 \end{pmatrix} \begin{pmatrix} 4.62 & 0 \\ 0 & 0.38 \end{pmatrix}^{1/2} \begin{pmatrix} -0.38 & 0.92 \\ 0.92 & 0.38 \end{pmatrix} = \begin{pmatrix} 0.84 & -0.54 \\ -0.54 & 1.95 \end{pmatrix}.$$

One then applies the above formula that linearly transforms Y into X.

EXERCISE 4.2. *Prove that if $X \sim N_p(\mu, \Sigma)$, then the variable $U = (X - \mu)^\top \Sigma^{-1}(X - \mu)$ has a χ_p^2 distribution.*

For a random vector $X \sim N_p(\mu, \Sigma)$ such that $\Sigma > 0$, the p-dimensional random vector

$$(Y_1, \ldots, Y_p)^\top = Y = \Sigma^{-1/2}(X - \mu)$$

has a multivariate normal distribution with mean vector $EY = 0_p$ and covariance matrix $\text{Var}(Y) = \mathcal{I}_p$, see Härdle & Simar (2003, theorem 4.5).

The linear transformation $\Sigma^{-1/2}(X - \mu)$ is called the Mahalanobis transformation.

Hence, the random variable

$$U = (X - \mu)^\top \Sigma^{-1}(X - \mu) = Y^\top Y = \sum_{i=1}^p Y_i^2$$

is a sum of squares of independent random variables with standard normal distribution and therefore it has the χ_p^2 distribution.

EXERCISE 4.3. *Suppose that X has mean zero and covariance $\Sigma = \left(\begin{smallmatrix} 1 & 0 \\ 0 & 2 \end{smallmatrix}\right)$. Let $Y = X_1 + X_2$. Write Y as a linear transformation, i.e., find the transformation matrix $\mathcal{A}$. Then compute $Var(Y)$.*

Clearly,

$$Y = X_1 + X_2 = \mathcal{A}X = (1,1) \begin{pmatrix} X_1 \\ X_2 \end{pmatrix}$$

and $Var(\mathcal{A}X) = E\{(\mathcal{A}X - E\mathcal{A}X)(\mathcal{A}X - E\mathcal{A}X)^\top\} = \mathcal{A}\{E(X - EX)(X - EX)^\top\}\mathcal{A}^\top = \mathcal{A}\,Var(X)\mathcal{A}^\top$.

Hence,

$$Var(Y) = \mathcal{A}\Sigma\mathcal{A}^\top = (1,1)\,\Sigma \begin{pmatrix} 1 \\ 1 \end{pmatrix} = (1,1) \begin{pmatrix} 1 & 0 \\ 0 & 2 \end{pmatrix} \begin{pmatrix} 1 \\ 1 \end{pmatrix} = 3$$

Another possibility is to write

$$Var(Y) = Var\,(X_1 + X_2) = Var(X_1) + 2\,\mathrm{Cov}(X_1, X_2) + Var(X_2) = 3.$$

EXERCISE 4.4. *Calculate the mean and the variance of the estimator $\hat{\beta} = (\mathcal{X}^\top\mathcal{X})^{-1}\mathcal{X}^\top Y$ in a linear model $Y = \mathcal{X}\beta + \varepsilon$, $E\varepsilon = 0_n$, $Var(\varepsilon) = \sigma^2\mathcal{I}_n$.*

The estimate $\hat{\beta} = (\mathcal{X}^\top\mathcal{X})^{-1}\mathcal{X}^\top Y$ of the unknown parameter β in the linear model has been derived in Exercise 3.23. It follows that

$$E\hat{\beta} = (\mathcal{X}^\top\mathcal{X})^{-1}\mathcal{X}^\top EY = (\mathcal{X}^\top\mathcal{X})^{-1}\mathcal{X}^\top(\mathcal{X}\beta + E\varepsilon) = \beta$$

since we assume that $E\varepsilon = 0_n$.

For the variance we have

$$\begin{aligned}
Var\,\hat{\beta} &= Var\{(\mathcal{X}^\top\mathcal{X})^{-1}\mathcal{X}^\top Y\} \\
&= (\mathcal{X}^\top\mathcal{X})^{-1}\mathcal{X}^\top\,Var(Y)\mathcal{X}(\mathcal{X}^\top\mathcal{X})^{-1} \\
&= (\mathcal{X}^\top\mathcal{X})^{-1}\mathcal{X}^\top\sigma^2\mathcal{I}_n\mathcal{X}(\mathcal{X}^\top\mathcal{X})^{-1} \\
&= \sigma^2(\mathcal{X}^\top\mathcal{X})^{-1},
\end{aligned}$$

where we used the assumption $Var(Y) = Var(\varepsilon) = \sigma^2\mathcal{I}_n$.

EXERCISE 4.5. *Compute the conditional moments $E(X_2 \mid x_1)$ and $E(X_1 \mid x_2)$ for the two-dimensional pdf*

$$f(x_1, x_2) = \begin{cases} \frac{1}{2}x_1 + \frac{3}{2}x_2 & 0 \le x_1, x_2 \le 1 \\ 0 & otherwise \end{cases}$$

The marginal densities of X_1 and X_2, for $0 \leq x_1, x_2 \leq 1$, are

$$f_{X_1}(x_1) = \int_0^1 f(x_1, x_2) dx_2 = \left[\frac{1}{2} x_1 x_2 + \frac{3}{4} x_2^2 \right]_0^1 = \frac{1}{2} x_1 + \frac{3}{4}$$

and

$$f_{X_2}(x_2) = \int_0^1 f(x_1, x_2) dx_1 = \left[\frac{1}{4} x_1^2 + \frac{3}{2} x_1 x_2 \right]_0^1 = \frac{1}{4} + \frac{3}{2} x_2.$$

Now, the conditional expectations, for $0 \leq x_1, x_2 \leq 1$, can be calculated as follows

$$E(X_2 | X_1 = x_1) = \int_0^1 x_2 f(x_2 | x_1) dx_2$$

$$= \int_0^1 x_2 \frac{f(x_1, x_2)}{f_{X_1}(x_1)} dx_2$$

$$= \int_0^1 x_2 \left(\frac{\frac{1}{2} x_1 + \frac{3}{2} x_2}{\frac{1}{2} x_1 + \frac{3}{4}} \right) dx_2$$

$$= \left[\frac{\frac{x_1 x_2^2}{4} + \frac{x_2^3}{2}}{\frac{3}{4} + \frac{x_1}{2}} \right]_0^1$$

$$= \frac{x_1 + 2}{3 + 2 x_1}$$

and

$$E(X_1 | X_2 = x_2) = \int_0^1 x_1 f(x_1 | x_2) dx_1$$

$$= \int_0^1 x_1 \frac{f(x_1, x_2)}{f_{X_2}(x_2)} dx_1$$

$$= \int x_1 \left(\frac{\frac{1}{2} x_1 + \frac{3}{2} x_2}{\frac{3}{2} x_2 + \frac{1}{4}} \right) dx_1$$

$$= \left[\frac{\frac{x_1^3}{6} + \frac{3 x_1^2 x_2}{4}}{\frac{1}{4} + \frac{3 x_2}{2}} \right]_0^1$$

$$= \frac{2 + 9 x_2}{3 + 18 x_2}.$$

EXERCISE 4.6. *Prove that* $EX_2 = E\{E(X_2 | X_1)\}$, *where* $E(X_2 | X_1)$ *is the conditional expectation of* X_2 *given* X_2.

Since $E(X_2 | X_1 = x_1)$ is a function of x_1, it is clear that $E(X_2 | X_1)$ is a random vector (function of random vector X_1).

Assume that the random vector $X = (X_1, X_2)^\top$ has the density $f(x_1, x_2)$. Then

$$E\{E(X_2|X_1)\} = \int \left\{ \int x_2 f(x_2|x_1) dx_2 \right\} f(x_1) dx_1$$

$$= \int \left\{ \int x_2 \frac{f(x_2, x_1)}{f(x_1)} dx_2 \right\} f(x_1) dx_1 = \int \int x_2 f(x_2, x_1) dx_2 dx_1$$

$$= EX_2.$$

EXERCISE 4.7. *Prove that*

$$\mathrm{Var}(X_2) = E\{\mathrm{Var}(X_2|X_1)\} + \mathrm{Var}\{E(X_2|X_1)\}. \tag{4.1}$$

Hint: Note that $\mathrm{Var}\{E(X_2|X_1)\} = E\{E(X_2|X_1) E(X_2^\top|X_1)\} - E(X_2) E(X_2^\top)$ *and that* $E\{\mathrm{Var}(X_2|X_1)\} = E\{E(X_2 X_2^\top|X_1) - E(X_2|X_1) E(X_2^\top|X_1)\}$.

Let us start with the right-hand side of the relation (4.1):

$$E\{\mathrm{Var}(X_2|X_1)\} + \mathrm{Var}\{E(X_2|X_1)\}$$
$$= E\{E(X_2 X_2^\top|X_1) - E(X_2|X_1)E(X_2^\top|X_1)\} + E\{E(X_2|X_1)E(X_2^\top|X_1)\}$$
$$\quad - E(X_2)E(X_2^\top)$$
$$= E(X_2 X_2^\top) - E(X_2)E(X_2^\top)$$
$$= \mathrm{Var}(X_2).$$

EXERCISE 4.8. *Compute the pdf of the random vector* $Y = \mathcal{A}X$ *with* $\mathcal{A} = \begin{pmatrix} 1 & 1 \\ 1 & -1 \end{pmatrix}$ *for the random vector* X *with the pdf:*

$$f_X(x) = f_X(x_1, x_2) = \begin{cases} \frac{1}{2}x_1 + \frac{3}{2}x_2 & 0 \le x_1, x_2 \le 1 \\ 0 & \text{otherwise.} \end{cases}$$

The pdf of Y is given by

$$f_Y(y) = \mathrm{abs}(|\mathcal{J}|) f_X\{u(y)\},$$

where $u(.)$ is the inverse transformation, i.e., $X = u(Y)$, and where $\mathcal{J}$ is the Jacobian of $u(.)$. In this case, $X = u(Y) = \mathcal{A}^{-1}Y = \mathcal{J}Y$.

We solve $y_1 = x_1 + x_2$ and $y_2 = x_1 - x_2$ for x_1 and x_2:

$$x_1 = u_1(y_1, y_2) = (y_1 + y_2)/2$$
$$x_2 = u_2(y_1, y_2) = (y_1 - y_2)/2$$

and it follows that the Jacobian of $u(.)$ is

$$J = \begin{pmatrix} \frac{\partial u_1(y)}{\partial y_1} & \frac{\partial u_1(y)}{\partial y_2} \\ \frac{\partial u_2(y)}{\partial y_1} & \frac{\partial u_2(y)}{\partial y_2} \end{pmatrix} = \begin{pmatrix} \frac{1}{2} & \frac{1}{2} \\ \frac{1}{2} & -\frac{1}{2} \end{pmatrix}.$$

Next, $|J| = -\frac{1}{2}$ and $\text{abs}(|J|) = \frac{1}{2}$ and we obtain the density of the transformed random vector Y,

$$f_Y(y) = \frac{1}{2} f_X \{u(y)\} = \frac{1}{2} f_X \left\{ \begin{pmatrix} \frac{1}{2} & \frac{1}{2} \\ \frac{1}{2} & -\frac{1}{2} \end{pmatrix} \begin{pmatrix} y_1 \\ y_2 \end{pmatrix} \right\}$$

$$= \frac{1}{2} f_X \left\{ \frac{1}{2}(y_1 + y_2), \frac{1}{2}(y_1 - y_2) \right\}$$

for $0 \le u_1(y_1, y_2), u_2(y_1, y_2) \le 1$ and $f_Y(y) = 0$ otherwise.

Plugging in the pdf of X, we obtain

$$f_Y(y) = \begin{cases} \frac{1}{2} \left[\frac{1}{2} \{ \frac{1}{2}(y_1 + y_2) \} + \frac{3}{2} \{ \frac{1}{2}(y_1 - y_2) \} \right] & 0 \le y_1 \pm y_2 \le 2, \\ 0 & \text{otherwise} \end{cases}$$

and, using simple algebra to determine the region for which the pdf $f_Y(y)$ is greater than zero, we have finally

$$f_Y(y) = \begin{cases} \frac{1}{2}y_1 - \frac{1}{4}y_2 & 0 \le y_1 \le 2, \ |y_2| \le 1 - |1 - y_1| \\ 0 & \text{otherwise.} \end{cases}$$

EXERCISE 4.9. *Show that the function*

$$f_Y(y) = \begin{cases} \frac{1}{2}y_1 - \frac{1}{4}y_2 & 0 \le y_1 \le 2, \ |y_2| \le 1 - |1 - y_1| \\ 0 & \text{otherwise} \end{cases}$$

is a probability density function.

The area for which the above function is non-zero is plotted in Figure 4.1.

In order to verify that $f_Y(y)$ is a two-dimensional pdf, we have to check that it is nonnegative and that it integrates to 1.

It is easy to see that the function $f_Y(y)$ is nonnegative inside the square plotted in Figure 4.1 since $y_1 \ge 0$ and $y_1 \ge y_2$ implies that $y_1/2 - y_2/4 > 0$.

It remains to verify that the function $f_Y(y)$ integrates to one by calculating the integral

$$\int f_Y(y) dy$$

for which we easily obtain the following:

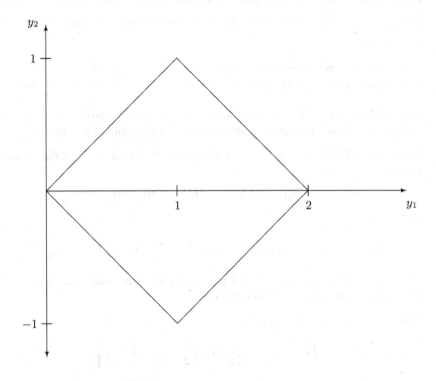

Fig. 4.1. The support of the pdf $f_Y(y_1, y_2)$ given in Exercise 4.9.

$$\iint f_Y(y_1, y_2) dy_2, y_1 = \int\limits_0^1 \int\limits_{-y_1}^{y_1} f_Y(y) dy_2 dy_1 + \int\limits_1^2 \int\limits_{y_1-2}^{2-y_1} f_Y(y) dy_2 dy_1$$

$$= \int\limits_0^1 \int\limits_{-y_1}^{y_1} \frac{1}{2}y_1 - \frac{1}{4}y_2 dy_2 dy_1 + \int\limits_1^2 \int\limits_{y_1-2}^{2-y_1} \frac{1}{2}y_1 - \frac{1}{4}y_2 dy_2 dy_1$$

$$= \int\limits_0^1 \left[\frac{1}{2}y_1 y_2 - \frac{1}{8}y_2^2\right]_{-y_1}^{y_1} dy_1 + \int\limits_1^2 \left[\frac{1}{2}y_1 y_2 - \frac{1}{8}y_2^2\right]_{y_1-2}^{2-y_1} dy_1$$

$$= \int\limits_0^1 y_1^2 dy_1 + \int\limits_1^2 -y_1^2 + 2y_1 dy_1$$

$$= \left[\frac{1}{3}y_1^3\right]_0^1 + \left[-\frac{1}{3}y_1^3 + y_1^2\right]_1^2 = \frac{1}{3} + \frac{2}{3} = 1.$$

EXERCISE 4.10. *Determine the distribution of the random vector* $Y = \mathcal{A}X$ *with* $\mathcal{A} = \begin{pmatrix} 1 & 1 \\ 1 & -1 \end{pmatrix}$, *where* $X = (X_1, X_2)^\top$ *has two-dimensional standard normal distribution.*

Show that the transformed random variables Y_1 *and* Y_2 *are independent. Give a geometrical interpretation of this result based on iso-distance curves.*

The random vector Y has a two-dimensional normal distribution since it is defined as a linear transformation of normally distributed random vector.

The normal distribution is fully determined by its mean and covariance matrix for which we have

$$EY = E\mathcal{A}X = \mathcal{A}EX = \mathcal{A}0_2 = 0_2$$

and

$$\mathrm{Var}(Y) = \mathrm{Var}(\mathcal{A}X) = \mathcal{A}\,\mathrm{Var}(X)\mathcal{A}^\top = \mathcal{A}\mathcal{I}_2\mathcal{A}^\top = \mathcal{A}\mathcal{A}^\top = \begin{pmatrix} 2 & 0 \\ 0 & 2 \end{pmatrix}.$$

Thus Y_1 and Y_2 are uncorrelated and, for jointly normal random variables, zero covariance implies independence.

The density of the random vector X,

$$f_X(x_1, x_2) = \frac{1}{2\pi} \exp\left\{\frac{1}{2}(x_1, x_2)\begin{pmatrix} x_1 \\ x_2 \end{pmatrix}\right\},$$

is obviously constant on circles with center in $(0,0)^\top$ since its value changes only when the value of the quadratic form $(x_1, x_2)\begin{pmatrix} x_1 \\ x_2 \end{pmatrix} = x_1^2 + x_2^2$ changes. We remark that a circle with diameter r is defined as a set of points $x = (x_1, x_2)^\top \in \mathbb{R}^2$ satisfying the equation $x_1^2 + x_2^2 = r^2$.

The density of the transformed random vector Y is also constant on the circles, but the distribution is more spread out. The transformation $Y = \mathcal{A}X$ corresponds to the rotation and then multiplication by factor $\sqrt{2}$.

EXERCISE 4.11. *Consider the Cauchy distribution which has no finite moment, so that the CLT cannot be applied. Simulate the distribution of* $\bar{x}$ *(for different* n's*). What can you expect for* $n \to \infty$? *Hint: The Cauchy distribution can be simulated by the quotient of two independent standard normally distributed random variables.*

For the Cauchy distribution, the distribution of $\bar{x}$ is the same as the distribution of X_1. Thus, the sample mean cannot be used for statistical inference.

In the simulations, you can observe that increasing the sample size doesn't improve the behavior of the sample mean as an estimator of the expected value.

EXERCISE 4.12. *A European car company has tested a new model and reports the consumption of gasoline (X_1) and oil (X_2). The expected consumption of gasoline is 8 liters per 100 km (μ_1) and the expected consumption of oil is 1 liter per 10.000 km (μ_2). The measured consumption of gasoline is 8.1 liters per 100 km ($\overline{x}_1$) and the measured consumption of oil is 1.1 liters per 10,000 km ($\overline{x}_2$). The asymptotic distribution of*

$$\sqrt{n}\left\{\begin{pmatrix}\overline{x}_1\\\overline{x}_2\end{pmatrix} - \begin{pmatrix}\mu_1\\\mu_2\end{pmatrix}\right\} \text{ is } N\left(\begin{pmatrix}0\\0\end{pmatrix}, \begin{pmatrix}0.1 & 0.05\\0.05 & 0.1\end{pmatrix}\right).$$

For the American market the basic measuring units are miles (1 mile $\approx$ 1.6 km) and gallons (1 gallon $\approx$ 3.8 liter). The consumptions of gasoline (Y_1) and oil (Y_2) are usually reported in miles per gallon. Can you express $\overline{y}_1$ and $\overline{y}_2$ in terms of $\overline{x}_1$ and $\overline{x}_2$? Recompute the asymptotic distribution for the American market!

The transformation of "liters per 100 km" to "miles per gallon" is given by the function

x liters per 100 km $= 1.6x/380$ gallons per mile $= 380/(1.6x)$ miles per gallon.

Similarly, we transform the oil consumption

x liters per 10000 km $= 38000/(1.6x)$ miles per gallon.

Thus, the transformation is given by the functions

$$f_1(x) = 380/(1.6x)$$
$$f_2(x) = 38000/(1.6x).$$

According to Härdle & Simar (2003, theorem 4.11), the asymptotic distribution is

$$\sqrt{n}\left\{\begin{pmatrix}f_1(\overline{x}_1)\\f_2(\overline{x}_2)\end{pmatrix} - \begin{pmatrix}f_1(\mu_1)\\f_2(\mu_2)\end{pmatrix}\right\} \sim N\left(\begin{pmatrix}0\\0\end{pmatrix}, \mathcal{D}^{\top}\begin{pmatrix}0.1 & 0.05\\0.05 & 0.1\end{pmatrix}\mathcal{D}\right),$$

where

$$D = \left(\frac{\partial f_j}{\partial x_i}\right)(x)\Big|_{x=\mu}$$

is the matrix of all partial derivatives.

In our example,

$$\mathcal{D} = \begin{pmatrix}-\frac{380}{1.6x_1^2} & 0\\0 & -\frac{38000}{1.6x_2^2}\end{pmatrix}\Bigg|_{x=\mu}$$

$$\doteq \begin{pmatrix}-\frac{380}{1.6\overline{x}_1^2} & 0\\0 & -\frac{38000}{1.6x_2^2}\end{pmatrix}$$

$$\doteq \begin{pmatrix}-3.62 & 0\\0 & -19628.10\end{pmatrix}.$$

Hence, the variance of the transformed random variable Y is given by

$$\Sigma_Y = \mathcal{D}^\top \begin{pmatrix} 0.1 & 0.05 \\ 0.05 & 0.1 \end{pmatrix} \mathcal{D}$$

$$= \begin{pmatrix} -3.62 & 0 \\ 0 & -19628.10 \end{pmatrix} \begin{pmatrix} 0.1 & 0.05 \\ 0.05 & 0.1 \end{pmatrix} \begin{pmatrix} -3.62 & 0 \\ 0 & -19628.10 \end{pmatrix}$$

$$= \begin{pmatrix} 1.31 & 3552.69 \\ 3552.69 & 38526230.96 \end{pmatrix}.$$

The average fuel consumption, transformed to American units of measurements is $\bar{y}_1 = 29.32$ miles per gallon and the transformed oil consumption is $\bar{y}_2 = 19628.10$. The asymptotic distribution is

$$\sqrt{n}\left\{ \begin{pmatrix} \bar{y}_1 \\ \bar{y}_2 \end{pmatrix} - \begin{pmatrix} f_1(\mu_1) \\ f_2(\mu_2) \end{pmatrix} \right\} \sim N\left(\begin{pmatrix} 0 \\ 0 \end{pmatrix}, \begin{pmatrix} 1.31 & 3552.69 \\ 3552.69 & 38526230.96 \end{pmatrix} \right).$$

EXERCISE 4.13. *Consider the pdf $f_X(x_1, x_2) = e^{-(x_1+x_2)}$, $x_1, x_2 > 0$ and let $U_1 = X_1 + X_2$ and $U_2 = X_1 - X_2$. Compute $f(u_1, u_2)$.*

For linear transformation

$$U = \mathcal{A}X = \begin{pmatrix} 1 & 1 \\ 1 & -1 \end{pmatrix} X,$$

the inverse transformation is $X = \mathcal{A}^{-1}U$, the Jacobian of the inverse transformation is $\mathcal{J} = \mathcal{A}^{-1}$ and, hence, the density of the transformed random vector is

$$f_U(u) = \text{abs}(|\mathcal{A}|^{-1})f_X(\mathcal{A}^{-1}u).$$

We have

$$|\mathcal{A}| = -2, \qquad \mathcal{A}^{-1} = -\frac{1}{2}\begin{pmatrix} -1 & -1 \\ -1 & 1 \end{pmatrix}$$

and it follows immediately that

$$f_U(u) = \frac{1}{2}f_X\left\{ \frac{1}{2}\begin{pmatrix} 1 & 1 \\ 1 & -1 \end{pmatrix} \begin{pmatrix} u_1 \\ u_2 \end{pmatrix} \right\}$$

$$= \frac{1}{2}\exp\left[-\left\{ \frac{1}{2}(u_1 + u_2) + \frac{1}{2}(u_1 - u_2) \right\} \right]$$

$$= \frac{1}{2}\exp(-u_1).$$

The support of the distribution has to be investigated carefully. The density of the random variable U_1 is nonzero only for $u_1 > 0$ since it is the sum of two positive random variables. The limits on U_2 are the following:

$$U_2 = X_1 - X_2 < X_1 + X_2 = U_1,$$
$$U_2 = -(X_1 - X_2) > -(X_2 + X_1) = -U_1.$$

We conclude that the pdf of the transformed random vector U is

$$f_U(u) = \begin{cases} \frac{1}{2}\exp(-u_1) & u_1 > 0, u_2 < |u_1|, \\ 0 & \text{otherwise.} \end{cases}$$

EXERCISE 4.14. *Consider the functions*

$$\begin{aligned} f_1(x_1, x_2) &= 4x_1 x_2 \exp(-x_1^2) & x_1, x_2 > 0, \\ f_2(x_1, x_2) &= 1 & 0 < x_1, x_2 < 1 \text{ and } x_1 + x_2 < 1 \\ f_3(x_1, x_2) &= \tfrac{1}{2}\exp(-x_1) & x_1 > |x_2|. \end{aligned}$$

Check whether they are pdfs and then compute $E(X)$, $\text{Var}(X)$, $E(X_1|X_2)$, $E(X_2|X_1)$, $\text{Var}(X_1|X_2)$, *and* $\text{Var}(X_2|X_1)$.

It is easy to see that the first function,

$$f_1(x_1, x_2) = 4x_1 x_2 \exp\{-x_1^2\}, \qquad x_1, x_2 > 0,$$

is not a probability density function. For any value of x_1, we can choose x_2 such that $f_1(x_1, x_2)$ is arbitrarily large on an infinite interval. Hence, it is clear that

$$\int_0^{+\infty} \int_0^{+\infty} f_1(x_1, x_2)dx_2 dx_1 = +\infty$$

and therefore the function $f_1(x_1, x_2)$ cannot be a pdf.

The second function,

$$f_2(x_1, x_2) = 1, \qquad 0 < x_1, x_2 < 1 \text{ and } x_1 + x_2 < 1,$$

is nonnegative and it obviously integrates to one. Hence, it is a probability density function. Notice that the function is symmetric in x_1 and x_2, it follows that $EX_1 = EX_2$ and $\text{Var } X_1 = \text{Var } X_2$.

For the expected value, we have

$$\begin{aligned} EX_1 &= \int_0^1 \int_0^{1-x_1} x_1 dx_2 dx_1 \\ &= \int_0^1 x_1(1 - x_1)dx_1 \\ &= \left[\frac{1}{2}x_1^2 - \frac{1}{3}x_1^3\right]_0^1 \\ &= \frac{1}{6}. \end{aligned}$$

We have already observed that $EX_1 = EX_2$ and, thus,

$$EX = \left(\frac{1}{6}, \frac{1}{6}\right)^{\top}.$$

The variances, $Var\, X_1 = Var\, X_2$, can be calculated as follows

$$\begin{aligned}
Var\, X_1 &= EX_1^2 - (EX_1^2) \\
&= \int_0^1 \int_0^{1-x_1} x_1^2 dx_2 dx_1 - \frac{1}{36} \\
&= \int_0^1 x_1^2(1 - x_1)dx_1 - \frac{1}{36} \\
&= \left[\frac{1}{3}x_1^3 - \frac{1}{4}x_1^4\right]_0^1 - \frac{1}{36} \\
&= \frac{1}{12} - \frac{1}{36} \\
&= \frac{1}{18}.
\end{aligned}$$

The covariance, $\mathrm{Cov}(X_1, X_2)$ is equal to

$$\begin{aligned}
\mathrm{Cov}(X_1, X_2) &= EX_1 X_2 - EX_1 EX_2 \\
&= \int_0^1 \int_0^{1-x_1} x_1 x_2 dx_2 dx_1 - \frac{1}{36} \\
&= \int_0^1 x_1 \left[\frac{x_2^2}{2}\right]_0^{1-x_1} dx_1 - \frac{1}{36} \\
&= \int_0^1 \frac{1}{2}x_1(1 - x_1)^2 dx_1 - \frac{1}{36} \\
&= \frac{1}{2}\int_0^1 x_1 - 2x_1^2 + x_1^3 dx_1 - \frac{1}{36} \\
&= \frac{1}{2}\left[\frac{1}{2}x_1^1 - \frac{2}{3}x_1^3 + \frac{1}{4}x_1^4\right]_0^1 - \frac{1}{36} \\
&= \frac{1}{4} - \frac{1}{3} + \frac{1}{8} - \frac{1}{36} \\
&= \frac{18 - 24 + 9 - 2}{72} \\
&= \frac{1}{72}.
\end{aligned}$$

The covariance matrix is

$$\mathrm{Var}(X) = \begin{pmatrix} \frac{1}{18} & \frac{1}{72} \\ \frac{1}{72} & \frac{1}{18} \end{pmatrix}.$$

The conditional expectations could be calculated by evaluating the appropriate integrals. However, in this case, the solution can be seen immediately. Clearly, the conditional distribution of X_2 given $X_1 = x_1$ is uniform on $(0, 1 - x_1)$. The expected value of uniform distribution is its center, i.e., $E(X_2|X_1 = x_1) = (1 - x_1)/2$. Due to the symmetry of the distribution, we have also that $E(X_1|X_2 = x_2) = (1 - x_2)/2$.

The conditional variances are also variances of uniform distributions:

$$Var(X_2|X_1 = x_1) = E(X_2^2|X_1 = x_1) - \{E(X_2|X_1 = x_1)\}^2$$
$$= \int_0^{1-x_1} \frac{x_2^2}{1 - x_1} dx_2 - \left(\frac{(1 - x_1)}{2}\right)^2$$
$$= \frac{1}{1 - x_1} \left[\frac{1}{3}x_2^3\right]_0^{1-x_1} - \frac{(1 - x_1)^2}{4}$$
$$= \frac{(1 - x_1)^2}{3} - \frac{(1 - x_1)^2}{4}$$
$$= \frac{(1 - x_1)^2}{12}.$$

Due to the symmetry, we have also that

$$Var(X_1|X_2 = x_2) = \frac{(1 - x_2)^2}{12}.$$

For the third function,

$$f_3(x_1, x_2) = \frac{1}{2} \exp\{-x_1\} \qquad x_1 > |x_2|,$$

we again start by verifying that it is a pdf. We have

$$\int_0^{+\infty} \int_{-x_1}^{x_1} f_3(x_1, x_2) dx_2 dx_1 = \int_0^{+\infty} \int_{-x_1}^{x_1} \frac{1}{2} \exp\{-x_1\} dx_2 dx_1$$
$$= \int_0^{+\infty} x_1 \exp\{-x_1\} dx_1$$
$$= 1.$$

Here, it is helpful to notice that the value of $f_3(x_1, x_2)$ is, for any value of x_1 symmetric around zero in x_2 and that the value of the pdf does not depend on x_2.

Notice that the conditional expected value of X_2 is finite since X_2 has bounded support for each value of X_1. From the symmetry, it follows that $E(X_2|X_1 = x_1) = 0$, this in turn implies that $EX_2 = E\{E(X_2|X_1)\} = 0$.

The fact that the value of the pdf does not depend on x_2 implies that the conditional distribution of X_2 given X_1 is uniform on the interval $(-x_1, x_1)$.

Looking at the above calculations for the variance of the uniform distribution, we can immediately write:

$$Var(X_2|X_1 = x_1) = \frac{(2x_1)^2}{12} = \frac{x_1^2}{3}.$$

In order to calculate the moments of X_1, we have to evaluate some integrals:

$$
\begin{aligned}
EX_1 &= \int_0^{+\infty} \int_{-x_1}^{x_1} \frac{1}{2} x_1 \exp\{-x_1\} dx_2 dx_1 \\
&= \int_0^{+\infty} x_1^2 \exp\{-x_1\} dx_1 \\
&= [x_1^2 \exp -x_1]_0^{+\infty} + \int_0^{+\infty} 2x_1 \exp\{-x_1\} dx_1 \\
&= [2x_1 \exp -x_1]_0^{+\infty} + 2 \int_0^{+\infty} \exp\{-x_1\} dx_1 \\
&= 2[-\exp\{-x_1\}]_0^{+\infty} = 2.
\end{aligned}
$$

Hence, the vector of expected values is $EX = (2,0)^\top$.

The variance of X_1 can be calculated similarly as the expected value

$$
\begin{aligned}
Var\, X_1 &= EX_1^2 - (EX_1)^2 \\
&= \int_0^{+\infty} \int_{-x_1}^{x_1} \frac{1}{2} x_1^2 \exp\{-x_1\} dx_2 dx_1 - 4 \\
&= \int_0^{+\infty} x_1^3 \exp\{-x_1\} dx_1 - 4 \\
&= \int_0^{+\infty} 3x_1^2 \exp\{-x_1\} dx_1 - 4 \\
&= 3EX_1 - 4 = 2.
\end{aligned}
$$

Now it is easy to calculate also the unconditional variance of X_2 since

$$Var(X_2) = E\{Var(X_2|X_1)\} + Var\{E(X_2|X_1)\} = E\left(\frac{X_1^2}{3}\right) = 2.$$

Notice that the symmetry of the pdf in x_2 implies that also the distribution of the random variable $X_1 X_2$ is symmetric around 0 and, hence, its expected value $EX_1 X_2 = 0$. It follows that

$$Cov(X_1, X_2) = EX_1 X_2 - EX_1 EX_2 = 0.$$

The variance matrix of the random vector X is $Var\, X = \begin{pmatrix} 2 & 0 \\ 0 & 2 \end{pmatrix}$.

It remains to investigate the conditional moments of X_1 given $X_2 = x_2$. The conditional density of X_1 given X_2 is

$$f_{X_1|X_2=x_2}(x_1) = \frac{f_3(x_1, x_2)}{f_{X_2}(x_2)} = \frac{\exp(-x_1)}{\int_{|x_2|}^{+\infty} \exp(-x_1)dx_1}$$

$$= \frac{\exp(-x_1)}{[-\exp(-x_1)]_{|x_2|}^{+\infty}} = \frac{\exp(-x_1)}{\exp(-|x_2|)},$$

for $x_1 > |x_2|$ and 0 otherwise.

The conditional expectation of X_1 can be calculated as

$$E(X_1|X_2 = x_2) = \int_{|x_2|}^{+\infty} x_1 f_{X_1|X_2=x_2}(x_1)dx_1$$

$$= \int_{|x_2|}^{+\infty} x_1 \frac{\exp(-x_1)}{\exp(-|x_2|)}dx_1$$

$$= \frac{1}{\exp(-|x_2|)} \int_{|x_2|}^{+\infty} x_1 \exp(-x_1)dx_1$$

$$= \frac{1}{\exp(-|x_2|)} \left\{ [x_1 \exp(-x_1)]_{|x_2|}^{+\infty} + \int_{|x_2|}^{+\infty} \exp(-x_1)dx_1 \right\}$$

$$= \frac{1}{\exp(-|x_2|)} \left\{ |x_2| \exp(-|x_2|) + \exp(-|x_2|) \right\}$$

$$= |x_2| + 1.$$

Finally, the conditional variance of X_1 given $X_2 = x_2$ is

$$Var(X_1|X_2 = x_2) = E(X_1^2|X_2 = x_2) - \{E(X_1|X_2 = x_2)\}^2$$

$$= \int_{|x_2|}^{+\infty} x_1^2 \frac{\exp(-x_1)}{\exp(-|x_2|)}dx_1 - (|x_2| + 1)^2$$

$$= \frac{1}{\exp(-|x_2|)} \int_{|x_2|}^{+\infty} x_1^2 \exp(-x_1)dx_1 - (|x_2| + 1)^2$$

$$= \frac{1}{\exp(-|x_2|)} \left\{ [-x_1^2 \exp(-x_1)]_{|x_2|}^{+\infty} + 2\int_{|x_2|}^{+\infty} x_1 \exp(-x_1)dx_1 \right\}$$

$$- (|x_2| + 1)^2$$

$$= \frac{1}{\exp(-|x_2|)} \left[|x_2|^2 \exp(-|x_2|) + 2\{|x_2| \exp(-|x_2|) \right.$$

$$\left. + \exp(-|x_2|)\} \right] - (|x_2| + 1)^2$$

$$= |x_2|^2 + 2|x_2| + 2 - (|x_2| + 1)^2$$

$$= 1.$$

EXERCISE 4.15. *Consider the pdf*

$$f(x_1, x_2) = \frac{3}{4}x_1^{-\frac{1}{2}}, \qquad 0 < x_1 < x_2 < 1.$$

Compute $P(X_1 < 0.25), P(X_2 < 0.25)$ *and* $P(X_2 < 0.25|X_1 < 0.25)$.

The probabilities can be expressed as integrals of the pdf as follows

$$
\begin{aligned}
P(X_1 < 0.25) &= \int_0^{0.25} \int_{x_1}^1 f(x_1, x_2) dx_2 dx_1 \\
&= \int_0^{0.25} \int_{x_1}^1 \frac{3}{4}x_1^{-\frac{1}{2}} dx_2 dx_1 \\
&= \frac{3}{4} \int_0^{0.25} x_1^{-\frac{1}{2}} [x_2]_{x_1}^1 dx_1 \\
&= \frac{3}{4} \int_0^{0.25} x_1^{-\frac{1}{2}} (1 - x_1) dx_1 \\
&= \frac{3}{4} \int_0^{0.25} x_1^{-\frac{1}{2}} - x_1^{\frac{1}{2}} dx_1 \\
&= \frac{3}{4} \left[2x_1^{\frac{1}{2}} - \frac{2}{3}x_1^{\frac{3}{2}} \right]_0^{0.25} \\
&= \frac{3}{4} \left(1 - \frac{1}{12} \right) = \frac{33}{48}.
\end{aligned}
$$

Similarly,

$$
\begin{aligned}
P(X_2 < 0.25) &= \int_0^{0.25} \int_0^{x_2} f(x_1, x_2) dx_1 dx_2 \\
&= \int_0^{0.25} \int_0^{x_2} \frac{3}{4}x_1^{-\frac{1}{2}} dx_1 dx_2 \\
&= \frac{3}{4} \int_0^{0.25} \left[2x_1^{\frac{1}{2}} \right]_0^{x_2} dx_2 \\
&= \int_0^{0.25} \frac{3}{2}x_2^{\frac{1}{2}} dx_2 \\
&= \left[x_2^{\frac{3}{2}} \right]_0^{0.25} = \frac{1}{8}.
\end{aligned}
$$

The conditional probability is defined as

$$P(X_2 < 0.25|X_1 < 0.25) = \frac{P(X_1 < 0.25, X_2 < 0.25)}{P(X_1 < 0.25)}.$$

It remains to calculate the probability in the numerator. Noticing that $P(X_1 > X_2) = 0$, we can write

$$P(X_2 < 0.25 | X_1 < 0.25) = \frac{P(X_1 < 0.25, X_2 < 0.25)}{P(X_1 < 0.25)}$$

$$= \frac{P(X_2 < 0.25)}{P(X_1 < 0.25)} = \frac{6}{33}.$$

EXERCISE 4.16. *Consider the pdf*

$$f(x_1, x_2) = \frac{1}{2\pi}, \qquad 0 < x_1 < 2\pi, 0 < x_2 < 1.$$

Let $U_1 = \sin X_1 \sqrt{-2 \log X_2}$ and $U_2 = \cos X_1 \sqrt{-2 \log X_2}$. Compute $f(u_1, u_2)$.

Notice that

$$U_1^2 + U_2^2 = -2 \log X_2 (\sin^2 X_1 + \cos^2 X_1) = -2 \log X_2$$

and

$$\frac{U_1}{U_2} = \frac{\sin X_1}{\cos X_1} = \tan X_1.$$

Hence, the inverse transformation is

$$X_1 = \arctan \frac{U_1}{U_2},$$

$$X_2 = \exp \left\{ -\frac{1}{2}(U_1^2 + U_2^2) \right\}.$$

Here, it is important to notice that this is not one-to-one transformation! The calculation has to be carried out very carefully.

In order to obtain a one-to-one transformation, we consider the conditional pdfs

$$f_{X|X_1 \in (\pi/2, 3\pi/2)}(x_1, x_2) = \frac{1}{\pi} I\{x_1 \in (\pi/2, 3\pi/2)\}$$

$$f_{X|X_1 \notin (\pi/2, 3\pi/2)}(x_1, x_2) = \frac{1}{\pi} I\{x_1 \notin (\pi/2, 3\pi/2)\}$$

which allow us to rewrite the pdf $f(.)$ as

$$f(x) = \begin{cases} \frac{1}{2} f_{X|X_1 \in (\pi/2, 3\pi/2)}(x) & \text{for } x_1 \in (\pi/2, 3\pi/2) \\ \frac{1}{2} f_{X|X_1 \notin (\pi/2, 3\pi/2)}(x) & \text{for } x_1 \notin (\pi/2, 3\pi/2) \end{cases}$$

since

$$\int_{x \in I} f(x) dx = P(X \in I)$$

$$= P\{X \in I | X_1 \in (\pi/2, 3\pi/2)\} P\{X_1 \in (\pi/2, 3\pi/2)\}$$
$$+ P\{X \in I | X_1 \notin (\pi/2, 3\pi/2)\} P\{X_1 \notin (\pi/2, 3\pi/2)\}$$
$$= \int_{x \in I} \left\{ f_{X|X_1 \in (\pi/2, 3\pi/2)}(x) \frac{1}{2} + f_{X|X_1 \notin (\pi/2, 3\pi/2)}(x) \frac{1}{2} \right\} dx$$
$$= \int_{x \in I} \frac{1}{2} \left\{ f_{X|X_1 \in (\pi/2, 3\pi/2)}(x) + f_{X|X_1 \notin (\pi/2, 3\pi/2)}(x) \right\} dx.$$

We transform each of the conditional pdfs separately and then combine the results into the pdf of the transformed random vector U. For the conditional pdf $f_{X|X_1 \in (\pi/2, 3\pi/2)}$, the Jacobian of the inverse transformation is given by

$$\mathcal{J} = \begin{pmatrix} \frac{u_2}{u_2^2 + u_1^2} & -\frac{u_1}{u_2^2 + u_1^2} \\ -u_1 \exp\left\{-\frac{1}{2}(u_1^2 + u_2^2)\right\} & -u_2 \exp\left\{-\frac{1}{2}(u_1^2 + u_2^2)\right\} \end{pmatrix}.$$

Plugging into the formula for the pdf of the transformed random variable, we obtain

$$f_{U|U_2<0}(u) = \text{abs}\,|\mathcal{J}| f_{X|X_1 \in (\pi/2, 3\pi/2)}(f_1(u), f_2(u))$$
$$= \text{abs} \left\{ \left(\frac{u_1^2}{u_2^2 + u_1^2} + \frac{u_2^2}{u_2^2 + u_1^2} \right) \exp\left\{ -\frac{1}{2}(u_1^2 + u_2^2) \right\} \right\} \frac{1}{\pi}$$
$$= \frac{1}{\pi} \exp\left\{ -\frac{1}{2}(u_1^2 + u_2^2) \right\}$$

for $u_1 \in \mathbb{R}, u_2 < 0$ and $f_{U|U_2<0}(u) = 0$ otherwise.

Similarly, it can be shown that

$$f_{U|U_2>0}(u) = \frac{1}{\pi} \exp\left\{ -\frac{1}{2}(u_1^2 + u_2^2) \right\}$$

for $u_1 \in \mathbb{R}, u_2 > 0$ and $f_{U|U_2<0}(u) = 0$ otherwise.

Combining the conditional pdfs $f_{U|U_2>0}(.)$ and $f_{U|U_2<0}(.)$, we obtain the (marginal) pdf of the transformed two-dimensional random vector

$$f_U(u) = f_{U|U_2<0}(u) P(U_2 < 0) + f_{U|U_2>0}(u) P(U_2 > 0)$$
$$= \frac{1}{2} \left\{ f_{U|U_2<0}(u) I(U_2 < 0) + f_{U|U_2>0}(u) I(U_2 > 0) \right\}$$
$$= \frac{1}{2} \left[\frac{1}{\pi} \exp\left\{ -\frac{1}{2}(u_1^2 + u_2^2) \right\} I(u_2 < 0) \right.$$
$$\left. + \frac{1}{\pi} \exp\left\{ -\frac{1}{2}(u_1^2 + u_2^2) \right\} I(u_2 > 0) \right]$$
$$= \frac{1}{2\pi} \exp\left\{ -\frac{1}{2}(u_1^2 + u_2^2) \right\}$$

for $u_1, u_2 \in \mathbb{R}$.

Notice that the pdf $f_U(.)$ defines a two-dimensional multinormal distribution with zero mean and identity variance matrix. This transformation is at the heart of the Box-Muller method to generate standard normal (pseudo) random numbers.

EXERCISE 4.17. *Consider* $f(x_1, x_2, x_3) = k(x_1 + x_2 x_3);\ 0 < x_1, x_2, x_3 < 1.$

a) *Determine* k *so that* f *is a valid pdf of* $(X_1, X_2, X_3) = X.$

b) *Compute the* (3×3) *matrix* Σ_X.

c) *Compute the* (2×2) *matrix of the conditional variance of* (X_2, X_3) *given* $X_1 = x_1.$

Ad a) We have to determine k for which

$$\int_0^1 \int_0^1 \int_0^1 f(x_1, x_2, x_3)\, dx_1\, dx_2\, dx_3 = 1.$$

Evaluating the integral leads to:

$$\int_0^1 \int_0^1 \int_0^1 f(x_1, x_2, x_3)\, dx_1\, dx_2\, dx_3$$
$$= \int_0^1 \int_0^1 \int_0^1 k(x_1 + x_2 x_3)\, dx_1\, dx_2\, dx_3$$
$$= k \int_0^1 \int_0^1 \left[\frac{1}{2}x_1^2 + x_1 x_2 x_3\right]_0^1 dx_2\, dx_3$$
$$= k \int_0^1 \int_0^1 \left(\frac{1}{2} + x_2 x_3\right) dx_2\, dx_3$$
$$= k \int_0^1 \left[\frac{1}{2}x_2 + \frac{1}{2}x_2^2 x_3\right]_0^1 dx_3$$
$$= k \int_0^1 \left(\frac{1}{2} + \frac{1}{2}x_3\right) dx_3$$
$$= k \left[\frac{1}{2}x_3 + \frac{1}{4}x_3^2\right]_0^1$$
$$= \frac{3}{4}k.$$

It follows that $k = 4/3$.

Ad b) For the expected values of X_i, $i = 1, \ldots, 3$, we have that

$$EX_1 = \int_0^1 \int_0^1 \int_0^1 x_1 f(x_1, x_2, x_3)\, dx_1\, dx_2\, dx_3$$

$$= \frac{4}{3} \int_0^1 \int_0^1 \int_0^1 (x_1^2 + x_1 x_2 x_3)\, dx_1\, dx_2\, dx_3$$

$$= \frac{11}{18}$$

$$EX_2 = \frac{4}{3} \int_0^1 \int_0^1 \int_0^1 (x_1 x_2 + x_2^2 x_3)\, dx_1\, dx_2\, dx_3$$

$$= \frac{5}{9}.$$

The pdf is symmetric in x_2 and x_3, $EX_2 = EX_3$ and, hence,

$$EX = \left(\frac{11}{18}, \frac{5}{9}, \frac{5}{9} \right)^\top.$$

In order to compute the covariance matrix of the three-dimensional random vector X, one has to compute the variances and covariances of its components:

$$\Sigma = \begin{pmatrix} Var(X_1) & Cov(X_1, X_2) & Cov(X_1, X_3) \\ Cov(X_2, X_1) & Var(X_2) & Cov(X_2, X_3) \\ Cov(X_3, X_1) & Cov(X_3, X_2) & Var(X_3) \end{pmatrix}.$$

We have

$$EX_1^2 = \frac{4}{3} \int_0^1 \int_0^1 \int_0^1 (x_1^3 + x_1^2 x_2 x_3)\, dx_1\, dx_2\, dx_3$$

$$= \frac{4}{9},$$

$$EX_2^2 = \frac{4}{3} \int_0^1 \int_0^1 \int_0^1 (x_1 x_2^2 + x_2^3 x_3)\, dx_1\, dx_2\, dx_3$$

$$= \frac{7}{18},$$

$$EX_3^2 = EX_2^2.$$

Now, we can compute the covariances using the formula $Cov(X_1, X_2) = EX_1 X_2 - EX_1 EX_2$ as

$$\Sigma = \begin{pmatrix} \frac{23}{324} & -\frac{1}{162} & -\frac{1}{162} \\ -\frac{1}{162} & \frac{13}{162} & \frac{13}{162} \\ -\frac{1}{162} & \frac{1}{162} & \frac{13}{162} \end{pmatrix}.$$

Ad c) The conditional density of $(X_2, X_3)^\top$ given $X_1 = x_1$ can be expressed as a ratio of the joint density of $(X_1, X_2, X_3)^\top$ and the marginal density of X_1.

The marginal density of X_1 is

$$f_{X_1}(x_1) = \int_0^1 \int_0^1 f(x_1, x_2, x_3) dx_2 dx_3$$

$$= \int_0^1 \int_0^1 \frac{4}{3}(x_1 + x_2 x_3) dx_2 dx_3$$

$$= \frac{4}{3} \int_0^1 \left[x_1 x_2 + \frac{1}{2} x_2^2 x_3 \right]_0^1 dx_3$$

$$= \frac{4}{3} \int_0^1 x_1 + \frac{1}{2} x_3 dx_3$$

$$= \frac{4}{3} \left[x_1 x_3 + \frac{1}{4} x_3^2 \right]_0^1$$

$$= \frac{4}{3} \left(x_1 + \frac{1}{4} \right).$$

It follows that the conditional density of X_2 and X_3 is

$$f(x_2, x_3 | x_1) = \frac{x_1 + x_2 x_3}{x_1 + \frac{1}{4}}.$$

Let us now compute the conditional moments $E(X_2 | X_1 = x_1) = E(X_3 | X_1 = x_1)$, $E(X_2^2 | X_1 = x_1) = E(X_3^2 | X_1 = x_1)$, and $E(X_2 X_3 | X_1 = x_1)$.

$$E(X_2 | X_1 = x_1) = \frac{1}{x_1 + \frac{1}{4}} \int_0^1 \int_0^1 (x_1 x_2 + x_2^2 x_3) \, dx_2 \, dx_3$$

$$= \frac{6x_1 + 2}{12x_1 + 3},$$

$$E(X_2^2 | X_1 = x_1) = \frac{1}{x_1 + \frac{1}{4}} \int_0^1 \int_0^1 (x_1 x_2^2 + x_2^3 x_3) \, dx_2 \, dx_3$$

$$= \frac{8x_1 + 3}{24x_1 + 6}.$$

Now we can compute the conditional variances of X_2 and X_3 :

$$Var(X_2 | X_1 = x_1) = E(X_2^2 | X_1 = x_1) - [E(X_2 | X_1 = x_1)]^2$$

$$= \frac{8x_1 + 3}{2(12x_1 + 3)} - \frac{36x_1^2 + 24x_1 + 4}{(12x_1 + 3)^2}$$

$$= \frac{96x_1^2 + 60x_1 + 9}{2(12x_1 + 3)^2} - \frac{72x_1^2 + 48x_1 + 8}{2(12x_1 + 3)^2}$$

$$= \frac{24x_1^2 + 12x_1 + 1}{2(12x_1 + 3)^2}.$$

Next, we have to compute $E(X_2 X_3 | X_1 = x_1)$:

$$E(X_2X_3|X_1 = x_1) = \frac{1}{x_1 + \frac{1}{4}} \int_0^1 \int_0^1 (x_1x_2x_3 + x_2^2x_3^2) \, dx_2 \, dx_3$$

$$= \frac{9x_1 + 4}{36x_1 + 9}.$$

Now, the conditional covariance can be expressed as:

$$Cov(X_2, X_3|X_1 = x_1) = E(X_2X_3|X_1 = x_1) - E(X_2|X_1 = x_1)E(X_3|X_1 = x_1)$$

$$= \frac{9x_1 + 4}{3(12x_1 + 3)} - \frac{36x_1^2 + 24x_1 + 4}{(12x_1^2 + 3)^2}$$

$$= \frac{108x_1^2 + 75x_1 + 12}{3(12x_1 + 3)^2} - \frac{108x_1^2 + 72x_1 + 12}{3(12x_1 + 3)^2}$$

$$= \frac{x_1}{(12x_1 + 3)^2}.$$

Summarizing the above results, the conditional covariance matrix is given by:

$$Var\left(\begin{pmatrix} X_2 \\ X_3 \end{pmatrix} \middle| X_1 = x_1\right) = \frac{1}{2(12x_1 + 3)^2} \begin{pmatrix} 24x_1^2 + 12x_1 + 1 & 2x_1 \\ 2x_1 & 24x_1^2 + 12x_1 + 1 \end{pmatrix}$$

EXERCISE 4.18. Let $X \sim N_2\left(\begin{pmatrix} 1 \\ 2 \end{pmatrix}, \begin{pmatrix} 2 & a \\ a & 2 \end{pmatrix}\right)$.

a) Represent the contour ellipses for $a = 0$; $-\frac{1}{2}$; $+\frac{1}{2}$; 1.

b) For $a = \frac{1}{2}$ find the regions of X centered on μ which cover the area of the true parameter with probability 0.90 and 0.95.

Ad a) The eigenvalues λ_1, λ_2 of the covariance matrix Σ are obtained as a solution of the equation $|\Sigma - \lambda I_2| = 0$. The eigenvectors γ_1, γ_2 are solutions of

$$\Sigma\gamma_i = \lambda_i\gamma_i, \quad i = 1, 2.$$

The contour ellipse has principal axes in the direction of γ_1 and γ_2 and it can be represented as follows:

$$E_d = \left\{x \in \mathbb{R}^2 \mid (x - \mu)^\top \Sigma^{-1}(x - \mu) = d^2\right\}.$$

The half-lengths of the axes of the ellipse E_d are $d\lambda_i^{1/2}$, where $i = 1, 2$.

i) For $a = 0$, we obtain

$$X \sim N_2\left(\begin{pmatrix} 1 \\ 2 \end{pmatrix}, \begin{pmatrix} 2 & 0 \\ 0 & 2 \end{pmatrix}\right)$$

and we have

$$\begin{vmatrix} 2-\lambda & 0 \\ 0 & 2-\lambda \end{vmatrix} = (2-\lambda)^2 = 0.$$

Hence,

$$\lambda_1 = \lambda_2 = 2,$$
$$\gamma_1 = (1,0)^\top,$$
$$\gamma_2 = (0,1)^\top$$

and

$$d^2 = (x_1 - 1, x_2 - 2) \begin{pmatrix} \frac{1}{2} & 0 \\ 0 & \frac{1}{2} \end{pmatrix} \begin{pmatrix} x_1 - 1 \\ x_2 - 2 \end{pmatrix} = \frac{(x_1 - 1)^2 + (x_2 - 2)^2}{2}.$$

The contour ellipse is centered in $(1,2)^\top$, its principal axes are in the direction $(1,0)^\top$, $(0,1)^\top$ and it can be represented as:

$$E_d = \left\{ x \in \mathbb{R}^2 \mid \frac{(x_1 - 1)^2 + (x_2 - 2)^2}{2} = d^2 \right\}.$$

The half-lengths of both axes are equal to $d\sqrt{2}$.

ii) For $a = -1/2$, we have

$$X \sim N_2 \left(\begin{pmatrix} 1 \\ 2 \end{pmatrix}, \begin{pmatrix} 2 & -\frac{1}{2} \\ -\frac{1}{2} & 2 \end{pmatrix} \right)$$

and from the equation

$$\begin{vmatrix} 2-\lambda & -\frac{1}{2} \\ -\frac{1}{2} & 2-\lambda \end{vmatrix} = (2-\lambda)^2 - \frac{1}{4} = 0.$$

it follows that $\lambda_1 = 5/2$, $\lambda_2 = 3/2$ and $\gamma_1 = \frac{1}{\sqrt{2}}(1,-1)^\top$, $\gamma_2 = \frac{1}{\sqrt{2}}(1,1)^\top$.

$$d^2 = (x_1 - 1, x_2 - 2) \begin{pmatrix} \frac{8}{15} & \frac{2}{15} \\ \frac{2}{15} & \frac{8}{15} \end{pmatrix} \begin{pmatrix} x_1 - 1 \\ x_2 - 2 \end{pmatrix}$$
$$= \frac{4}{15}(2x_1^2 + 2x_2^2 - 6x_1 - 9x_2 + x_1 x_2 + 12)$$

The contour ellipse is centered in $(1,2)^\top$, its principal axes are in directions of $(1,-1)^\top$, $(1,1)^\top$ and it can be represented as:

$$E_d = \left\{ x \in \mathbb{R}^2 \mid \frac{4}{15}(2x_1^2 + 2x_2^2 - 6x_1 - 9x_2 + x_1 x_2 + 12) = d^2 \right\}.$$

The half-lengths of its axes are equal to $d\sqrt{5/2}$ and $d\sqrt{3/2}$.

iii) For $a = 1/2$, we have

$$X \sim N_2 \left(\begin{pmatrix} 1 \\ 2 \end{pmatrix}, \begin{pmatrix} 2 & \frac{1}{2} \\ \frac{1}{2} & 2 \end{pmatrix} \right)$$

and from the equation

$$\begin{vmatrix} 2 - \lambda & \frac{1}{2} \\ \frac{1}{2} & 2 - \lambda \end{vmatrix} = (2 - \lambda)^2 - \frac{1}{4} = 0$$

it follows that $\lambda_1 = \frac{5}{2}, \lambda_2 = \frac{3}{2}$ and $\gamma_1 = \frac{1}{\sqrt{2}}(1,1)^\top$, $\gamma_2 = \frac{1}{\sqrt{2}}(1,-1)^\top$.

$$d^2 = (x_1 - 1, x_2 - 2) \begin{pmatrix} \frac{8}{15} & -\frac{2}{15} \\ -\frac{2}{15} & \frac{8}{15} \end{pmatrix} \begin{pmatrix} x_1 - 1 \\ x_2 - 2 \end{pmatrix}$$

$$= \frac{4}{15}(2x_1^2 + 2x_2^2 - 2x_1 - 7x_2 - x_1 x_2 + 8)$$

The contour ellipse is centered in $(1,2)^\top$, its principal axes are in directions of $(1,1)^\top$, $(1,-1)^\top$ and it can be represented as:

$$E_d = \left\{ x \in \mathbb{R}^2 \mid \frac{4}{15}(2x_1^2 + 2x_2^2 - 2x_1 - 7x_2 - x_1 x_2 + 8) = d^2 \right\}.$$

The half-lengths of its axes are $d\sqrt{5/2}$ and $d\sqrt{3/2}$.

iv) For $a = 1$ we have

$$X \sim N_2 \left(\begin{pmatrix} 1 \\ 2 \end{pmatrix}, \begin{pmatrix} 2 & 1 \\ 1 & 2 \end{pmatrix} \right)$$

and from the equation

$$\begin{vmatrix} 2 - \lambda & 1 \\ 1 & 2 - \lambda \end{vmatrix} = (2 - \lambda)^2 - 1 = 0$$

it follows that $\lambda_1 = 3, \lambda_2 = 1$ and $\gamma_1 = \frac{1}{\sqrt{2}}(1,1)^\top$, $\gamma_2 = \frac{1}{\sqrt{2}}(1,-1)^\top$.

$$d^2 = (x_1 - 1, x_2 - 2) \begin{pmatrix} \frac{2}{3} & -\frac{1}{3} \\ -\frac{1}{3} & \frac{2}{3} \end{pmatrix} \begin{pmatrix} x_1 - 1 \\ x_2 - 2 \end{pmatrix} = \frac{2}{3}(x_1^2 + x_2^2 - 3x_2 - x_1 x_2 + 3)$$

The contour ellipse is centered in $(1,2)^\top$, its principal axes are in the direction of $(1,1)^\top$, $(1,-1)^\top$ and the ellipse can be represented as:

$$E_d = \left\{ x \in \mathbb{R}^2 \mid \frac{2}{3}(x_1^2 + x_2^2 - 3x_2 - x_1 x_2 + 3) = d^2 \right\}.$$

The half-lengths of its axes are $d\sqrt{3}$ and d.

Ad b) We know that the random variable $U = (X - \mu)^\top \Sigma^{-1}(X - \mu)$ has a χ_2^2 distribution. The definition of critical value says that $P(U \leq \chi_{0.90;2}^2) = 0.90$ and $P(U \leq \chi_{0.95;2}^2) = 0.95$. This implies that the wanted regions for X can be written as

$$\left\{ x \in \mathbb{R}^2 \mid \frac{4}{15}(2x_1^2 + 2x_2^2 - 2x_1 - 7x_2 - x_1 x_2 + 8) \le \chi^2_{0.90;2} = 4.61 \right\}$$

covering realizations of X with probability 0.90 and

$$\left\{ x \in \mathbb{R}^2 \mid \frac{4}{15}(2x_1^2 + 2x_2^2 - 2x_1 - 7x_2 - x_1 x_2 + 8) \le \chi^2_{0.95;2} = 5.99 \right\}$$

containing future realizations of X with probability 0.95. The regions are ellipses corresponding to $d^2_{0.90} = \chi^2_{0.90;2} = 4.61$ and $d^2_{0.95} = \chi^2_{0.95;2} = 5.99$.

EXERCISE 4.19. *Consider the pdf*

$$f(x_1, x_2) = \frac{1}{8x_2} \exp\left\{ -\left(\frac{x_1}{2x_2} + \frac{x_2}{4} \right) \right\}, \qquad x_1, x_2 > 0.$$

Compute $f(x_2)$ and $f(x_1|x_2)$. Also give the best (MSE) approximation of X_1 by a function of X_2. Compute the variance of the error of the approximation.

The marginal distribution of x_2 can be calculated by "integrating out" x_1 from the joint pdf $f(x_1, x_2)$:

$$
\begin{aligned}
f_{X_2}(x_2) &= \int_0^{+\infty} f(x_1, x_2) dx_1 \\
&= -\frac{1}{4} \exp\left\{ -\frac{x_2}{4} \right\} \int_0^{+\infty} -\frac{1}{2x_2} \exp\left\{ -\frac{x_1}{2x_2} \right\} dx_1 \\
&= -\frac{1}{4} \exp\left\{ -\frac{x_2}{4} \right\} \left[\exp\left\{ -\frac{x_2}{4} \right\} \right]_0^{+\infty} \\
&= \frac{1}{4} \exp\left\{ -\frac{x_2}{4} \right\},
\end{aligned}
$$

for $x_2 > 0$, in other words, the distribution of X_2 is exponential with expected value $EX_2 = 4$.

The conditional distribution $f(x_1|x_2)$ is calculated as a ratio of the joint pdf $f(x_1, x_2)$ and the marginal pdf $f_{X_2}(x_2)$:

$$
\begin{aligned}
f_{X_1|X_2=x_2}(x_1) &= \frac{f(x_1, x_2)}{f_{X_2}(x_2)} \\
&= \frac{1}{2x_2} \exp\left(-\frac{x_1}{2x_2} \right),
\end{aligned}
$$

for $x_1, x_2 > 0$. Note that this is just the exponential distribution with expected value $2x_2$.

The best approximation of X_1 by X_2, from the point of view of MSE, is the conditional expectation $E(X_1, X_2 = x_2)$. We have already remarked that the conditional expected value is $E(X_1|X_2 = x_2) = 2x_2$.

The variance of the "error of approximation" is the variance of X_1 around its expected value, i.e., the conditional variance of X_1 given $X_2 = x_2$. From the properties of the exponential distribution, we can immediately say that it is equal to $Var(X_1|X_2 = x_2) = 4x_2^2$.

EXERCISE 4.20. *Prove Theorem 4.5 in Härdle & Simar (2003), i.e., that the linear transformation of a p-variate normally distributed random variable $Y = \mathcal{A}X + b$ ($\mathcal{A}$ is square and nonsingular) has again a p-variate normal distribution.*

The multinormal distribution has pdf

$$f_X(x) = |2\pi\Sigma|^{-1/2} \exp\left\{-\frac{1}{2}(x-\mu)^\top \Sigma^{-1}(x-\mu)\right\}.$$

For the linear transformation, $Y = \mathcal{A}X + b$, the inverse transformation is $X = \mathcal{A}^{-1}Y - b$, the Jacobian of the inverse transformation is $\mathcal{J} = \mathcal{A}^{-1}$ and the density of the transformed random vector is

$$f_Y(y) = \mathrm{abs}(|\mathcal{A}|^{-1})f_X\{\mathcal{A}^{-1}(y-b)\}.$$

From the assumption, that $\mathcal{A}$ is square and nonsingular, we know that the inverse matrix $\mathcal{A}^{-1}$ exists and we can write the pdf of the transformed random vector as

$f_Y(y)$

$$= |2\pi\Sigma|^{-1/2}\,\mathrm{abs}(|\mathcal{A}|^{-1})\exp\left[-\frac{1}{2}\{\mathcal{A}^{-1}(y-b)-\mu\}^\top \Sigma^{-1}\{\mathcal{A}^{-1}(y-b)-\mu\}\right]$$

$$= |2\pi\mathcal{A}\Sigma\mathcal{A}^\top|^{-1/2}\exp\left[-\frac{1}{2}\{y-(b+\mathcal{A}\mu)\}^\top (\mathcal{A}^{-1})^\top \Sigma^{-1}\mathcal{A}^{-1}\{y-(b+\mathcal{A}\mu)\}\right]$$

$$= |2\pi\mathcal{A}\Sigma\mathcal{A}^\top|^{-1/2}\exp\left[-\frac{1}{2}\{y-(b+\mathcal{A}\mu)\}^\top \left(\mathcal{A}\Sigma\mathcal{A}^\top\right)^{-1}\{y-(b+\mathcal{A}\mu)\}\right].$$

This is the probability density function of a p-variate multinormal distribution with mean $EY = \mathcal{A}\mu + b$ and variance matrix $\mathrm{Var}(Y) = \mathcal{A}\Sigma\mathcal{A}^\top$ and we conclude that

$$\mathcal{A}X + b = Y \sim N_p(\mathcal{A}\mu + b, \mathcal{A}\Sigma\mathcal{A}^\top).$$

Theory of the Multinormal

> ...while the individual man is an insoluble puzzle, in the aggregate
> he becomes a mathematical certainty.
> Sherlock Holmes in "The Sign of Four"

In the preceeding chapter we realized the importance of the multivariate normal distribution, its geometry and connection with elliptic dependence structures. The multivariate normal comes into play in many applications and statistical tests. It is therefore important to know how this distribution behaves when we apply conditioning or linear or nonlinear transformation.

It is also of interest to check whether partitioned random vectors are still normally distributed and how the multinormal distribution is popping out of theoretical concepts. It is stable under linear transforms, zero correlation corresponds to independence, the marginals and all the conditionals are also multivariate normal variates, etc. The mathematical properties of the multinormal make analyses much simpler. We consider here best linear approximations, partial correlation (expressed via partitioned matrices), and conditioning on parts of a multinormal random vector.

In order to better explain the basic properties of the multivariate normal distribution, we start by introducing several theorems.

Theorem 5.1 says that a subvector of a multivariate normal vector has again multivariate normal distribution, and it shows how to calculate its orthogonal (independent) complement.

THEOREM 5.1. *Let $X = \begin{pmatrix} X_1 \\ X_2 \end{pmatrix} \sim N_p(\mu, \Sigma)$, $X_1 \in \mathbb{R}^r$, and $X_2 \in \mathbb{R}^{p-r}$. Define $X_{2.1} = X_2 - \Sigma_{21}\Sigma_{11}^{-1}X_1$ from the partitioned covariance matrix*

$$\Sigma = \begin{pmatrix} \Sigma_{11} & \Sigma_{12} \\ \Sigma_{21} & \Sigma_{22} \end{pmatrix}.$$

Then

$$X_1 \sim N_r(\mu_1, \Sigma_{11}), \qquad (5.1)$$

$$X_{2.1} \sim N_{p-r}(\mu_{2.1}, \Sigma_{22.1}) \qquad (5.2)$$

are independent with

$$\mu_{2.1} = \mu_2 - \Sigma_{21}\Sigma_{11}^{-1}\mu_1, \quad \Sigma_{22.1} = \Sigma_{22} - \Sigma_{21}\Sigma_{11}^{-1}\Sigma_{12}. \qquad (5.3)$$

Theorem 5.2 says that linear transformation of a multivariate normal vector also has multivariate normal distribution. The mean and the variance matrix of the linearly transformed random vector actually follow from the results presented in previous chapters.

THEOREM 5.2. *If $X \sim N_p(\mu, \Sigma)$, $\mathcal{A}(q \times p)$, $c \in \mathbb{R}^q$, and* rank$(\mathcal{A}) = q \leq p$, *then $Y = \mathcal{A}X + c$ is a q-variate normal, i.e.,*

$$Y \sim N_q(\mathcal{A}\mu + c, \mathcal{A}\Sigma\mathcal{A}^\top).$$

Theorem 5.3 gives the formula for conditional distribution, which is also multivariate normal.

THEOREM 5.3. *The conditional distribution of X_2 given $X_1 = x_1$ is normal with mean $\mu_2 + \Sigma_{21}\Sigma_{11}^{-1}(x_1 - \mu_1)$ and covariance $\Sigma_{22.1}$, i.e.,*

$$(X_2 \mid X_1 = x_1) \sim N_{p-r}(\mu_2 + \Sigma_{21}\Sigma_{11}^{-1}(x_1 - \mu_1), \Sigma_{22.1}). \qquad (5.4)$$

Using Theorem 5.1, we can say that the conditional distribution $(X_2 \mid X_1 = x_1)$ and the random vector X_1 are independent.

Apart from the multivariate normal distribution, we mention the Wishart and the Hotelling distributions, which can be seen as generalizations of the one-dimensional χ^2 and t-distribution, respectively.

For a data matrix $\mathcal{X}(n \times p)$, containing n independent observations of the centered normal vector $X \sim N_p(0, \Sigma)$, the estimated covariance matrix is proportional to $\mathcal{X}^\top\mathcal{X}$. The distribution of the random matrix $\mathcal{M}(p \times p) = \mathcal{X}^\top\mathcal{X} = \sum_{i=1}^n x_i x_i^\top$ is the so-called Wishart distribution $W_p(\Sigma, n)$, which proves to be very useful in the analysis of estimated covariance matrices.

Suppose that the random vector $Y \sim N_p(0, \mathcal{I})$ is independent of the random matrix $\mathcal{M} \sim W_p(\mathcal{I}, n)$. Then the random variable $n\, Y^\top\mathcal{M}^{-1}Y$ has Hotelling T^2 (p, n) distribution. The Hotelling T^2 (p, n) is closely related to the F-distribution:

$$T^2(p, n) = \frac{np}{n - p + 1} \; F_{p, n-p+1}.$$

EXERCISE 5.1. *Consider* $X \sim N_2(\mu, \Sigma)$ *with* $\mu = (2,2)^\top$ *and* $\Sigma = \begin{pmatrix} 1 & 0 \\ 0 & 1 \end{pmatrix}$
and the matrices $\mathcal{A} = \begin{pmatrix} 1 \\ 1 \end{pmatrix}^\top$, $\mathcal{B} = \begin{pmatrix} 1 \\ -1 \end{pmatrix}^\top$. *Show that* $\mathcal{A}X$ *and* $\mathcal{B}X$ *are independent.*

Since $X \sim N_2(\mu, \Sigma)$ is multivariate normal, Theorem 5.2 implies that also both $\mathcal{A}X$ and $\mathcal{B}X$ are normal. More precisely, $\mathcal{A}X \sim N(\mathcal{A}\mu, \mathcal{A}\Sigma\mathcal{A}^\top) = N(4,2)$ and $\mathcal{B}X \sim N(\mathcal{B}\mu, \mathcal{B}\Sigma\mathcal{B}^\top) = N(0,2)$.

However, in order to show the independence, we have to study the joint distribution of $(\mathcal{A}X, \mathcal{B}X)^\top$. Theorem 5.2 implies that

$$\begin{pmatrix} \mathcal{A}X \\ \mathcal{B}X \end{pmatrix} = \begin{pmatrix} \mathcal{A} \\ \mathcal{B} \end{pmatrix} X \sim N_2\left(\begin{pmatrix} 4 \\ 0 \end{pmatrix}, \begin{pmatrix} 2 & 0 \\ 0 & 2 \end{pmatrix} \right)$$

With this diagonal structure of the covariance matrix, the joint pdf of $(\mathcal{A}X, \mathcal{B}X)$ can be factorized as follows:

$$
\begin{aligned}
f(x_1, x_2) &= \frac{1}{4\pi} \exp\left\{ -\frac{1}{2} \cdot (x_1 - 4, x_2) \begin{pmatrix} \frac{1}{2} & 0 \\ 0 & \frac{1}{2} \end{pmatrix} \begin{pmatrix} x_1 - 4 \\ x_2 \end{pmatrix} \right\} \\
&= \frac{1}{4\pi} \exp\left\{ -\frac{(x_1 - 4)^2 + x_2^2}{4} \right\} \\
&= \frac{1}{2\sqrt{\pi}} \exp\left\{ -\frac{(x_1 - 4)^2}{4} \right\} \frac{1}{2\sqrt{\pi}} \exp\left\{ -\frac{x_2^2}{4} \right\} \\
&= f_{\mathcal{A}X}(x_1) f_{\mathcal{B}X}(x_2),
\end{aligned}
$$

i.e., as the product of the marginal densities of ($\mathcal{A}X$ and $\mathcal{B}X$). This factorization, following from the diagonal structure of the variance matrix of multivariate normal distribution, proves the independence of the random variables $\mathcal{A}X$ and $\mathcal{B}X$, see also Exercise 4.10.

EXERCISE 5.2. *Prove that if* $X_1 \sim N_r(\mu_1, \Sigma_{11})$ *and* $(X_2|X_1 = x_1) \sim N_{p-r}(\mathcal{A}x_1 + b, \Omega)$ *where* Ω *does not depend on* x_1, *then* $X = \begin{pmatrix} X_1 \\ X_2 \end{pmatrix} \sim N_p(\mu, \Sigma)$, *where*

$$\mu = \begin{pmatrix} \mu_1 \\ \mathcal{A}\mu_1 + b \end{pmatrix} \quad and \quad \Sigma = \begin{pmatrix} \Sigma_{11} & \Sigma_{11}\mathcal{A}^\top \\ \mathcal{A}\Sigma_{11} & \Omega + \mathcal{A}\Sigma_{11}\mathcal{A}^\top \end{pmatrix}.$$

The conditional distribution of $(X_2|X_1 = x_1)$ can be written as $X_2 = \mathcal{A}x_1 + b + X_3$, where $X_3 \sim N(0, \Omega)$ is independent of X_1. Hence, the marginal distribution of the random vector X_2 is the same as the distribution of $\mathcal{A}X_1 + b + X_3$. Now, according to Theorem 5.2, the random vector

$$X = \begin{pmatrix} X_1 \\ X_2 \end{pmatrix} = \begin{pmatrix} \mathcal{I}_p & 0_p 0_p^\top \\ \mathcal{A} & \mathcal{I}_p \end{pmatrix} \begin{pmatrix} X_1 \\ X_3 \end{pmatrix} + \begin{pmatrix} 0 \\ b \end{pmatrix}$$

has multivariate normal distribution.

It remains to calculate $E(X)$ and $\text{Var}\,X$:

$$\begin{aligned}
E(X_2) &= E\{E(X_2|X_1)\} = E\{\mathcal{A}X_1 + b + X_3\} \\
&= \mathcal{A}\mu_1 + b, \\
\text{Var}(X_2) &= E\{\text{Var}(X_2|X_1)\} + Var\{E(X_2|X_1)\} \\
&= E\{\Omega\} + \text{Var}\{\mathcal{A}X_1 + b\} \\
&= \Omega + \mathcal{A}\Sigma_{11}\mathcal{A}^\top, \\
\text{Cov}(X_1, X_2) &= E\{(X_1 - EX_1)(X_2 - EX_2)^\top\} = \\
&= E\{(X_1 - \mu_1)(\mathcal{A}X_1 + b - \mathcal{A}\mu_1 - b)^\top\} = \\
&= E\{(X_1 - \mu_1)(X_1 - \mu_1)^\top \mathcal{A}^\top\} = \Sigma_{11}\mathcal{A}^\top.
\end{aligned}$$

Since $X_1 \sim N_r(\mu_1, \Sigma_{11})$, it follows that

$$X = \begin{pmatrix} X_1 \\ X_2 \end{pmatrix} \sim N_p\left(\begin{pmatrix} \mu_1 \\ \mathcal{A}\mu_1 + b \end{pmatrix} \begin{pmatrix} \Sigma_{11} & \Sigma_{11}\mathcal{A}^\top \\ \mathcal{A}\Sigma_{11} & \Omega + \mathcal{A}\Sigma_{11}\mathcal{A}^\top \end{pmatrix} \right).$$

EXERCISE 5.3. *Let $\mathcal{X}(n \times p)$ be a data matrix from a $N_p(\mu, \Sigma)$ distribution. Show that $n\mathcal{S} = \mathcal{X}^\top \mathcal{H}\mathcal{X}$ is distributed as $W_p(\Sigma, n - 1)$.*

In order to arrive at the Wishart distribution, we have to consider transformations of $\mathcal{X}$ that will allow us to write $\mathcal{S}$ in terms of independent centered identically distributed multivariate normal observations.

The centering matrix $\mathcal{H}(n \times n)$ is idempotent, see Exercise 3.16, and rank $(\mathcal{H}) = \text{tr}(\mathcal{H}) = n(1 - 1/n) = n - 1$. Thus, the spectral decomposition of $\mathcal{H}$ can be written as $\mathcal{H} = \Gamma \mathcal{I}_{n-1}\Gamma^\top$.

Define the data matrix $\mathcal{Y} = \Gamma^\top \mathcal{X} = (\gamma_i \mathcal{X}_j)_{i=1,\ldots,n-1; j=1,\ldots,p} = (y_{ij})_{i;j}$, where γ_i denotes the ith eigenvector of $\mathcal{H}$ and $\mathcal{X}_j$ is the jth column of matrix $\mathcal{X}$.

We start by rewriting the spectral decomposition of the centering matrix:

$$\begin{aligned}
\mathcal{H} &= \Gamma \mathcal{I}_{n-1}\Gamma^\top \\
\Gamma^\top \mathcal{H}\Gamma &= \mathcal{I}_{n-1} \\
\Gamma^\top (\mathcal{I}_n - n^{-1}1_n 1_n^\top)\Gamma &= \mathcal{I}_{n-1} \\
\Gamma^\top \Gamma - n^{-1}\Gamma^\top 1_n 1_n^\top \Gamma &= \mathcal{I}_{n-1} \\
n^{-1}\Gamma^\top 1_n 1_n^\top \Gamma &= 0_{n-1}0_{n-1}^\top.
\end{aligned}$$

The above equality means that $\Gamma^\top 1_n = 0_{n-1}$ which in turn implies, for any $j = 1,\ldots,p$ and $i = 1,\ldots,n - 1$, that

$$Ey_{ij} = E\gamma_i^\top \mathcal{X}_j = \gamma_i^\top E\mathcal{X}_j = \mu_j \gamma_i^\top 1_n = 0,$$

i.e., the expected value of every element of matrix $\mathcal{Y}$ is zero.

Next, for any $j, k = 1, \ldots, p$ and $i = 1, \ldots, n-1$, we can write

$$\text{Cov}(y_{ij}, y_{ik}) = \text{Cov}(\gamma_i^\top \mathcal{X}_j, \gamma_i^\top \mathcal{X}_k) = \sigma_{jk} \gamma_i^\top \gamma_i = \sigma_{jk}$$

and it follows that all rows of the random matrix $\mathcal{Y}$ have the same variance matrix Σ. Furthermore, the rows of the matrix $\mathcal{Y}$ are independent since, for any $i, h = 1, \ldots, n-1, i \neq h$ and $j, k = 1, \ldots, p$, we have

$$\text{Cov}(y_{ij}, y_{hk}) = \text{Cov}(\gamma_i^\top \mathcal{X}_j, \gamma_h^\top \mathcal{X}_j) = \sigma_{jk} \gamma_i^\top \gamma_h = 0.$$

From Theorem 5.2 and from the normality of $\mathcal{X}$ it follows that the distribution of $\mathcal{Y}$ is also multivariate normal.

Now we can write

$$n\mathcal{S} = \mathcal{X}^\top \mathcal{H} \mathcal{X} = \mathcal{X}^\top \Gamma \Gamma^\top \mathcal{X} = \mathcal{Y}^\top \mathcal{Y},$$

where the $n-1$ rows of the matrix $\mathcal{Y}$ are independent observations of multivariate normally distributed random vector $Y \sim N_p(0, \Sigma)$. From the definition of the Wishart distribution, it is now straightforward that $n\mathcal{S} \sim W_p(\Sigma, n-1)$.

EXERCISE 5.4. *Let*

$$X \sim N_2 \left(\begin{pmatrix} 1 \\ 2 \end{pmatrix}, \begin{pmatrix} 2 & 1 \\ 1 & 2 \end{pmatrix} \right)$$

and

$$Y \mid X \sim N_2 \left(\begin{pmatrix} X_1 \\ X_1 + X_2 \end{pmatrix}, \begin{pmatrix} 1 & 0 \\ 0 & 1 \end{pmatrix} \right).$$

a) Determine the distribution of $Y_2 \mid Y_1$.

b) Determine the distribution of $W = X - Y$.

We start by computing the joint distribution of the vector $(X_1, X_2, Y_1, Y_2)^\top$ from the marginal distribution of X and the conditional distribution $Y \mid X$. Exercise 5.2, where

$$\mathcal{A} = \begin{pmatrix} 1 & 0 \\ 1 & 1 \end{pmatrix}, \quad b = \begin{pmatrix} 0 \\ 0 \end{pmatrix}, \quad \Omega = \begin{pmatrix} 1 & 0 \\ 0 & 1 \end{pmatrix},$$

provides the following result:

$$\begin{pmatrix} X \\ Y \end{pmatrix} \sim N_4 \left(\begin{pmatrix} 1 \\ 2 \\ 1 \\ 3 \end{pmatrix}, \begin{pmatrix} 2 & 1 & 2 & 3 \\ 1 & 2 & 1 & 3 \\ 2 & 1 & 3 & 3 \\ 3 & 3 & 3 & 7 \end{pmatrix} \right).$$

In particular, the marginal distribution of Y is

$$Y \sim N_2 \left(\begin{pmatrix} 1 \\ 3 \end{pmatrix}, \begin{pmatrix} 3 & 3 \\ 3 & 7 \end{pmatrix} \right).$$

Now we are ready to solve our problem.

a) The conditional distribution of Y_2 given Y_1 is normal

$$Y_2 \,|\, Y_1 = N(Y_1 + 2, 4)$$

by Theorem 5.3.

b) It is clear that W can be written as a linear transformation $W = X - Y = \mathcal{B}(X_1, X_2, Y_1, Y_2)^\top$, where

$$\mathcal{B} = \begin{pmatrix} 1 & 0 & -1 & 0 \\ 0 & 1 & 0 & -1 \end{pmatrix}.$$

Using Theorem 5.2, we obtain

$$W \sim N_2 \left(\begin{pmatrix} 0 \\ -1 \end{pmatrix}, \begin{pmatrix} 1 & 0 \\ 0 & 3 \end{pmatrix} \right).$$

EXERCISE 5.5. *Consider* $\begin{pmatrix} X \\ Y \\ Z \end{pmatrix} \sim N_3(\mu, \Sigma)$. *Compute μ and Σ knowing that*

$$Y \,|\, Z \sim N_1(-Z, 1) \tag{5.5}$$
$$\mu_{Z|Y} = -\frac{1}{3} - \frac{1}{3} Y \tag{5.6}$$
$$X \,|\, Y, Z \sim N_1(2 + 2Y + 3Z, 1). \tag{5.7}$$

Determine the conditional distributions of $X \,|\, Y$ and of $X \,|\, Y + Z$.

Since we know the conditional distribution $Y|Z \sim N_1(-Z, 1)$, we can apply Theorem 5.3:

$$\mu_{Y|Z} = \mu_Y + \sigma_{YZ}\sigma_{ZZ}^{-1}(Z - \mu_Z) = -Z \tag{5.8}$$
$$\Sigma_{YY.Z} = \sigma_{YY} - \sigma_{YZ}\sigma_{ZZ}^{-1}\sigma_{YZ} = 1 \tag{5.9}$$

By calculating the expected value and the variance of both sides of (5.8) we get:

$$\mu_Y = -\mu_Z$$
$$\sigma_{YZ} = \sigma_{ZZ}.$$

The equation (5.9) now implies:

$$\sigma_{YY} = 1 + \sigma_{YZ} = 1 + \sigma_{ZZ}.$$

Now we are ready to use (5.6). Theorem 5.3 allows to express the expected value $\mu_{Z|Y}$ of the conditional distribution $Z|Y$ as

$$\mu_{Z|Y} = \mu_Z + \sigma_{ZY}\sigma_{YY}^{-1}(Y - \mu_Y) = -\frac{1}{3} - \frac{1}{3}Y$$

Again, by calculating the expected value and the variance of both sides of the above formula, we obtain:

$$-3\mu_Z = 1 + \mu_Y$$
$$\sigma_{ZY} = \frac{1}{3}\sigma_{YY}.$$

For the expected values of Y and Z we now have the system of equations:

$$\mu_Y = -\mu_Z$$
$$-3\mu_Z = 1 + \mu_Y$$

so that $\mu_Z = -\frac{1}{2}$ and $\mu_Y = \frac{1}{2}$.

The equations for the covariances are:

$$\sigma_{YY} = 1 + \sigma_{YZ}$$
$$\sigma_{ZY} = \frac{1}{3}\sigma_{YY}$$
$$\sigma_{YZ} = \sigma_{ZZ}$$

and it is easy to calculate $\sigma_{YY} = \frac{3}{2}$, $\sigma_{YZ} = \frac{1}{2}$ and $\sigma_{ZZ} = \frac{1}{2}$.

Thus, we have derived the distribution of the vector

$$\binom{Y}{Z} \sim N_2\left(\binom{\frac{1}{2}}{-\frac{1}{2}}, \begin{pmatrix} \frac{3}{2} & -\frac{1}{2} \\ -\frac{1}{2} & \frac{1}{2} \end{pmatrix}\right)$$

and, since we know that $X|(Y, Z) \sim N_1(2 + 2Y + 3Z, 1)$, it is straightforward to apply the result derived in Exercise 5.2 with $\Omega = 1$, $\mathcal{A} = (2, 3)$, and $b = 2$. We obtain

$$\begin{pmatrix} X \\ Y \\ Z \end{pmatrix} \sim N_3\left(\begin{pmatrix} \frac{3}{2} \\ \frac{1}{2} \\ -\frac{1}{2} \end{pmatrix}, \begin{pmatrix} \frac{45}{2} & \frac{11}{2} & \frac{5}{2} \\ \frac{11}{2} & \frac{3}{2} & \frac{1}{2} \\ \frac{5}{2} & \frac{1}{2} & \frac{1}{2} \end{pmatrix}\right).$$

The distribution of $X|Y$ can now be found easily by applying Theorem 5.3:

$$\mu_{X|Y} = \mu_X + \sigma_{XY}\sigma_{YY}^{-1}(Y - \mu_Y) = \frac{3}{2} + \frac{11}{2}\frac{2}{3}\left(Y - \frac{1}{2}\right) = \frac{11}{3}Y + \frac{10}{3}$$

$$\sigma_{X|Y} = \sigma_{XX} - \sigma_{XY}\sigma_{YY}^{-1}\sigma_{XY} = \frac{45}{2} - \frac{11}{2}\frac{2}{3}\frac{11}{2} = \frac{7}{3}.$$

Hence, the conditional distribution is $X|Y \sim N_1(\frac{11}{3}Y + \frac{10}{3}, \frac{7}{3})$.

To determine the conditional distribution of $X|Y + Z$ we have to determine the joint distribution of $X + Y$:

$$\mu_{Y+Z} = \mu_Y + \mu_Z = \frac{1}{2} - \frac{1}{2} = 0$$

$$\sigma_{Y+Z,Y+Z} = \sigma_{YY} + \sigma_{ZZ} + 2\sigma_{YZ} = \frac{3}{2} + \frac{1}{2} + 2\frac{1}{2} = 3$$

$$\sigma_{X,Y+Z} = \sigma_{XY} + \sigma_{XZ} = \frac{11}{2} + \frac{5}{2} = 8.$$

Now we can use Theorem 5.3 again and write

$$\mu_{X|Y+Z} = \mu_X + \sigma_{X,Y+Z}\sigma_{Y+Z,Y+Z}^{-1}(Y + Z - \mu_{Y+Z})$$
$$= \frac{3}{2} + \frac{8}{3}(Y + Z)$$

$$\sigma_{X|Y+Z} = \sigma_{XX} - \sigma_{X,Y+Z}\sigma_{Y+Z,Y+Z}^{-1}\sigma_{X,Y+Z} = \frac{45}{2} - \frac{64}{3} = \frac{7}{6}$$

so that $X|(Y + Z) \sim N_1\left(\frac{8}{3}(Y + Z) + \frac{3}{2}, \frac{7}{6}\right)$

EXERCISE 5.6. *Knowing that*

$$Z \sim N_1(0, 1)$$
$$Y \mid Z \sim N_1(1 + Z, 1)$$
$$X \mid Y, Z \sim N_1(1 - Y, 1)$$

a) find the distribution of $\begin{pmatrix} X \\ Y \\ Z \end{pmatrix}$ *and of* $Y \mid (X, Z)$.

b) find the distribution of

$$\begin{pmatrix} U \\ V \end{pmatrix} = \begin{pmatrix} 1 + Z \\ 1 - Y \end{pmatrix}.$$

c) compute $E(Y \mid U = 2)$.

a) The distribution of the random vector $\begin{pmatrix} X \\ Y \\ Z \end{pmatrix}$ can be derived easily by applying the result derived in Exercise 5.2 repeatedly. In the first step, we find the distribution of

$$\begin{pmatrix} Y \\ Z \end{pmatrix} \sim N_2\left(\begin{pmatrix} 1 \\ 0 \end{pmatrix}, \begin{pmatrix} 2 & 1 \\ 1 & 1 \end{pmatrix}\right)$$

and, applying the same procedure with $b = 1$, $A = (-1, 0)$, and $\Omega = 1$, we can combine the known distributions of $\begin{pmatrix} Y \\ Z \end{pmatrix}$ and $X \mid (Y, Z)$ to obtain

$$\begin{pmatrix} X \\ Y \\ Z \end{pmatrix} \sim N_3 \left(\begin{pmatrix} 0 \\ 1 \\ 0 \end{pmatrix}, \begin{pmatrix} 3 & -2 & -1 \\ -2 & 2 & 1 \\ -1 & 1 & 1 \end{pmatrix} \right).$$

The conditional distribution $Y \mid (X, Z)$ can be derived using Theorem 5.3. The moments of the resulting normal distribution are

$$\mu_{Y|(X,Z)} = 1 + (-2, 1) \begin{pmatrix} 3 & -1 \\ -1 & 1 \end{pmatrix}^{-1} \left(\begin{pmatrix} X \\ Z \end{pmatrix} - \begin{pmatrix} 0 \\ 0 \end{pmatrix} \right)$$

$$= 1 + (-2, 1) \frac{1}{2} \begin{pmatrix} 1 & 1 \\ 1 & 3 \end{pmatrix} \begin{pmatrix} X \\ Z \end{pmatrix}$$

$$= 1 + (0, 1) \begin{pmatrix} X \\ Z \end{pmatrix}$$

$$= 1 + Z$$

and

$$\sigma^2_{Y|(X,Z)} = 2 - (-2, 1) \begin{pmatrix} 3 & -1 \\ -1 & 1 \end{pmatrix}^{-1} \begin{pmatrix} -2 \\ 1 \end{pmatrix}$$

$$= 2 - (-2, 1) \frac{1}{2} \begin{pmatrix} 1 & 1 \\ 1 & 3 \end{pmatrix} \begin{pmatrix} -2 \\ 1 \end{pmatrix}$$

$$= 2 - \frac{3}{2} = \frac{1}{2}.$$

Hence, we arrive to the conditional distribution $Y \mid (X, Z) \sim N_1 \left(1 + Z, \frac{1}{2} \right)$.

b) The distribution of $\begin{pmatrix} U \\ V \end{pmatrix} = \begin{pmatrix} 1 + Z \\ 1 - Y \end{pmatrix}$ is obviously normal since it is a linear transformation of normally distributed random vector, see Theorem 5.2. The distribution of $(U, V)^\top$ can be deduced by calculating the corresponding first and second moments:

$$\mu_U = E(1 + Z) = 1 + EZ = 1$$
$$\mu_V = E(1 - Y) = 0$$
$$\sigma^2_U = \sigma^2_Z = 1$$
$$\sigma^2_V = \sigma^2_Y = 2$$
$$\sigma_{UV} = -\sigma_{Z,Y} = -1$$

and it follows that the distribution of $(U, V)^\top$ is

$$\begin{pmatrix} U \\ V \end{pmatrix} = \begin{pmatrix} 1 + Z \\ 1 - Y \end{pmatrix} \sim N_2 \left(\begin{pmatrix} 1 \\ 0 \end{pmatrix}, \begin{pmatrix} 1 & -1 \\ -1 & 2 \end{pmatrix} \right).$$

c) The conditional distribution of $(Y|U = 2)$ is the same as the conditional distribution of $(Y|Z + 1 = 2)$, i.e., $(Y|Z = 1)$. We know that $Y \mid Z \sim N_1(1 + Z, 1)$ and thus, the conditional distribution of $(Y|U = 2)$ is

$$(Y|U = 2) \sim N_1(1 + 1, 1) = N_1(2, 1).$$

EXERCISE 5.7. *Suppose* $\begin{pmatrix} X \\ Y \end{pmatrix} \sim N_2(\mu, \Sigma)$ *with* Σ *positive definite. Is it possible that*

a) $\mu_{X|Y} = 3Y^2$,

b) $\sigma_{XX|Y} = 2 + Y^2$,

c) $\mu_{X|Y} = 3 - Y$, *and*

d) $\sigma_{XX|Y} = 5$?

Using Theorem 5.3, we see that c) and d) are, in principle, possible (the conditional mean is a linear function of the condition and the conditional variance is constant).

Parts a) and b) are not possible since the resulting conditional means and variances in Theorem 5.3 do not contain any quadratic term.

EXERCISE 5.8. *Let* $X \sim N_3 \left(\begin{pmatrix} 1 \\ 2 \\ 3 \end{pmatrix}, \begin{pmatrix} 11 & -6 & 2 \\ -6 & 10 & -4 \\ 2 & -4 & 6 \end{pmatrix} \right)$.

a) *Find the best linear approximation of* X_3 *by a linear function of* X_1 *and* X_2 *and compute the multiple correlation coefficient between* X_3 *and* (X_1, X_2).

b) *Let* $Z_1 = X_2 - X_3$, $Z_2 = X_2 + X_3$ *and* $(Z_3 \mid Z_1, Z_2) \sim N_1(Z_1 + Z_2, 10)$.
Compute the distribution of $\begin{pmatrix} Z_1 \\ Z_2 \\ Z_3 \end{pmatrix}$.

a) The best linear approximation of X_3 by a linear function of X_1 and X_2 is given by the conditional expected value calculated according to Theorem 5.3:

$$\mu_{X_3|(X_1,X_2)} = 3 + (2, -4) \begin{pmatrix} 11 & -6 \\ -6 & 10 \end{pmatrix}^{-1} \begin{pmatrix} X_1 - 1 \\ X_2 - 2 \end{pmatrix}$$

$$= 3 + (2, -4) \frac{1}{74} \begin{pmatrix} 10 & 6 \\ 6 & 11 \end{pmatrix} \begin{pmatrix} X_1 - 1 \\ X_2 - 2 \end{pmatrix}$$

$$= 3 + \frac{1}{74}(-4, 32) \begin{pmatrix} X_1 - 1 \\ X_2 - 2 \end{pmatrix}$$

$$= 3 + \frac{1}{74}(-4, 32) \begin{pmatrix} X_1 - 1 \\ X_2 - 2 \end{pmatrix}$$

$$= \frac{145}{37} - \frac{2}{37}X_1 - \frac{16}{37}X_2,$$

The multiple correlation coefficient, $\rho_{3.12}$, between X_3 and (X_1, X_2) is defined as the correlation between X_3 and its best linear approximation based on X_1 and X_2, i.e.,

$$\rho_{3.12}^2 = \frac{\mathrm{Cov}(X_3, -\frac{2}{37}X_1 - \frac{16}{37}X_2)}{\sqrt{Var(X_3)\,Var(-\frac{2}{37}X_1 - \frac{16}{37}X_2)}}$$

$$= -\frac{\mathrm{Cov}(X_3, X_1 + 8X_2)}{\sqrt{Var(X_3)\,Var(X_1 + 8X_2)}}$$

$$= -\frac{\mathrm{Cov}(X_3, X_1) + 8\,\mathrm{Cov}(X_3, X_2)}{\sqrt{Var(X_3)\{Var(X_1) + 64\,Var(X_2) + 16\,\mathrm{Cov}(X_1, X_2)\}}}$$

$$= -\frac{2 - 32}{\sqrt{6(11 + 640 - 96)}}$$

$$= \frac{1}{\sqrt{37}} \doteq 0.1644.$$

b) The random vector $\begin{pmatrix} Z_1 \\ Z_2 \end{pmatrix}$ can be calculated as a linear transformation of the random vector X as

$$\begin{pmatrix} Z_1 \\ Z_2 \end{pmatrix} = \mathcal{A}X + b,$$

where $\mathcal{A} = \begin{pmatrix} 0 & 1 & -1 \\ 0 & 1 & 1 \end{pmatrix}$ and $b = 0$. According to Theorem 5.2, the vector $(Z_1, Z_2)^\top$ is normally distributed with the expected value

$$\mu_{12} = \mathcal{A}\mu = \begin{pmatrix} 0 & 1 & -1 \\ 0 & 1 & 1 \end{pmatrix} \begin{pmatrix} 1 \\ 2 \\ 3 \end{pmatrix} = \begin{pmatrix} -1 \\ 5 \end{pmatrix}$$

and the variance matrix

$$\Sigma_{12} = \mathcal{A}\Sigma\mathcal{A}^\top = \begin{pmatrix} 0 & 1 & -1 \\ 0 & 1 & 1 \end{pmatrix} \begin{pmatrix} 11 & -6 & 2 \\ -6 & 10 & -4 \\ 2 & -4 & 6 \end{pmatrix} \begin{pmatrix} 0 & 0 \\ 1 & 1 \\ -1 & 1 \end{pmatrix} = \begin{pmatrix} 24 & 4 \\ 4 & 8 \end{pmatrix}.$$

Since we know $(Z_3|Z_1Z_2) \sim N(Z_1 + Z_2, 10)$, we can apply the result derived in Exercise 5.2 with $\mathcal{A} = \begin{pmatrix} 1 \\ 1 \end{pmatrix}$, $b = 0$, and $\Omega = 10$. Then

$$Z = \begin{pmatrix} Z_1 \\ Z_2 \\ Z_3 \end{pmatrix} \sim N_3(\mu_Z, \Sigma_Z),$$

where

$$\mu_Z = \begin{pmatrix} \mu_{Z_{12}} \\ \mathcal{A}\mu_{Z_{12}} + b \end{pmatrix} = \begin{pmatrix} -1 \\ 5 \\ 4 \end{pmatrix}$$

and

$$\Sigma_Z = \begin{pmatrix} \Sigma_{12} & \Sigma_{12}\mathcal{A}^\top \\ \mathcal{A}\Sigma_{12} & \Omega + \mathcal{A}\Sigma_{12}\mathcal{A}^\top \end{pmatrix} = \begin{pmatrix} 24 & 4 & 28 \\ 4 & 8 & 12 \\ 28 & 12 & 50 \end{pmatrix}.$$

EXERCISE 5.9. *Let $(X, Y, Z)^\top$ be a tri-variate normal r.v. with*

$$Y \mid Z \sim N_1(2Z, 24)$$
$$Z \mid X \sim N_1(2X + 3, 14)$$
$$X \sim N_1(1, 4)$$
$$\text{and } \rho_{XY} = 0.5.$$

Find the distribution of $(X, Y, Z)^\top$ and compute the partial correlation between X and Y for fixed Z. Do you think it is reasonable to approximate X by a linear function of Y and Z?

Using the known marginal distribution $X \sim N_1(\mu_X, \sigma_X^2) \sim N(1, 4)$ and the conditional distribution $Z|X \sim N_1(\mathcal{A}X + b, \Omega) \sim N(2X + 3, 14)$, the method explained in Exercise 5.2 leads to

$$\begin{pmatrix} X \\ Z \end{pmatrix} \sim N_2\left(\begin{pmatrix} \mu_X \\ \mathcal{A}\mu_X + b \end{pmatrix}, \begin{pmatrix} \sigma_X^2 & \mathcal{A}\sigma_X^2 \\ \mathcal{A}\sigma_X^2 & \mathcal{A}\sigma_X^2\mathcal{A} + \Omega \end{pmatrix} \right)$$

$$\sim N_2\left(\begin{pmatrix} 1 \\ 2 + 3 \end{pmatrix}, \begin{pmatrix} 4 & 8 \\ 8 & 16 + 14 \end{pmatrix} \right)$$

$$\sim N_2\left(\begin{pmatrix} 1 \\ 5 \end{pmatrix}, \begin{pmatrix} 4 & 8 \\ 8 & 30 \end{pmatrix} \right).$$

Clearly, the marginal distribution of the random variable Z is $Z \sim N(5, 30)$ and the same rule can be used to determine the joint distribution of $(Y, Z)^\top$ from the conditional distribution $Y|Z \sim N(\mathcal{C}Z + d, \Phi) \sim N(2Z, 24)$:

$$\begin{pmatrix} Y \\ Z \end{pmatrix} \sim N_2 \left(\begin{pmatrix} \mathcal{C}\mu_Z + d \\ \mu_Z \end{pmatrix}, \begin{pmatrix} \mathcal{C}\sigma_Z^2\mathcal{C} + \Phi\,\mathcal{C}\sigma_Z^2 \\ \mathcal{C}\sigma_Z^2 \quad \sigma_Z^2 \end{pmatrix} \right)$$

$$\sim N_2 \left(\begin{pmatrix} 10 \\ 5 \end{pmatrix}, \begin{pmatrix} 120 + 24 & 60 \\ 60 & 30 \end{pmatrix} \right)$$

$$\sim N_2 \left(\begin{pmatrix} 10 \\ 5 \end{pmatrix}, \begin{pmatrix} 144 & 60 \\ 60 & 30 \end{pmatrix} \right)$$

Finally, the correlation ρ_{XY} of X and Y allows us to calculate the covariance σ_{XY} of X and Y:

$$\sigma_{XY} = \rho_{XY}\sqrt{\sigma_{XX}\sigma_{YY}}$$
$$= \frac{1}{2}\sqrt{4 \cdot 144} = 12$$

and the joint distribution of the random vector $(X, Y, Z)^\top$ is thus

$$\begin{pmatrix} X \\ Y \\ Z \end{pmatrix} \sim N_3 \left(\begin{pmatrix} 1 \\ 10 \\ 5 \end{pmatrix}, \begin{pmatrix} 4 & 12 & 8 \\ 12 & 144 & 60 \\ 8 & 60 & 30 \end{pmatrix} \right).$$

The partial correlation coefficient, $\rho_{XY|Z}$, of X and Y for fixed Z can be written in terms of simple correlation coefficients as

$$\rho_{XY|Z} = \frac{\rho_{XY} - \rho_{XZ}\rho_{YZ}}{\sqrt{(1 - \rho_{XZ}^2)(1 - \rho_{YZ}^2)}}.$$

Plugging in the appropriate elements of the covariance matrix, we obtain

$$\rho_{XY|Z} = \frac{\frac{\sigma_{XY}}{\sqrt{\sigma_{XX}\sigma_{YY}}} - \frac{\sigma_{XZ}\sigma_{YZ}}{\sqrt{\sigma_{XX}\sigma_{YY}\sigma_{ZZ}^2}}}{\sqrt{(1 - \frac{\sigma_{XZ}^2}{\sigma_{XX}\sigma_{ZZ}})(1 - \frac{\sigma_{YZ}^2}{\sigma_{YY}\sigma_{ZZ}})}}$$

$$= \frac{\frac{12}{\sqrt{4 \cdot 144}} - \frac{8 \cdot 60}{\sqrt{4 \cdot 144 \cdot 30^2}}}{\sqrt{(1 - \frac{8^2}{4 \cdot 30})(1 - \frac{60^2}{144 \cdot 30})}}$$

$$= \frac{\frac{1}{2} - \frac{2}{3}}{\sqrt{(\frac{56}{120})(\frac{1}{6})}} = -\frac{\frac{1}{6}}{\sqrt{\frac{7}{90}}} = -\frac{1}{2}\sqrt{\frac{2}{7}} \doteq -0.2673.$$

The best linear approximation of X in terms of Y and Z is given by the conditional expectation $\mu_{X|YZ}$ which, using Theorem 5.3, can be calculated as

$$\mu_{X|YZ} = \mu_X + (12, 8) \begin{pmatrix} 144 & 60 \\ 60 & 30 \end{pmatrix}^{-1} \begin{pmatrix} Y - \mu_Y \\ Z - \mu_Z \end{pmatrix}$$

$$= 1 + (12, 8) \frac{1}{720} \begin{pmatrix} 30 & -60 \\ -60 & 144 \end{pmatrix} \begin{pmatrix} Y - 10 \\ Z - 5 \end{pmatrix}$$

$$= 1 + \frac{1}{720}(-120, 432) \begin{pmatrix} Y - 10 \\ Z - 5 \end{pmatrix}$$

$$= 1 + \frac{1}{720}(-120, 432) \begin{pmatrix} Y - 10 \\ Z - 5 \end{pmatrix}$$

$$= \frac{7}{4} - \frac{1}{6}Y + \frac{3}{5}Z.$$

Such a linear approximation seems to make a good sense, the quality of the linear approximation can be assessed via the multiple correlation coefficient:

$$\rho_{X;(Y,Z)} = \frac{5\sigma_{XY} - 18\sigma_{XZ}}{\sqrt{\sigma_X X(25\sigma_{YY} + 324\sigma_{ZZ} - 180\sigma_{YZ})}}$$

$$= \frac{60 - 144}{\sqrt{4(3600 + 9720 - 10800)}}$$

$$= \frac{-84}{2\sqrt{2520}} = -\frac{7}{\sqrt{70}} = -\sqrt{\frac{7}{10}} \doteq -0.8367$$

suggesting quite a strong relationship between X and $(Y, Z)^\top$.

EXERCISE 5.10. Let $X \sim N_4 \left(\begin{pmatrix} 1 \\ 2 \\ 3 \\ 4 \end{pmatrix}, \begin{pmatrix} 4 & 1 & 2 & 4 \\ 1 & 4 & 2 & 1 \\ 2 & 2 & 16 & 1 \\ 4 & 1 & 1 & 9 \end{pmatrix} \right)$.

a) Give the best linear approximation of X_2 as a function of (X_1, X_4) and evaluate the quality of the approximation.

b) Give the best linear approximation of X_2 as a function of (X_1, X_3, X_4) and compare your answer with part a).

a) The best linear approximation of X_2 in terms of X_1 and X_4 is the conditional expectation, $\mu_{2|14}$, given as:

$$\mu_{2|14} = \mu_2 + (\sigma_{21} \; \sigma_{24}) \begin{pmatrix} \sigma_{11} & \sigma_{14} \\ \sigma_{14} & \sigma_{44} \end{pmatrix}^{-1} \begin{pmatrix} X_1 - \mu_1 \\ X_4 - \mu_4 \end{pmatrix}$$

$$= 2 + (1,1) \begin{pmatrix} 4 & 4 \\ 4 & 9 \end{pmatrix}^{-1} \begin{pmatrix} X_1 - 1 \\ X_4 - 4 \end{pmatrix}$$

$$= 2 + (1,1) \frac{1}{20} \begin{pmatrix} 9 & -4 \\ -4 & 4 \end{pmatrix} \begin{pmatrix} X_1 - 1 \\ X_4 - 4 \end{pmatrix}$$

$$= 2 + \frac{1}{20} (5,0) \begin{pmatrix} X_1 - 1 \\ X_4 - 4 \end{pmatrix}$$

$$= \frac{7}{4} + \frac{1}{4} X_1.$$

b) To determine the best linear approximation of X_2 as a function of (X_1, X_2, X_3), we use the same procedure so that

$$\mu_{2|134} = \mu_2 + (\sigma_{21} \; \sigma_{23} \; \sigma_{24}) \begin{pmatrix} \sigma_{11} & \sigma_{13} & \sigma_{14} \\ \sigma_{31} & \sigma_{33} & \sigma_{34} \\ \sigma_{41} & \sigma_{43} & \sigma_{44} \end{pmatrix}^{-1} \begin{pmatrix} X_1 - \mu_1 \\ X_3 - \mu_3 \\ X_4 - \mu_4 \end{pmatrix}$$

$$= 2 + (1,2,1) \begin{pmatrix} 4 & 2 & 4 \\ 2 & 16 & 1 \\ 4 & 1 & 9 \end{pmatrix}^{-1} \begin{pmatrix} X_1 - 1 \\ X_3 - 3 \\ X_4 - 4 \end{pmatrix}$$

$$= 2 + (1,2,1) \frac{1}{296} \begin{pmatrix} 143 & -14 & -62 \\ -14 & 20 & 4 \\ -62 & 4 & 60 \end{pmatrix} \begin{pmatrix} X_1 - 1 \\ X_3 - 3 \\ X_4 - 4 \end{pmatrix}$$

$$= 2 + \frac{1}{296} (53, 30, 6) \begin{pmatrix} X_1 - 1 \\ X_3 - 3 \\ X_4 - 4 \end{pmatrix}$$

$$= \frac{425}{296} + \frac{53}{296} X_1 + \frac{15}{148} X_3 + \frac{3}{148} X_4.$$

This exercise demonstrates that the variable X_4, which was not important for the prediction of X_2 based on X_1 and X_4, can enter the formula for the conditional expected value when another explanatory variable, X_3, is added. In multivariate analyses, such dependencies occur very often.

EXERCISE 5.11. *Prove Theorem 5.2.*

As in Theorem 5.2, let us assume that $X \sim N_p(\mu, \Sigma)$, $\mathcal{A}(q \times p)$, $c \in \mathbb{R}^q$ and rank$(\mathcal{A}) = q \le p$. Our goal is to calculate the distribution of the random vector $Y = \mathcal{A}X + c$.

Recall that the pdf of $X \sim N_p(\mu, \Sigma)$ is

$$f_X(x) = |2\pi \Sigma|^{-1/2} \exp\left\{-\frac{1}{2}(x-\mu)^\top \Sigma^{-1}(x-\mu)\right\}. \tag{5.10}$$

We start by considering the linear transformation

$$Z = \begin{pmatrix} \mathcal{A} \\ \mathcal{B} \end{pmatrix} X + \begin{pmatrix} c \\ 0_{p-q} \end{pmatrix} = \mathcal{D}X + e,$$

where $\mathcal{B}$ contains in its rows $p - q$ arbitrary linearly independent vectors orthogonal to the rows of the matrix $\mathcal{A}$. Hence, the matrix $\mathcal{D} = \begin{pmatrix} \mathcal{A} \\ \mathcal{B} \end{pmatrix}$ has full rank and the density of Z can be expressed as:

$$f_Z(z)$$
$$= \text{abs} \, |\mathcal{D}|^{-1} f_X\{\mathcal{D}^{-1}(z-e)\}$$
$$= (|\mathcal{D}|^2)^{-1/2}|2\pi \Sigma|^{-1/2} \exp\left\{-\frac{1}{2}\{\mathcal{D}^{-1}(z-e)-\mu\}^\top \Sigma^{-1}\{\mathcal{D}^{-1}(z-e)-\mu\}\right\}$$
$$= |2\pi \mathcal{D}\Sigma\mathcal{D}|^{-1/2} \exp\left\{-\frac{1}{2}(z-e-\mathcal{D}\mu)^\top (\mathcal{D}^{-1})^\top \Sigma^{-1}\mathcal{D}^{-1}(z-e-\mathcal{D}\mu)\right\}$$
$$= |2\pi \mathcal{D}\Sigma\mathcal{D}|^{-1/2} \exp\left\{-\frac{1}{2}\{z-(\mathcal{D}\mu+e)\}^\top (\mathcal{D}\Sigma\mathcal{D}^\top)^{-1}\{z-(\mathcal{D}\mu+e)\}\right\}.$$

Notice that the above formula is exactly the density of the p-dimensional normal distribution $N_p(\mathcal{D}\mu + e, \mathcal{D}\Sigma\mathcal{D}^\top)$.

More precisely, we can write that

$$Z \sim N_p(\mathcal{D}\mu + e, \mathcal{D}\Sigma\mathcal{D}^\top)$$
$$\sim N_p\left(\begin{pmatrix} \mathcal{A} \\ \mathcal{B} \end{pmatrix}\mu + e, \begin{pmatrix} \mathcal{A} \\ \mathcal{B} \end{pmatrix}\Sigma(\mathcal{A}^\top, \mathcal{B}^\top)\right)$$
$$\sim N_p\left(\begin{pmatrix} \mathcal{A}\mu + c \\ \mathcal{B}\mu \end{pmatrix}, \begin{pmatrix} \mathcal{A}\Sigma\mathcal{A}^\top & \mathcal{A}\Sigma\mathcal{B}^\top \\ \mathcal{B}\Sigma\mathcal{A}^\top & \mathcal{B}\Sigma\mathcal{B}^\top \end{pmatrix}\right)$$

Noticing that the first part of the random vector Z is exactly the random vector Y and applying Theorem 5.1 we have that the distribution of $Y = \mathcal{A}X + c$ is q-variate normal, i.e.,

$$Y \sim N_q(\mathcal{A}\mu + c, \mathcal{A}\Sigma\mathcal{A}^\top).$$

EXERCISE 5.12. *Let* $X = \begin{pmatrix} X_1 \\ X_2 \end{pmatrix} \sim N_p(\mu, \Sigma)$, $\Sigma = \begin{pmatrix} \Sigma_{11} & \Sigma_{12} \\ \Sigma_{21} & \Sigma_{22} \end{pmatrix}$. *Prove that* $\Sigma_{12} = 0$ *if and only if* X_1 *is independent of* X_2.

We already know, from the previous chapters, that independence implies zero covariance since, for X_1 and X_2 independent, we have

$$\text{Cov}(X_1, X_2) = EX_1EX_2 - EX_1X_2 = EX_1EX_2 - EX_1EX_2 = 0.$$

It remains to show that, for normally distributed random vectors, zero covariance implies independence.

Applying Theorem 5.1 with the given covariance matrix

$$\Sigma = \begin{pmatrix} \Sigma_{11} & 0 \\ 0 & \Sigma_{22} \end{pmatrix}$$

we obtain that $X_{2.1} = X_2 + 0\Sigma_{11}^{-1}\mu_1 = X_2$ and from Theorem 5.1 we immediately have that $X_2 = X_{2.1}$ and X_1 are independent.

EXERCISE 5.13. *Show that if $X \sim N_p(\mu, \Sigma)$ and given some matrices $\mathcal{A}$ and $\mathcal{B}$, then $\mathcal{A}X$ and $\mathcal{B}X$ are independent if and only if $\mathcal{A}\Sigma\mathcal{B}^\top = 0$.*

Let us define the random vector

$$Z = \begin{pmatrix} \mathcal{A} \\ \mathcal{B} \end{pmatrix} X = \begin{pmatrix} \mathcal{A}X \\ \mathcal{B}X \end{pmatrix}.$$

Using the result of the previous Exercise 5.12, where $X_1 = \mathcal{A}X$ and $X_2 = \mathcal{B}X$, it is clear that the multivariate random vectors $\mathcal{A}X$ and $\mathcal{B}X$ are independent if and only if their covariance matrix $\mathcal{A}\Sigma\mathcal{B}^\top$ is equal to zero.

6

Theory of Estimation

No, no; I never guess. It is a shocking habit—destructive to the logical
faculty.
Sherlock Holmes in "The Sign of Four"

The basic objective of statistics is to understand and model the underlying
processes that generate the data. This involves statistical inference, where we
extract information contained in a sample by applying a model. In general,
we assume an i.i.d. random sample $\{x_i\}_{i=1}^n$ from which we extract unknown
characteristics of its distribution. In parametric statistics these are condensed
in a p-variate vector θ characterizing the unknown properties of the population pdf $f(x, \theta)$: this could be the mean, the covariance matrix, kurtosis, or
something else.

The aim is to estimate θ from the sample $\mathcal{X}$ through estimators $\hat{\theta}$ that are
functions of the sample: $\hat{\theta} = b(\mathcal{X})$. When an estimator is proposed, we must
derive its sampling distribution to analyze its properties: are they related to
the unknown characteristic it is supposed to estimate?

Let the symbol $\mathcal{X}(n \times p)$ denote the data matrix containing p-dimensional
observations, $x_i \sim f(., \theta)$, $i = 1, \ldots, n$, in each row. The maximum likelihood
estimator (MLE) of θ is defined as

$$\hat{\theta} = \arg \max_\theta L(\mathcal{X}; \theta) = \arg \max_\theta \ell(\mathcal{X}; \theta),$$

where $L(\mathcal{X}; \theta) = \prod_{i=1}^n f(x_i; \theta)$ is the likelihood function, i.e., the joint density
of the observations $x_i \sim f(., \theta)$ considered as a function of θ and $\ell(\mathcal{X}; \theta) =
\log L(\mathcal{X}; \theta)$ is the log-likelihood function.

The score function $s(\mathcal{X}; \theta)$ is the derivative of the log-likelihood function w.r.t.
$\theta \in \mathbb{R}^k$

$$s(\mathcal{X}; \theta) = \frac{\partial}{\partial \theta} \ell(\mathcal{X}; \theta) = \frac{1}{L(\mathcal{X}; \theta)} \frac{\partial}{\partial \theta} L(\mathcal{X}; \theta).$$

The covariance matrix

$$\mathcal{F}_n = E\{s(\mathcal{X}; \theta)s(\mathcal{X}; \theta)^\top\} = \text{Var}\{s(\mathcal{X}; \theta)\} = -E\left\{\frac{\partial^2}{\partial\theta\partial\theta^\top}\ell(\mathcal{X}; \theta)\right\}$$

is called the Fisher information matrix.

The importance of the Fisher information matrix is explained by the following Cramer-Rao theorem, which gives the lower bound for the variance matrix for any unbiased estimator of θ. An unbiased estimator with the variance equal to $\mathcal{F}_n^{-1}$ is called a minimum variance unbiased estimator.

THEOREM 6.1 (Cramer-Rao). *If $\hat\theta = t = t(\mathcal{X})$ is an unbiased estimator for θ, then under regularity conditions*

$$\text{Var}(t) \geq \mathcal{F}_n^{-1}.$$

Another important result says that the MLE is asymptotically unbiased, efficient (minimum variance), and normally distributed.

THEOREM 6.2. *Suppose that the sample $\{x_i\}_{i=1}^n$ is i.i.d. If $\widehat\theta$ is the MLE for $\theta \in \mathbb{R}^k$ then under some regularity conditions, as $n \to \infty$:*

$$\sqrt{n}(\widehat\theta - \theta) \xrightarrow{\mathcal{L}} N_k(0, \mathcal{F}_1^{-1}),$$

where $\mathcal{F}_1$ denotes the Fisher information for sample size $n = 1$.

Whenever we are not able to calculate the exact distribution of the MLE $\widehat\theta$, Theorem 6.2 gives us a very useful and simple approximation.

In this chapter we present calculation of the Fisher information matrix for several examples. We also discuss and calculate Cramer-Rao lower bounds for these situations. We will illustrate the estimation for multivariate normal pdf in detail and discuss constrained estimation.

EXERCISE 6.1. *Consider a uniform distribution on the interval $[0, \theta]$. What is the MLE of θ? (Hint: the maximization here cannot be performed by means of derivatives. Here the support of x depends on θ!)*

The density of the uniform distribution on the interval $[0, \theta]$ is

$$f(x) = \begin{cases} \frac{1}{\theta} & \text{if } x \in [0, \theta], \\ 0 & \text{else.} \end{cases}$$

Assuming that we have n independent and identically distributed (iid) random variables, $X_1, \ldots, X_n$, from this distribution, the likelihood function

$$L(X_1, \ldots, X_n; \theta) = \begin{cases} \theta^{-n} & \text{if } X_1, \ldots, X_n \in [0, \theta], \\ 0 & \text{else.} \end{cases}$$

The maximum of the likelihood is achieved by choosing θ as small as possible such that $0 \le X_1, \ldots, X_n \le \theta$. The maximum likelihood estimator,

$$\hat{\theta} = \arg\max_\theta L(X_1, \ldots, X_n; \theta),$$

can thus be written as $\hat{\theta} = \max_{i=1,\ldots,n} X_i$.

EXERCISE 6.2. *Consider an iid sample of size n from a bivariate population with pdf* $f(x_1, x_2) = \frac{1}{\theta_1 \theta_2} exp\left\{ - \left(\frac{x_1}{\theta_1} + \frac{x_2}{\theta_2} \right) \right\}$, $x_1, x_2 > 0$. *Compute the MLE of* $\theta = (\theta_1, \theta_2)^\top$. *Find the Cramer-Rao lower bound. Is it possible to derive a minimum variance unbiased estimator of* θ?

The function $f(.)$ is a probability density function (pdf) only if $\theta_1, \theta_2 > 0$.

Let $\mathcal{X}(n \times 2) = (x_{ij})$ denote the data matrix containing in its rows the n independent bivariate observations from the given pdf.

The marginal densities can be calculated by integrating the bivariate pdf:

$$f_1(x_1) = \int_0^{+\infty} f(x_1, x_2) dx_2 = \frac{1}{\theta_1} \exp(-x_1/\theta_1),$$

$$f_2(x_2) = \int_0^{+\infty} f(x_1, x_2) dx_1 = \frac{1}{\theta_2} \exp(-x_2/\theta_2).$$

Notice that $f(x_1, x_2) = f_1(x_1) f_2(x_2)$. Thus, the marginal distributions are independent.

The expected values, μ_1 and μ_2, of the marginal distributions are

$$\mu_1 = \int_0^{+\infty} x_1 \frac{1}{\theta_1} \exp(-x_1/\theta_1) dx_1$$

$$= \left[x_1 e^{-\frac{x_1}{\theta_1}} \right]_0^{+\infty} - \int_0^{+\infty} \exp(-x_1/\theta_1) dx_1$$

$$= -\left[\theta_1 \exp(-x_1/\theta_1) \right]_0^{+\infty} = \theta_1$$

and $\mu_2 = \theta_2$ since the marginal distributions are identical. Similarly, the variances are

$$\sigma_1^2 = E(X^2) + \mu_1^2$$

$$= \int_0^{+\infty} x_1^2 \frac{1}{\theta_1} \exp(-x_1/\theta_1) dx_1 + \mu_1^2$$

$$= \theta_1^2$$

and $\sigma_2^2 = \theta_2^2$.

After writing down the log-likelihood function, $\ell(\mathcal{X}; \theta_1, \theta_2)$, where the pdf $f(.)$ is thought of as a function of the (unknown) parameters θ_1 and θ_2,

$$\ell(\mathcal{X}; \theta_1, \theta_2) = \log \prod_{i=1}^{n} f(x_{i1}, x_{i2}; \theta_1, \theta_2)$$

$$= \log \prod_{i=1}^{n} \frac{1}{\theta_1 \theta_2} e^{-\left(\frac{x_{i1}}{\theta_1} + \frac{x_{i2}}{\theta_2}\right)}$$

$$= n \log \frac{1}{\theta_1} + n \log \frac{1}{\theta_2} - \sum_{i=1}^{n} \frac{x_{i1}}{\theta_1} - \sum_{i=1}^{n} \frac{x_{i2}}{\theta_2},$$

the MLE of θ_1 and θ_2 are obtained by solving the system of equations

$$\frac{\partial \ell(\mathcal{X}; \theta_1, \theta_2)}{\partial \theta_1} = -\frac{n}{\theta_1} + \sum_{i=1}^{n} \frac{x_{i1}}{\theta_1^2} = 0$$

and

$$\frac{\partial \ell(\mathcal{X}; \theta_1, \theta_2)}{\partial \theta_2} = -\frac{n}{\theta_2} + \sum_{i=1}^{n} \frac{x_{i2}}{\theta_2^2} = 0.$$

It follows that the MLEs are the sample means $\widehat{\theta}_1 = \overline{x}_1$ and $\widehat{\theta}_2 = \overline{x}_2$.

The Cramer-Rao lower bound for the variance of any unbiased estimator for θ is $\mathcal{F}_n^{-1}$, the inverse of the Fisher information matrix $\mathcal{F}_n = E\{s(\mathcal{X}; \theta)s(\mathcal{X}; \theta)^{\top}\} = \text{Var } s(\mathcal{X}; \theta)$, where $s(\mathcal{X}; \theta) = \frac{\partial}{\partial \theta}\ell(\mathcal{X}; \theta)$ is the so-called score function.

In this exercise, the score function is

$$s(\mathcal{X}; \theta) = \begin{pmatrix} -\frac{n}{\theta_1} + \sum_{i=1}^{n} \frac{x_{i1}}{\theta_1^2} \\ -\frac{n}{\theta_2} + \sum_{i=1}^{n} \frac{x_{i2}}{\theta_2^2} \end{pmatrix}$$

Since the observations are iid, the Fisher information matrix $\mathcal{F}_n = n\mathcal{F}_1$ and from the Fisher information matrix calculated as if $n = 1$,

$$\mathcal{F}_1 = \text{Var} \begin{pmatrix} -\frac{1}{\theta_1} + \frac{x_{11}}{\theta_1^2} \\ -\frac{1}{\theta_2} + \frac{x_{12}}{\theta_2^2} \end{pmatrix} = \begin{pmatrix} \frac{1}{\theta_1^2} & 0 \\ 0 & \frac{1}{\theta_2^2} \end{pmatrix}$$

we easily obtain the Cramer-Rao lower bound:

$$\mathcal{F}_n^{-1} = \frac{1}{n}\mathcal{F}_1^{-1} = \frac{1}{n}\begin{pmatrix} \theta_1^2 & 0 \\ 0 & \theta_2^2 \end{pmatrix}.$$

Calculating the expected values and variances of the maximum likelihood estimators:

$$E(\widehat{\theta}_1) = E\frac{1}{n}\sum_{i=1}^{n} x_{i1} = \mu_1,$$

$$E(\widehat{\theta}_2) = \mu_2,$$

$$Var(\widehat{\theta}_1) = Var\,\frac{1}{n}\sum_{i=1}^{n} x_{i1} = \frac{1}{n}\,Var\,x_{i1} = \frac{1}{n}\theta_1^2,$$

$$Var(\widehat{\theta}_2) = \frac{1}{n}\theta_2^2,$$

we can see that the estimators $\widehat{\theta}_1$ and $\widehat{\theta}_2$ achieve the Cramer-Rao lower bound and, hence, $\widehat{\theta} = (\widehat{\theta}_1, \widehat{\theta}_2)^\top$ is the minimum variance unbiased estimator of the parameter θ.

EXERCISE 6.3. *Consider a sample $\{x_i\}_{i=1}^{n}$ from $N_p(\theta, \mathcal{I}_p)$, where $\theta \in \mathbb{R}^p$ is the mean vector parameter. Show that the MLE of θ is the minimum variance estimator.*

The log-likelihood is in this case

$$\ell(\mathcal{X};\theta) = \sum_{i=1}^{n} \log\{f(x_i;\theta)\}$$

$$= \log(2\pi)^{-np/2} - \frac{1}{2}\sum_{i=1}^{n}(x_i - \theta)^\top(x_i - \theta)$$

$$= \log(2\pi)^{-np/2} - \frac{1}{2}\sum_{i=1}^{n}\{(x_i - \overline{x})^\top(x_i - \overline{x}) + (\overline{x} - \theta)^\top(\overline{x} - \theta)$$

$$+ 2(\overline{x} - \theta)^\top(x_i - \overline{x})\}$$

$$= \log(2\pi)^{-np/2} - \frac{1}{2}\sum_{i=1}^{n}(x_i - \overline{x})^\top(x_i - \overline{x}) - \frac{n}{2}(\overline{x} - \theta)^\top(\overline{x} - \theta)$$

The last term is the only part depending on θ and it is obviously maximized for $\theta = \overline{x}$. Thus $\widehat{\theta} = \overline{x}$ is the MLE of θ for this family of pdfs $f(x, \theta)$.

It follows that the score function is

$$s(\mathcal{X};\theta) = \frac{\partial}{\partial\theta}\ell(\mathcal{X};\theta)$$

$$= -\frac{n}{2}\frac{\partial}{\partial\theta}(\overline{x} - \theta)^\top(\overline{x} - \theta)$$

$$= n(\overline{x} - \theta).$$

We obtain the Fisher information matrix as the variance of the score function:

$$\mathcal{F}_n = Var\{s(\mathcal{X};\theta)\} = Var\{n(\overline{x} - \theta)\} = n^2\,Var\,\overline{x} = n\,\mathcal{I}_p$$

and the Cramer-Rao lower bound for this case is

$$\mathcal{F}_n^{-1} = \frac{1}{n}\mathcal{I}_p. \tag{6.1}$$

We know that the mean and the variance of $\hat{\theta} = \bar{x}$ are:

$$E\,\bar{x} = \theta,$$

$$Var\ \bar{x} = \frac{1}{n}\mathcal{I}_p.$$

Hence, the MLE estimator is unbiased and its variance attains the Cramer-Rao lower bound, see (6.1). Thus it is the minimum variance unbiased estimator.

EXERCISE 6.4. *We know from Exercise 6.3 that the MLE of parameter θ based on observations from the multinormal distribution $N_p(\theta, \mathcal{I}_p)$ has the Fisher information $\mathcal{F}_1 = \mathcal{I}_p$. This leads to the asymptotic distribution*

$$\sqrt{n}(\bar{x} - \theta) \xrightarrow{\mathcal{L}} N_p(0, \mathcal{I}_p),$$

see also Theorem 6.2. Can you derive an analogous result for the square $\bar{x}^2$?

One possibility is to consider $\bar{x}^2$ as a transformation of the statistics $\bar{x}$. In this way, with transformation $f(x) = (x_1^2, \ldots, x_p^2)$, we immediately obtain that the matrix of partial derivatives is

$$\mathcal{D} = \left(\frac{\partial f_j}{\partial x_i}\right)(x)\bigg|_{x=\theta} = \mathrm{diag}(2\theta_1, \ldots, 2\theta_p)$$

and that the asymptotic distribution of the transformed asymptotically normal statistics is

$$\sqrt{n}(\bar{x}^2 - \theta^2) \xrightarrow{\mathcal{L}} N_p(0, \mathcal{D}^\top \mathcal{I}_p \mathcal{D}) = N_p(0, 4\,\mathrm{diag}(\theta_1^2, \ldots, \theta_p^2)).$$

Second possibility, in this situation more straightforward, is to denote by $x \cdot y$ the componentwise product of vectors x and y and to write

$$\sqrt{n}(\bar{x}^2 - \theta^2) = (\bar{x} + \theta) \cdot \sqrt{n}(\bar{x} - \theta)$$
$$\xrightarrow{\mathcal{L}} 2\theta N_p(0, \mathcal{I}_p) = N_p(0, 4\theta^2 \mathcal{I}_p)$$

since $(\bar{x} + \theta) \xrightarrow{P} 2\theta$ and $\sqrt{n}(\bar{x} - \theta) \sim N_p(0, \mathcal{I}_p)$.

EXERCISE 6.5. *Consider an iid sample of size n from the bivariate population with pdf*

$$f(x_1, x_2) = \frac{1}{\theta_1^2 \theta_2} \frac{1}{x_2} \exp\left\{-\left(\frac{x_1}{\theta_1 x_2} + \frac{x_2}{\theta_1 \theta_2}\right)\right\}, \quad x_1, x_2 > 0.$$

Compute the MLE, $\hat{\theta}$, of the unknown parameter $\theta = (\theta_1, \theta_2)^\top$. Find the Cramer-Rao lower bound and the asymptotic variance of $\hat{\theta}$.

The estimator $\widehat{\theta}$ is the maximizer of the log-likelihood function

$$
\begin{aligned}
\ell(\mathcal{X}; \theta_1, \theta_2) &= \log \prod_{i=1}^{n} f(x_{i1}, x_{i2}; \theta_1, \theta_2) \\
&= \log \prod_{i=1}^{n} \frac{1}{\theta_1^2 \theta_2} \frac{1}{x_{i2}} e^{-\left(\frac{x_{i1}}{\theta_1 x_{i2}} + \frac{x_{i2}}{\theta_1 \theta_2}\right)} \\
&= n \log \frac{1}{\theta_1^2 \theta_2} + \sum_{i=1}^{n} \log \frac{1}{x_{i2}} - \sum_{i=1}^{n} \left(\frac{x_{i1}}{\theta_1 x_{i2}} + \frac{x_{i2}}{\theta_1 \theta_2}\right) \\
&= -n(2 \log \theta_1 + \log \theta_2) - \sum_{i=1}^{n} \log x_{i2} - \sum_{i=1}^{n} \left(\frac{x_{i1}}{\theta_1 x_{i2}} + \frac{x_{i2}}{\theta_1 \theta_2}\right).
\end{aligned}
$$

The MLE of θ can be found by solving the system of equations

$$
\frac{\partial \ell(\mathcal{X}; \theta_1, \theta_2)}{\partial \theta_1} = -\frac{2n}{\theta_1} + \sum_{i=1}^{n} \left(\frac{x_{i1}}{\theta_1^2 x_{i2}} + \frac{x_{i2}}{\theta_1^2 \theta_2}\right) = 0
$$

and

$$
\frac{\partial \ell(\mathcal{X}; \theta_1, \theta_2)}{\partial \theta_2} = -\frac{n}{\theta_2} + \sum_{i=1}^{n} \frac{x_{i2}}{\theta_1 \theta_2^2} = 0.
$$

From the second equation it follows that $\bar{x}_2 = \theta_1 \theta_2$. Plugging this into the first equation leads to the MLE

$$
\widehat{\theta}_1 = \frac{1}{n} \sum_{i=1}^{n} \frac{x_{i1}}{x_{i2}} \quad \text{and} \quad \widehat{\theta}_2 = \frac{\bar{x}_2}{\widehat{\theta}_1} = \frac{n \bar{x}_2}{\sum_{i=1}^{n} \frac{x_{i1}}{x_{i2}}}.
$$

From the score function,

$$
s(\mathcal{X}; \theta) = \begin{pmatrix} -\frac{2n}{\theta_1} + \sum_{i=1}^{n} \left(\frac{x_{i1}}{\theta_1^2 x_{i2}} + \frac{x_{i2}}{\theta_1^2 \theta_2}\right) \\ -\frac{n}{\theta_2} + \sum_{i=1}^{n} \frac{x_{i2}}{\theta_1 \theta_2^2} \end{pmatrix},
$$

we can express the Fisher information matrix

$$
\begin{aligned}
\mathcal{F}_n &= n \mathcal{F}_1 \\
&= n \operatorname{Var} \begin{pmatrix} -\frac{2}{\theta_1} + \left(\frac{x_{11}}{\theta_1^2 x_{12}} + \frac{x_{12}}{\theta_1^2 \theta_2}\right) \\ -\frac{1}{\theta_2} + \frac{x_{12}}{\theta_1 \theta_2^2} \end{pmatrix},
\end{aligned}
$$

where the variance matrix can be calculated from the moments similarly as in Exercise 6.2:

$$
\operatorname{Var}\left(\frac{x_{11}}{x_{12}}\right) = E\left(\frac{x_{11}^2}{x_{12}^2}\right) - \left\{E\left(\frac{x_{11}}{x_{12}}\right)\right\}^2 = \theta_1^2 \quad \text{and} \quad \operatorname{Var}(x_{12}) = \theta_2^2 \theta_1^2.
$$

The covariance, $\text{Cov}\left(\frac{x_{11}}{x_{12}}, x_{12}\right) = 0$ because the given density can be decomposed into a product of two independent parts. We obtain

$$\mathcal{F}_1 = \begin{pmatrix} \frac{2}{\theta_1^2} & \frac{1}{\theta_1 \theta_2} \\ \frac{1}{\theta_1 \theta_2} & \frac{1}{\theta_2^2} \end{pmatrix},$$

which leads to the Cramer-Rao lower bound

$$\mathcal{F}_n^{-1} = \frac{1}{n}\mathcal{F}_1^{-1} = \frac{\theta_1^2 \theta_2^2}{n} \begin{pmatrix} \frac{1}{\theta_2^2} & -\frac{1}{\theta_1 \theta_2} \\ -\frac{1}{\theta_1 \theta_2} & \frac{2}{\theta_1^2} \end{pmatrix} = \frac{1}{n}\begin{pmatrix} \theta_1^2 & -\theta_1\theta_2 \\ -\theta_1\theta_2 & 2\theta_2^2 \end{pmatrix}.$$

From Theorem 6.2, we can finally say that the maximum likelihood estimator $\widehat{\theta}$ is asymptotically multivariate normally distributed:

$$\sqrt{n}(\widehat{\theta} - \theta) \xrightarrow{\mathcal{L}} N_2(0_2, \mathcal{F}_1^{-1}).$$

EXERCISE 6.6. *Consider an iid sample $\{x_i\}_{i=1}^n$ from $N_p(\mu, \Sigma_0)$ where Σ_0 is known. Compute the Cramer-Rao lower bound for μ. Can you derive a minimum variance unbiased estimator for μ?*

For the case of n iid observations, we know that the Fisher information matrix $\mathcal{F}_n = n\mathcal{F}_1$. Hence, we start by writing down the likelihood, the log-likelihood and the score function for a "sample" containing only one observation $x_1 = (x_{11}, \ldots, x_{1p})$:

$$L(x_1; \mu) = \prod_{i=1}^{1} f(x_i, \mu) = |2\pi \Sigma_0|^{-1/2} \exp\left\{ -\frac{1}{2}(x_1 - \mu)^\top \Sigma_0^{-1}(x_1 - \mu) \right\}$$

$$\ell(x_1; \mu) = \log L(x_1; \mu) = -\frac{1}{2}\log|2\pi \Sigma_0| - \frac{1}{2}(x_1 - \mu)^\top \Sigma_0^{-1}(x_1 - \mu)$$

$$s(x_1; \mu) = \frac{\partial}{\partial \mu}\ell(x_1; \mu) = \Sigma_0^{-1}(x_1 - \mu).$$

Next, we calculate the Fisher information $\mathcal{F}_1$ as the variance matrix of the score function:

$$\mathcal{F}_1 = \text{Var}\,\Sigma_0^{-1}(x_1 - \mu) = \Sigma_0^{-1}\,\text{Var}(x_1)\Sigma_0^{-1} = \Sigma_0^{-1}\Sigma_0\Sigma_0^{-1} = \Sigma_0^{-1}$$

with inverse $\mathcal{F}_1^{-1} = \Sigma_0$. We thus obtain the Cramer-Rao lower bound is $\mathcal{F}_n^{-1} = \frac{1}{n}\mathcal{F}_1^{-1} = \frac{1}{n}\Sigma_0$.

Remember that for the sample mean, $\overline{x}_n$, we have that $E(\overline{x}_n) = \mu$ and $\text{Var}(\overline{x}_n) = \frac{1}{n}\Sigma_0$. In other words, the sample mean is an unbiased estimator achieving the Cramer-Rao lower bound, i.e., the minimum variance unbiased estimator. By maximizing the log-likelihood function, $\ell(\mathcal{X}; \mu)$, it can be shown that it is also the MLE.

EXERCISE 6.7. *Let $X \sim N_p(\mu, \Sigma)$ where Σ is unknown but we know that $\Sigma = diag(\sigma_{11}, \sigma_{22}, \ldots, \sigma_{pp})$. From an iid sample of size n, find the MLE of μ and of Σ.*

Let σ denote the vector of the unknown parameters $(\sigma_{11}, \sigma_{22}, \ldots, \sigma_{pp})^\top$. The likelihood and the log-likelihood, based on the data matrix $\mathcal{X}$ containing the n observations $x_1, \ldots, x_n$, are

$$
L(\mathcal{X}; \mu, \sigma) = \prod_{i=1}^{n} f(x_i; \mu, \sigma)
$$

$$
= \prod_{i=1}^{n} |2\pi \operatorname{diag}(\sigma)|^{-1/2} \exp\left\{ -\frac{1}{2}(x_i - \mu)^\top \operatorname{diag}(\sigma^{-1})(x_i - \mu) \right\}
$$

$$
= \left(2\pi \prod_{j=1}^{p} \sigma_{jj} \right)^{-n/2} \prod_{i=1}^{n} \exp\left\{ -\frac{1}{2}(x_i - \mu)^\top \operatorname{diag}(\sigma^{-1})(x_i - \mu) \right\},
$$

$$
\ell(\mathcal{X}; \mu, \sigma) = \log L(\mathcal{X}; \mu, \sigma)
$$

$$
= -\frac{n}{2} \log(2\pi) - \frac{n}{2} \sum_{j=1}^{p} \log \sigma_{jj} - \frac{1}{2} \sum_{i=1}^{n} (x_i - \mu)^\top \operatorname{diag}(\sigma^{-1})(x_i - \mu).
$$

In order to maximize this log-likelihood function, we first have to compute the partial derivatives

$$
\frac{\partial}{\partial \mu} \ell(\mathcal{X}; \mu, \sigma) = \operatorname{diag}(\sigma^{-1}) \sum_{i=1}^{n} (x_i - \mu)
$$

$$
\frac{\partial}{\partial \sigma} \ell(\mathcal{X}; \mu, \sigma) = -\frac{n}{2}\sigma^{-1} - \frac{1}{2} \frac{\partial}{\partial \sigma} \sum_{i=1}^{n} \operatorname{tr}\{(x_i - \mu)^\top \operatorname{diag}(\sigma^{-1})(x_i - \mu)\}
$$

$$
= -\frac{n}{2}\sigma^{-1} - \frac{1}{2} \frac{\partial}{\partial \sigma} \sum_{i=1}^{n} \operatorname{tr}\{(x_i - \mu)(x_i - \mu)^\top \operatorname{diag}(\sigma^{-1})\}
$$

$$
= -\frac{n}{2}\sigma^{-1} + \frac{1}{2} \sum_{i=1}^{n} \operatorname{diag}\{(x_i - \mu)(x_i - \mu)^\top\}\sigma^{-2}.
$$

Setting the partial derivatives equal to zero, we obtain the MLE

$$
0 = \operatorname{diag}(\sigma^{-1}) \sum_{i=1}^{n} (x_i - \widehat{\mu})
$$

$$
n\widehat{\mu} = \sum_{i=1}^{n} x_i
$$

$$
\widehat{\mu} = \frac{1}{n} \sum_{i=1}^{n} x_i
$$

and

$$0 = -\frac{n}{2}\widehat{\sigma}^{-1} + \frac{1}{2}\sum_{i=1}^{n}\mathrm{diag}\{(x_i - \mu)(x_i - \mu)^\top\}\widehat{\sigma}^{-2}$$

$$n\widehat{\sigma} = \sum_{i=1}^{n}\mathrm{diag}\{(x_i - \mu)(x_i - \mu)^\top\}$$

$$\widehat{\sigma} = \mathrm{diag}\left\{\frac{1}{n}\sum_{i=1}^{n}(x_i - \mu)(x_i - \mu)^\top\right\} = \mathrm{diag}(\mathcal{S})$$

where $\mathcal{S}$ is the empirical covariance matrix.

EXERCISE 6.8. *Reconsider the setup of the previous exercise with the diagonal covariance matrix* $\Sigma = diag(\sigma) = diag(\sigma_{11}, \sigma_{22}, \ldots, \sigma_{pp})$. *Derive the Cramer-Rao lower bound for the parameter* $\theta = (\mu_1, \ldots, \mu_p, \sigma_{11}, \ldots, \sigma_{pp})^\top$.

The score function $s(\mathcal{X}; \mu, \sigma)$ consists of the partial derivatives of the log-likelihood that were derived in the previous Exercise 6.7:

$$\frac{\partial}{\partial\mu}\ell(\mathcal{X};\mu,\sigma) = \mathrm{diag}(\sigma^{-1})\sum_{i=1}^{n}(x_i - \mu)$$

$$\frac{\partial}{\partial\sigma}\ell(\mathcal{X};\mu,\sigma) = -\frac{n}{2}\sigma^{-1} + \frac{1}{2}\sum_{i=1}^{n}\mathrm{diag}\{(x_i - \mu)(x_i - \mu)^\top\}\sigma^{-2}.$$

In this exercise, we will calculate the Fisher information matrix as

$$\mathcal{F}_n = -E\left\{\frac{\partial^2}{\partial\theta\partial\theta^\top}\ell(\mathcal{X};\theta)\right\}$$

$$= -E\begin{pmatrix} \frac{\partial^2}{\partial\mu\mu^\top}\ell(\mathcal{X};\mu,\sigma) & \frac{\partial^2}{\partial\mu\sigma^\top}\ell(\mathcal{X};\mu,\sigma) \\ \frac{\partial^2}{\partial\sigma\mu^\top}\ell(\mathcal{X};\mu,\sigma) & \frac{\partial^2}{\partial\sigma\sigma^\top}\ell(\mathcal{X};\mu,\sigma) \end{pmatrix}.$$

We split the calculation into three steps by evaluating each of the four sub-matrices separately, i.e.,

$$-E\frac{\partial^2}{\partial\mu\mu^\top}\ell(\mathcal{X};\mu,\sigma) = -E\frac{\partial}{\partial\mu^\top}\mathrm{diag}(\sigma^{-1})\sum_{i=1}^{n}(x_i - \mu)$$

$$= -\mathrm{diag}(\sigma^{-1})\sum_{i=1}^{n}E\frac{\partial}{\partial\mu^\top}(x_i - \mu)$$

$$= -\mathrm{diag}(\sigma^{-1})\sum_{i=1}^{n}E\,\mathrm{diag}-1_p$$

$$= \mathrm{diag}(n\sigma^{-1}) = n\Sigma,$$

$$-E\frac{\partial^2}{\partial\mu\sigma^\top}\ell(\mathcal{X};\mu,\sigma) = -E\frac{\partial}{\partial\sigma^\top}\,\mathrm{diag}(\sigma^{-1})\sum_{i=1}^{n}(x_i-\mu)$$

$$= \mathrm{diag}(\sigma^{-2})E\sum_{i=1}^{n}(x_i-\mu)$$

$$= 0_p0_p^\top,$$

$$-E\frac{\partial^2}{\partial\sigma\sigma^\top}\ell(\mathcal{X};\mu,\sigma) = -E\frac{\partial}{\partial\sigma^\top}\left\{-\frac{n}{2}\sigma^{-1}\right.$$

$$\left.+\frac{1}{2}\sum_{i=1}^{n}\mathrm{diag}\{(x_i-\mu)(x_i-\mu)^\top\}\sigma^{-2}\right\}$$

$$= -\frac{n}{2}\,\mathrm{diag}(\sigma^{-2})$$

$$+E\sum_{i=1}^{n}\mathrm{diag}\{(x_i-\mu)(x_i-\mu)^\top\}\,\mathrm{diag}(\sigma^{-3})$$

$$= -\frac{n}{2}\,\mathrm{diag}(\sigma^{-2})+\mathrm{diag}(\sigma^{-3})n\,\mathrm{diag}\,\sigma$$

$$= \frac{n}{2}\,\mathrm{diag}(\sigma^{-2}).$$

Due to its simple diagonal structure, we can now write directly the Cramer-Rao lower bound for the parameter θ as the inverse of the derived Fisher information matrix:

$$\mathcal{F}_n^{-1} = \begin{pmatrix} \frac{1}{n}\,\mathrm{diag}\,\sigma & 0_p0_p^\top \\ 0_p0_p^\top & \frac{2}{n}\,\mathrm{diag}(\sigma^2) \end{pmatrix} = \begin{pmatrix} \frac{1}{n}\Sigma & 0_p0_p^\top \\ 0_p0_p^\top & \frac{2}{n}\Sigma^2 \end{pmatrix}.$$

EXERCISE 6.9. *Prove that if $s = s(\mathcal{X};\theta)$ is the score function and if $\hat{\theta} = t = t(\mathcal{X},\theta)$ is any function of $\mathcal{X}$ and θ, then under certain regularity conditions*

$$E(st^\top) = \frac{\partial}{\partial\theta}E(t^\top) - E\left(\frac{\partial t^\top}{\partial\theta}\right). \tag{6.2}$$

Note that

$$s(\mathcal{X};\theta) = \frac{\partial}{\partial\theta}\ell(\mathcal{X};\theta) = \frac{1}{L(\mathcal{X};\theta)}\frac{\partial}{\partial\theta}L(\mathcal{X};\theta).$$

Next, assuming that the regularity conditions allow us to permute the integral and the derivative, we write

$$\frac{\partial}{\partial\theta}E(t^{\top}) = \frac{\partial}{\partial\theta}\int t^{\top}(\mathcal{X};\theta)L(\mathcal{X};\theta)d\mathcal{X}$$

$$= \int \left(\frac{\partial}{\partial\theta}t^{\top}(\mathcal{X};\theta)L(\mathcal{X};\theta)\right)d\mathcal{X}$$

$$= \int \left(L(\mathcal{X};\theta)\frac{\partial t^{\top}}{\partial\theta} + t^{\top}\frac{\partial}{\partial\theta}L(\mathcal{X};\theta)\right)d\mathcal{X}$$

$$= \int L(\mathcal{X};\theta)\frac{\partial t^{\top}}{\partial\theta}d\mathcal{X} + \int t^{\top}L(\mathcal{X};\theta)s(\mathcal{X};\theta)d\mathcal{X}$$

$$= E\left(\frac{\partial t^{\top}(\mathcal{X};\theta)}{\partial\theta}\right) + E(t^{\top}(\mathcal{X};\theta)s(\mathcal{X};\theta))$$

and rearranging terms proves the statement (6.2).

EXERCISE 6.10. *Prove that the score function has zero expectation.*

We start by writing down the expectation as an integral with respect to the appropriate probability density function, the likelihood, of all observations:

$$E\{s(\mathcal{X};\theta)\} = \int s(\mathcal{X};\theta)L(\mathcal{X};\theta)d\mathcal{X}.$$

Similarly as in the previous exercise, we assume that the regularity conditions are such that we can exchange the integral and the derivative in the following formulas:

$$E\{s(\mathcal{X};\theta)\} = \int \left\{\frac{1}{L(\mathcal{X};\theta)}\frac{\partial L(\mathcal{X};\theta)}{\partial\theta}\right\}L(\mathcal{X};\theta)d\mathcal{X}$$

$$= \int \frac{\partial}{\partial\theta}L(\mathcal{X};\theta)d\mathcal{X}$$

$$= \frac{\partial}{\partial\theta}\int L(\mathcal{X};\theta)d\mathcal{X} = \frac{\partial}{\partial\theta}1_p = 0_p.$$

7

Hypothesis Testing

Criminal cases are continually hinging upon that one point. A man is suspected of a crime months perhaps after it has been committed. His linen or clothes are examined, and brownish stains discovered upon them. Are they blood stains, or mud stains, or rust stains, or fruit stains, or what are they? That is a question which has puzzled many an expert, and why? Because there was no reliable test. Now we have the Sherlock Holmes' test, and there will no longer be any difficulty.
Sherlock Holmes in "Study in Scarlet"

A first step in data modeling and understanding is the estimation of parameters in a supposed model. The second step—and very important statistical work—is the inferential conclusion on a hypothesis of the involved parameters. The construction of tests for different kinds of hypotheses is at the heart of this chapter.

A likelihood ratio is the ratio of the likelihood calculated under the null, H_0, and the alternative, H_1. The null hypothesis involves the supposed values of the parameter, e.g., H_0: $\mu = 0$. The ratio of the two likelihoods measures the closeness of the two hypotheses H_0 and H_1. By taking logarithms, the likelihood ratio is transformed into a difference of log likelihoods. By Wilks' theorem, two times this difference converges to a χ^2 distribution. Large values of this test statistic indicate a deviance from the null H_0 and thus lead us to reject the null hypothesis.

Formally, we will consider two hypotheses:

$$H_0 : \theta \in \Omega_0,$$
$$H_1 : \theta \in \Omega_1,$$

where θ is a parameter of the distribution of $\{x_i\}_{i=1}^n$, $x_i \in \mathbb{R}^p$.

THEOREM 7.1 (Wilks' Theorem). *If $\Omega_1 \subset \mathbb{R}^q$ is a q-dimensional space and if $\Omega_0 \subset \Omega_1$ is an r-dimensional subspace, then under regularity conditions:*

$$\forall\, \theta \in \Omega_0 : -2\log\lambda = 2(\ell_1^* - \ell_0^*) \xrightarrow{\mathcal{L}} \chi_{q-r}^2 \quad as \quad n \to \infty,$$

where ℓ_j^*, $j = 1, 2$ are the maxima of the log-likelihood for each hypothesis.

We will learn how to apply Theorem 7.1 to construct likelihood ratio tests and how to build confidence regions. We focus on the parameters of the multivariate normal distribution, e.g., we study (simultaneous) confidence intervals for linear combinations of the mean vector. The presented exercises and solutions cover the questions of testing dice, comparing company outputs, and testing the profiles of citrate concentrations in plasma. Other applications contain the linear regression model for the bank notes data and prediction of the vocabulary score for eighth graders.

EXERCISE 7.1. *Suppose that X has pdf $f(x; \theta)$, $\theta \in \mathbb{R}^k$. Using Theorem 7.1, construct an asymptotic rejection region of size α for testing, the hypothesis $H_0 : \theta = \theta_0$ against alternative $H_1 : \theta \neq \theta_0$.*

Let $\ell(\mathcal{X}; \theta) = \sum_{i=1}^n \log f(x_i; \theta)$ be the log-likelihood function and $\ell_j^* = \max_{\theta \in \Omega_j} \ell(\mathcal{X}; \theta)$. We construct the log-likelihood test statistic:

$$-2\log\lambda = 2(\ell_1^* - \ell_0^*)$$

for which the rejection region can be expressed as:

$$R = \{\mathcal{X} : -2\log\lambda > \kappa\}$$

The critical value κ has to be determined so that, if H_0 is true, $P(-2\log\lambda > \kappa) = \alpha$.

In line with Theorem 7.1 we know that under H_0 the log-likelihood ratio test statistic $-2\log\lambda$ is asymptotically distributed as:

$$-2\log\lambda \xrightarrow{\mathcal{L}} \chi_{q-r}^2 \quad as \quad n \to \infty,$$

where $r = \dim\Omega_0$ and $q = \dim\Omega_1$ denote the dimensions of the parameter spaces under the null and the alternative hypothesis. Fixing the value of the k-dimensional parameter θ reduces the dimension of the parameter space by $q - r = k$ and it follows that the asymptotic rejection region of H_0 (vs. H_1) of size α is:

$$R = \{\mathcal{X} : -2\log\lambda > \chi_{1-\alpha;k}^2\}.$$

EXERCISE 7.2. *Use Theorem 7.1 to derive a test for testing the hypothesis that a dice is balanced, based on n tosses of that dice.*

The probability that the number 1 occurs x_1-times, 2 occurs x_2-times,..., and 6 occurs x_6-times, is given by the multinomial distribution:

$$P(x_1,\ldots,x_6) = \frac{n!}{x_1!\ldots x_6!}p_1^{x_1}\ldots p_6^{x_6}, \tag{7.1}$$

where $\sum_{i=1}^{6} x_i = n$ and p_i, $i = 1,\ldots,6$, is the probability of i in a single toss, $\sum_{i=1}^{6} p_i = 1$.

The null hypothesis, the balanced dice, is $H_0 : p_1 = \cdots = p_6 = \frac{1}{6}$ and we will test it against the alternative hypothesis $H_1 : \exists i, j \in \{1,\ldots,6\} : p_i \neq p_j$.

Let $X = (x_1,\ldots,x_6)^\top$ denote the observed frequencies. The likelihood and the log-likelihood functions are based on (7.1):

$$L(X;p_1,\ldots,p_6) = \frac{n!}{x_1!\ldots x_6!}p_1^{x_1}\ldots p_6^{x_6},$$

$$\ell(X;p_1,\ldots,p_6) = \log n! - \sum_{j=1}^{6} \log x_j! + \sum_{j=1}^{6} x_j \log p_j$$

$$= \log n! - \sum_{j=1}^{6} \log x_j! + \sum_{j=1}^{5} x_j \log p_j + x_6 \log\left(1 - \sum_{j=1}^{5} p_j\right).$$

Setting the derivative of the log-likelihood w.r.t. the unknown parameters equal to zero, we obtain that $x_j/p_j = x_6/p_6$, $j = 1,\ldots,5$. This entails that $\dim \Omega_1 = 5$. From the condition $\sum_{j=1}^{6} p_j = 1$ it follows that the MLE for each of the unknown parameters is $\hat{p}_j = x_j/n$, $j = 1,\ldots,6$ which implies that the maximum of the log-likelihood under the alternative hypothesis is

$$\ell_1^* = \log n! - \sum_{j=1}^{6} \log x_j! + \sum_{j=1}^{6} x_j \log\left(\frac{x_j}{n}\right).$$

Under the null hypothesis $\Omega_0 = \{(1/6, 1/6, \ldots, 1/6)\}$ with $\dim \Omega_0 = 0$, the maximum of the log-likelihood is, obviously,

$$\ell_0^* = \log n! - \sum_{j=1}^{6} \log x_j! + \sum_{j=1}^{6} x_j \log\left(\frac{1}{6}\right).$$

Thus, we have for the likelihood ratio statistics:

$$-2\log\lambda = 2(\ell_1^* - \ell_0^*) = 2\left(\sum_{i=1}^{6} x_i \log x_i - n \log n + n \log 6\right) \sim \chi_5^2,$$

where the degrees of freedom of the asymptotic χ^2 distribution were determined, according to Theorem 7.1, as $\dim \Omega_1 - \dim \Omega_0 = 5 - 0$.

The application of this result is straightforward: the observed frequencies $x_1, \ldots, x_6$ are used to calculate the value of the likelihood ratio test statistics λ which is then compared to the appropriate quantile of the χ_5^2 distribution: if λ is too large, we reject the null hypothesis in favor of the alternative.

For example, if the observed frequencies are $X = (10, 7, 8, 12, 13, 6)^{\top}$, we obtain the value of the likelihood ratio statistics $-2\log \lambda = 4.23$. This value is smaller than the 95% critical value of the asymptotic χ_5^2 distribution, $\chi_{0.95;5}^2 = 11.07$, and we do not reject the null hypothesis. The null hypothesis is not rejected since the observed values are consistent with a balanced dice.

EXERCISE 7.3. *In Exercise 6.5, we have considered the pdf*

$$f(x_1, x_2) = \frac{1}{\theta_1^2 \theta_2^2 x_2} e^{-\left(\frac{x_1}{\theta_1 x_2} + \frac{x_2}{\theta_1 \theta_2}\right)}, \quad for \; x_1, x_2 > 0.$$

Solve the problem of testing $H_0 : \theta^{\top} = (\theta_{01}, \theta_{02})$ from an iid sample $x_i = (x_{i1}, x_{i2})^{\top}$, $i = 1, \ldots, n$, for large number of observations n.

Both the log-likelihood function:

$$\ell(\mathcal{X}; \theta_1, \theta_2) = \log \prod_{i=1}^{n} f(x_{i1}, x_{i2}; \theta_1, \theta_2)$$

$$= -n(2\log\theta_1 + \log\theta_2) - \sum_{i=1}^{n}\log x_{i2} - \sum_{i=1}^{n}\left(\frac{x_{i1}}{\theta_1 x_{i2}} + \frac{x_{i2}}{\theta_1 \theta_2}\right)$$

and the MLEs maximizing the likelihood under the alternative hypothesis:

$$\widehat{\theta}_1 = \frac{1}{n}\sum_{i=1}^{n}\frac{x_{i1}}{x_{i2}} \quad and \quad \widehat{\theta}_2 = \frac{\overline{x}_2}{\widehat{\theta}_1} = \frac{n\overline{x}_2}{\sum_{i=1}^{n}\frac{x_{i1}}{x_{i2}}}$$

are given in Exercise 6.5

The likelihood ratio test statistic can be derived as follows:

$$
\begin{aligned}
-2\log\lambda &= 2(\ell_1^* - \ell_0^*) \\
&= 2\{\ell(\mathcal{X}; \widehat{\theta}_1, \widehat{\theta}_2) - \ell(\mathcal{X}; \theta_{01}, \theta_{02})\} \\
&= -2n(2\log\widehat{\theta}_1 + \log\widehat{\theta}_2) - 2\sum_{i=1}^{n}\left(\frac{x_{i1}}{\widehat{\theta}_1 x_{i2}} + \frac{x_{i2}}{\widehat{\theta}_1 \widehat{\theta}_2}\right) \\
&\quad + 2n(2\log\theta_{01} + \log\theta_{02}) + 2\sum_{i=1}^{n}\left(\frac{x_{i1}}{\theta_{01} x_{i2}} + \frac{x_{i2}}{\theta_{01}\theta_{02}}\right) \\
&= 2n\left(2\log\frac{\theta_{01}}{\widehat{\theta}_1} + \log\frac{\theta_{02}}{\widehat{\theta}_2}\right) - 4n + 2\sum_{i=1}^{n}\left(\frac{x_{i1}}{\theta_{01} x_{i2}} + \frac{x_{i2}}{\theta_{01}\theta_{02}}\right).
\end{aligned}
$$

Note that dim $\Omega_1 = 2$ and dim $\Omega_0 = 0$. The likelihood ratio test statistic has, under the null hypothesis, asymptotically χ^2 distribution with $2 - 0 = 2$ degrees of freedom.

EXERCISE 7.4. *Consider a $N_3(\mu, \Sigma)$ distribution. Formulate the hypothesis $H_0 : \mu_1 = \mu_2 = \mu_3$ in terms of $\mathcal{A}\mu = a$.*

One possibility is to select matrix

$$\mathcal{A}_1 = \begin{pmatrix} 1 & -1 & 0 \\ 0 & 1 & -1 \end{pmatrix}$$

and vector $a = (0, 0)^\top$.

Then, the equation $\mathcal{A}_1 \mu = a$ can be written as

$$\mathcal{A}_1 \begin{pmatrix} \mu_1 \\ \mu_2 \\ \mu_3 \end{pmatrix} = \begin{pmatrix} \mu_1 - \mu_2 \\ \mu_2 - \mu_3 \end{pmatrix} = \begin{pmatrix} 0 \\ 0 \end{pmatrix},$$

which implies conditions $\mu_1 - \mu_2 = 0$ and $\mu_2 - \mu_3 = 0$ from which we get $\mu_1 = \mu_2 = \mu_3$ as desired.

Notice that the hypothesis H_0 can be written in infinitely many ways, e.g., using matrices

$$\mathcal{A}_2 = \begin{pmatrix} 1 & -\frac{1}{2} & -\frac{1}{2} \\ 0 & 1 & -1 \end{pmatrix} \quad \text{or} \quad \mathcal{A}_3 = \begin{pmatrix} 1 & -1 & 0 \\ 1 & 0 & -1 \end{pmatrix}.$$

EXERCISE 7.5. *Simulate a normal sample with $\mu = \begin{pmatrix} 1 \\ 2 \end{pmatrix}$ and $\Sigma = \begin{pmatrix} 1 & 0.5 \\ 0.5 & 2 \end{pmatrix}$ and test $H_0 : 2\mu_1 - \mu_2 = 0.2$ first with Σ known and then with Σ unknown. Compare the results.*

In general, suppose that $X_1, \ldots, X_n$ is an iid random sample from a $N_p(\mu, \Sigma)$ population and consider the hypothesis:

$$H_0 : \mathcal{A}\mu = a, \ \Sigma \text{ known versus } H_1 : \text{ no constraints},$$

where $\mathcal{A}(q \times p)$, $q \leq p$, has linearly independent rows. Under H_0, we have that:

$$n(\mathcal{A}\bar{x} - a)^\top (\mathcal{A}\Sigma\mathcal{A}^\top)^{-1}(\mathcal{A}\bar{x} - a) \sim \chi_q^2, \tag{7.2}$$

and we reject H_0 if this test statistic is too large at the desired significance level.

The test statistics (7.2) cannot be calculated if the variance matrix Σ is not known. Replacing the unknown variance matrix Σ by its estimate S leads to the test:

$$(n-1)(\mathcal{A}\overline{x}-a)^{\top}(\mathcal{A}\mathcal{S}\mathcal{A}^{\top})^{-1}(\mathcal{A}\overline{x}-a) \sim T^2(q, n-1). \qquad (7.3)$$

The tests described in (7.2) and (7.3) can be applied in our situation with $a = 0.2$ and $\mathcal{A} = (2, -1)$ since the null hypothesis $H_0 : 2\mu_1 - \mu_2 = 0.2$ can be written as

$$H_0 : (2, -1) \begin{pmatrix} \mu_1 \\ \mu_2 \end{pmatrix} = 0.2.$$

First, applying the test (7.2) with the known variance matrix and $n = 100$ simulations, we obtain the test statistics 0.8486. Comparing this value with the appropriate critical value $\chi^2_{0.95;1} = 3.8415$ of the χ^2_1 distribution, we see that the observed values are at level 95% not significantly different from the assumed values.   Q SMStestsim

Performing the test (7.3), where the variance matrix Σ is replaced by the estimate $\mathcal{S}$, we obtain the test statistics 0.9819 which is again smaller than the 95% critical value of the Hotelling T^2 distribution, $T^2(0.95; 1, 99) = F_{0.95;1,99} = 3.9371$.

Notice that the tests (7.2) and (7.3) with the known and unknown variance matrix are very similar. The critical value for the test (7.3) is slightly larger since it has to reflect also the uncertainty coming from the estimation of the variance matrix. However, for large number of observations, both tests should provide very similar results.

EXERCISE 7.6. *Suppose that* $x_1, \ldots, x_n$ *is an iid random sample from a* $N_p(\mu, \Sigma)$ *population. Show that the maximum of the log-likelihood under* $H_0 : \mu = \mu_0$ *with unknown variance matrix* Σ *is*

$$\ell_0^* = \ell(\mathcal{X}; \mu_0, \mathcal{S} + dd^{\top}), \quad d = (\overline{x} - \mu_0).$$

From the likelihood function for parameters Σ and μ:

$$L(\mathcal{X}; \mu, \Sigma) = |2\pi\Sigma|^{-n/2} \exp\left\{ -\frac{1}{2} \sum_{i=1}^{n} (x_i - \mu)^{\top} \Sigma^{-1} (x_i - \mu) \right\}$$

we obtain the log-likelihood

$$\ell(\mathcal{X}; \mu, \Sigma) = -\frac{n}{2} \log|2\pi\Sigma| - \frac{1}{2} \sum_{i=1}^{n} (x_i - \mu)^{\top} \Sigma^{-1} (x_i - \mu). \qquad (7.4)$$

Notice that, in the definition of the multinormal pdf given in Exercise 4.20, we assume that the variance matrix Σ is positive definite.

Under the null hypothesis $H_0 : \mu = \mu_0$, we have to maximize (w.r.t. Σ) the expression

$$\ell_0(\mathcal{X}; \Sigma)$$

$$= -\frac{n}{2}\log|2\pi\Sigma| - \frac{1}{2}\sum_{i=1}^{n}(x_i - \mu_0)^{\top}\Sigma^{-1}(x_i - \mu_0)$$

$$= -\frac{np\log 2\pi}{2} - \frac{n}{2}\log|\Sigma| - \frac{1}{2}\sum_{i=1}^{n}\left\{\operatorname{tr}\Sigma^{-1}(x_i - \mu_0)(x_i - \mu_0)^{\top}\right\}$$

$$= -\frac{np\log 2\pi}{2} - \frac{n}{2}\log|\Sigma| - \frac{1}{2}\left\{\operatorname{tr}\Sigma^{-1}\sum_{i=1}^{n}(x_i - \mu_0)(x_i - \mu_0)^{\top}\right\}. \quad (7.5)$$

Let us now state two useful rules for matrix differentiation (Lütkepohl 1996, Harville 1997):

$$\frac{\partial \log|\mathcal{X}|}{\partial \mathcal{X}} = (X^{\top})^{-1} \quad \text{and} \quad \frac{\partial \operatorname{tr}\mathcal{X}^{\top}\mathcal{A}}{\partial \mathcal{X}} = \mathcal{A}$$

which are in turn applied to express the derivative of the log-likelihood (7.5) with respect to the unknown parameter Σ^{-1} as follows:

$$\frac{\partial \ell_0(\mathcal{X}; \Sigma)}{\partial(\Sigma^{-1})} = \frac{n}{2}\Sigma - \frac{1}{2}\sum_{i=1}^{n}(x_i - \mu_0)(x_i - \mu_0)^{\top}.$$

Setting the derivative equal to zero, we immediately obtain the MLE of the unknown parameter Σ as:

$$\widehat{\Sigma} = \frac{1}{n}\sum_{i=1}^{n}(x_i - \mu_0)(x_i - \mu_0)^{\top}$$

$$= \frac{1}{n}\sum_{i=1}^{n}\left\{(x_i - \overline{x})(x_i - \overline{x})^{\top} + (\overline{x} - \mu_0)(\overline{x} - \mu_0)^{\top} + 2(\overline{x} - \mu_0)(x_i - \overline{x})^{\top}\right\}$$

$$= \frac{1}{n}\sum_{i=1}^{n}(x_i - \overline{x})(x_i - \overline{x})^{\top} + \frac{1}{n}\sum_{i=1}^{n}(\overline{x} - \mu_0)(\overline{x} - \mu_0)^{\top} + 0$$

$$= \frac{1}{n}\sum_{i=1}^{n}(x_i - \overline{x})(x_i - \overline{x})^{\top} + (\overline{x} - \mu_0)(\overline{x} - \mu_0)^{\top}$$

$$= \mathcal{S} + dd^{\top},$$

where $\mathcal{S}$ is the empirical covariance matrix and $d = (\overline{x} - \mu_0)$. It is clear that the maximum of the log-likelihood under the null hypothesis is

$$\ell_0^* = \max_{\Sigma}\ell_0(\mathcal{X}; \Sigma) = \ell_0(\mathcal{X}; \widehat{\Sigma}) = \ell(\mathcal{X}; \mu_0, \widehat{\Sigma}) = \ell(\mathcal{X}; \mu_0, \mathcal{S} + dd^{\top}).$$

EXERCISE 7.7. *Suppose that $X_1, \ldots, X_n$ is an iid random sample from a $N_p(\mu, \Sigma)$ population and consider the test of the hypothesis*

$$H_0 : \mu = \mu_0, \ \Sigma \ unknown \ versus \ H_1 : \ no \ constraints.$$

Show that the likelihood ratio test statistic is equal to

$$-2 \log \lambda = 2(\ell_1^* - \ell_0^*) = n \log(1 + d^{\top} \mathcal{S}^{-1} d), \quad d = (\overline{x} - \mu_0).$$

The maximum of the likelihood under the null hypothesis, ℓ_0^*, was already derived in Exercise 7.6:

$$\ell_0^* = \ell(\mathcal{X}; \mu_0, \mathcal{S} + dd^{\top}), \quad d = (\overline{x} - \mu_0).$$

In order to calculate the maximum of the likelihood under the alternative hypothesis, we have to maximize (w.r.t. (μ, Σ)) the log-likelihood:

$$\ell(\mathcal{X}; \mu, \Sigma) = -\frac{n}{2} \log |2\pi \Sigma| - \frac{1}{2} \sum_{i=1}^{n} (x_i - \mu)^{\top} \Sigma^{-1}(x_i - \mu)$$

$$= -\frac{np \log 2\pi}{2} - \frac{n}{2} \log |\Sigma| - \frac{1}{2} \left\{ \operatorname{tr} \Sigma^{-1} \sum_{i=1}^{n} (x_i - \mu)(x_i - \mu)^{\top} \right\}.$$

Let us start by calculating the derivative of the likelihood w.r.t. the parameter μ:

$$\frac{\partial \ell(\mathcal{X}; \Sigma)}{\partial \mu} = -\frac{1}{2} \Sigma^{-1} \sum_{i=1}^{n} (x_i - \mu)$$

and we see the MLE $\widehat{\mu}$ is equal to the sample mean $\overline{x}$ for any matrix Σ^{-1}.

Let us now maximize the function $\ell(\mathcal{X}; \overline{x}, \Sigma)$ in terms of the parameter Σ. Similarly as in Exercise 7.6, we express the derivative of the log-likelihood $\ell(\mathcal{X}; \overline{x}, \Sigma)$ with respect to the unknown parameter Σ^{-1} as follows:

$$\frac{\partial \ell(\mathcal{X}; \Sigma)}{\partial \Sigma^{-1}} = \frac{n}{2} \Sigma - \frac{1}{2} \sum_{i=1}^{n} (x_i - \overline{x})(x_i - \overline{x})^{\top}.$$

Setting the derivative equal to zero, we immediately obtain that the MLE $\widehat{\Sigma}$ is equal to the sample variance matrix

$$\mathcal{S} = \frac{1}{n} \sum_{i=1}^{n} (x_i - \overline{x})(x_i - \overline{x})^{\top}$$

and the maximum of the log-likelihood under the alternative hypothesis is

$$\ell_1^* = \ell(\mathcal{X}; \overline{x}, \mathcal{S}).$$

Hence, using the rule for calculating the determinant derived in Exercise 2.8, the likelihood ratio test statistic can be written as

$$-2\lambda = 2(\ell_1^* - \ell_0^*)$$

$$= 2\{\ell(\mathcal{X}; \overline{x}, \mathcal{S}) - \ell(\mathcal{X}; \mu_0, \mathcal{S} + dd^\top)\}$$

$$= -n\log|\mathcal{S}| + n\log|\mathcal{S} + dd^\top| - \text{tr}\left\{\mathcal{S}^{-1}\sum_{i=1}^{n}(x_i - \overline{x})(x_i - \overline{x})^\top\right\}$$

$$+ \text{tr}\left\{(\mathcal{S} + dd^\top)^{-1}\sum_{i=1}^{n}(x_i - \mu_0)(x_i - \overline{\mu}_0)^\top\right\}$$

$$= n\log\{|S|(1 + d^\top \mathcal{S}^{-1}d)\} - n\log|\mathcal{S}|$$

$$= n\log(1 + d^\top \mathcal{S}^{-1}d).$$

EXERCISE 7.8. *In the U.S. companies data set in Table A.17, test the equality of means between the energy and manufacturing sectors taking the full vector of observations X_1 to X_6. Derive simultaneous confidence intervals for the differences.*

Assume that we have a random sample consisting of $X_{i1} \sim N_p(\mu_1, \Sigma)$, $i = 1, \cdots, n_1$, and $X_{j2} \sim N_p(\mu_2, \Sigma)$, $j = 1, \cdots, n_2$. The test of the equality of the means μ_1 and μ_2 can be formally written as

$$H_0 : \mu_1 = \mu_2, \text{ versus } H_1 : \text{ no constraints.}$$

Both samples provide the statistics $\overline{x}_k$ and $\mathcal{S}_k$, $k = 1, 2$. Let $\delta = \mu_1 - \mu_2$ and $n = n_1 + n_2$. We have

$$(\overline{x}_1 - \overline{x}_2) \sim N_p\left(\delta, \frac{n}{n_1 n_2}\Sigma\right) \text{ and } n_1 S_1 + n_2 S_2 \sim W_p(\Sigma, n_1 + n_2 - 2).$$

Let $\mathcal{S} = (n_1 + n_2)^{-1}(n_1 S_1 + n_2 S_2)$ be the weighted mean of $\mathcal{S}_1$ and $\mathcal{S}_2$. This leads to a test statistic with a Hotelling T^2-distribution:

$$\frac{n_1 n_2 (n - 2)}{n^2}\left\{(\overline{x}_1 - \overline{x}_2) - \delta\right\}^\top \mathcal{S}^{-1}\left\{(\overline{x}_1 - \overline{x}_2) - \delta\right\} \sim T^2(p, n - 2)$$

or

$$\left\{(\overline{x}_1 - \overline{x}_2) - \delta\right\}^\top \mathcal{S}^{-1}\left\{(\overline{x}_1 - \overline{x}_2) - \delta\right\} \sim \frac{pn^2}{(n - p - 1)n_1 n_2}F_{p, n-p-1}.$$

This result can be used to test the null hypothesis of equality of two means, $H_0 : \delta = 0$, or to construct a confidence region for $\delta \in \mathbb{R}^p$.

The rejection region of the test is given by:

$$\frac{n_1 n_2 (n - p - 1)}{pn^2}(\overline{x}_1 - \overline{x}_2)^\top \mathcal{S}^{-1}(\overline{x}_1 - \overline{x}_2) \geq F_{1-\alpha; p, n-p-1}. \tag{7.6}$$

A $(1 - \alpha)$ confidence region for δ is given by the ellipsoid centered at $(\overline{x}_1 - \overline{x}_2)$

$$\{\delta - (\overline{x}_1 - \overline{x}_2)\}^\top \mathcal{S}^{-1} \{\delta - (\overline{x}_1 - \overline{x}_2)\} \le \frac{pn^2}{(n - p - 1)(n_1 n_2)} F_{1-\alpha;p,n-p-1}.$$

The simultaneous confidence intervals for all linear combinations $a^\top \delta$ of the elements of δ are given by

$$a^\top \delta \in a^\top (\overline{x}_1 - \overline{x}_2) \pm \sqrt{\frac{pn^2}{(n - p - 1)(n_1 n_2)} F_{1-\alpha;p,n-p-1} a^\top \mathcal{S} a}.$$

In particular, we have at the $(1 - \alpha)$ level, for $j = 1, \ldots, p$,

$$\delta_j \in (\overline{x}_{1j} - \overline{x}_{2j}) \pm \sqrt{\frac{pn^2}{(n - p - 1)(n_1 n_2)} F_{1-\alpha;p,n-p-1} s_{jj}}. \qquad (7.7)$$

In the U.S. companies data set, we observe altogether 6 variables. We have $n_1 = 15$ observations from the energy sector and $n_2 = 10$ observations from the manufacturing sector.

The test statistic

$$\frac{n_1 n_2 (n - p - 1)}{pn^2} (\overline{x}_1 - \overline{x}_2)^\top \mathcal{S}^{-1} (\overline{x}_1 - \overline{x}_2) = 2.15$$

is smaller than the corresponding critical value $F_{1-\alpha;p,n-p-1} = F_{0.95;6,18} = 2.66$ and, hence, we do not reject the null hypothesis.

Let us now derive the simultaneous confidence interval for the difference of the means at level $1 - \alpha = 95\%$ by calculating the intervals

$$(\overline{x}_{1j} - \overline{x}_{2j}) \pm \sqrt{\frac{pn^2}{n_1 n_2 (n - p - 1)} F_{1-\alpha;p,n-p-1} s_{jj}}$$

for $j = 1, \ldots, p$.

We only have to take the mean and the variances of the variables into account since the covariances do not appear in the formula. At the 95% level we have the confidence intervals:

$$-7639 \le \delta_1 \le 7193$$
$$-9613 \le \delta_2 \le 4924$$
$$-2924 \le \delta_3 \le 2103$$
$$-295 \le \delta_4 \le 530$$
$$-527 \le \delta_5 \le 791$$
$$-102 \le \delta_6 \le 20.$$

We remark that all above confidence intervals contain zero which corresponds to not rejecting the null hypothesis. ◻ SMStestuscomp

EXERCISE 7.9. *Consider an iid sample of size $n = 5$ from a bivariate normal distribution*

$$X \sim N_2\left(\mu, \begin{pmatrix} 3 & \rho \\ \rho & 1 \end{pmatrix}\right)$$

where ρ is a known parameter. Suppose $\overline{x}^\top = (1,0)$. For what value of ρ would the hypothesis $H_0 : \mu^\top = (0,0)$ be rejected in favor of $H_1 : \mu^\top \neq (0,0)$ (at the 5% level)?

Since the variance matrix Σ is known, we can use the test statistic:

$$-2\log\lambda = n(\overline{x} - \mu_0)^\top \Sigma^{-1}(\overline{x} - \mu_0)$$

which has, under the null hypothesis, exactly a χ^2 distribution with $p = 2$ degrees of freedom.

Plugging in the observed values, we obtain

$$n(\overline{x} - \mu_0)^\top \Sigma^{-1}(\overline{x} - \mu_0) = 5(1,0)\begin{pmatrix} 3 & \rho \\ \rho & 1 \end{pmatrix}^{-1}\begin{pmatrix} 1 \\ 0 \end{pmatrix}$$

$$= 5(1,0)\frac{1}{3 - \rho^2}\begin{pmatrix} 1 & -\rho \\ -\rho & 3 \end{pmatrix}\begin{pmatrix} 1 \\ 0 \end{pmatrix}$$

$$= \frac{5}{3 - \rho^2}$$

and it follows that the null hypothesis is rejected if

$$\frac{5}{3 - \rho^2} > \chi^2_{0.95;2} = 5.99,$$

i.e., if $\mathrm{abs}(\rho) > \sqrt{3 - 5.99/5} = 1.471$.

At the same time, $\mathrm{abs}(\rho) < \sqrt{3}$ since the variance matrix must be positive definite (and the covariance $\rho = \pm\sqrt{3}$ if the correlation coefficient is equal to ± 1).

Hence, the null hypothesis is rejected for covariances ρ such that

$$\mathrm{abs}(\rho) \in \left(3 - \frac{5}{\chi^2_{0.95;2}}, \sqrt{3}\right) = (1.471, 1.732).$$

EXERCISE 7.10. *Consider $X \sim N_3(\mu, \Sigma)$. An iid sample of size $n = 10$ provides:*

$$\overline{x} = (1,0,2)^\top \quad and \quad S = \begin{pmatrix} 3 & 2 & 1 \\ 2 & 3 & 1 \\ 1 & 1 & 4 \end{pmatrix}.$$

a) *Knowing that the eigenvalues of S are integers, describe a 95% confidence region for μ.*

b) *Calculate the simultaneous confidence intervals for μ_1, μ_2 and μ_3.*

c) *Can we assert that μ_1 is an average of μ_2 and μ_3?*

a) The test statistic $(n-p)(\mu - \overline{x})^{\top} S^{-1}(\mu - \overline{x})$ has $F_{p,n-p}$ distribution. Comparison of the test statistic with the appropriate quantile of its distribution yields the following confidence region, covering the unknown parameter μ with probability $1 - \alpha$:

$$\left\{ \mu \in \mathbb{R}^p; |(\mu - \overline{x})^{\top} S^{-1}(\mu - \overline{x})| \leq \frac{p}{n-p} F_{1-a;p,n-p} \right\}.$$

In our case, we obtain

$$\left\{ \mu \in \mathbb{R}^3 |(\mu - \overline{x})^T S^{-1}(\mu - \overline{x}) \leqslant \frac{3}{7} F_{0,95;3,7} \right\} \tag{7.8}$$

Calculating the trace and the determinant of $\mathcal{S}$:

$$|\mathcal{S}| = 18 = \prod_{j=1}^{3} \lambda_j \quad \text{and} \quad \operatorname{tr}(\mathcal{S}) = 3 + 3 + 4 = 10 \sum_{j=1}^{3} \lambda_j$$

and searching for positive integers satisfying these two equations yields easily $\lambda = (\lambda_1.\lambda_2, \lambda_3)^{\top} = (6, 3, 1)^{\top}$.

Next, we can calculate the eigenvectors $\gamma_1, \gamma_2, \gamma_3$ by solving the three systems of equations $(\mathcal{S} - \lambda_i \mathcal{I}_3) = 0_3$, respectively:

$$\begin{pmatrix} -3 & 2 & 1 \\ 2 & -3 & 1 \\ 1 & 1 & -2 \end{pmatrix} \gamma_1 = 0_3, \quad \begin{pmatrix} 0 & 2 & 1 \\ 2 & 0 & 1 \\ 1 & 1 & 1 \end{pmatrix} \gamma_2 = 0_3, \quad \text{and} \quad \begin{pmatrix} 2 & 2 & 1 \\ 2 & 2 & 1 \\ 1 & 1 & 3 \end{pmatrix} \gamma_3 = 0_3$$

and it is easy to verify that $\gamma_1 = (1, 1, 1)^{\top}/\sqrt{3}$, $\gamma_2 = (1, 1, -2)^{\top}/\sqrt{6}$, and $\gamma_3 = (-1, 1, 0)^{\top}/\sqrt{2}$.

The confidence region (7.8) can now be described in words as a 3-dimensional ellipsoid with axes of lengths $\sqrt{\frac{3}{7} F_{0.95;3,7} \lambda_i}$, $i = 1, 2, 3$, oriented in the directions of the eigenvectors γ_1, γ_2, and γ_3, respectively.

b) Simultaneous confidence intervals for components of μ may be calculated using the formula:

$$\overline{x}_j - \sqrt{\frac{p}{n-p} F_{1-\alpha;p,n-p} s_{jj}} < \mu_j < \overline{x}_j + \sqrt{\frac{p}{n-p} F_{1-\alpha;p,n-p} s_{jj}}$$

In this particular case we have

$$\overline{x}_j - \sqrt{\frac{3}{7}F_{0.95;3,7}s_{jj}} < \mu_j < \overline{x}_j + \sqrt{\frac{3}{7}F_{0.95;3,7}s_{jj}}.$$

It should be noticed that these intervals define a rectangle inscribing the confidence ellipsoid (7.8) for μ given above. Calculations yield:

$$-1.364 < \mu_1 < 3.364$$
$$-2.364 < \mu_2 < 2.364$$
$$-0.729 < \mu_3 < 4.730.$$

c) The problem can be solved applying the test statistic:

$$(n-1)(\mathcal{A}\overline{x} - a)^\top (\mathcal{A}S\mathcal{A}^\top)^{-1}(\mathcal{A}\overline{x} - a) \sim T^2(q, n-1)$$

where $\mathcal{A} = (2, -1, -1)$. In this case, with the observed $\overline{x} = (1, 0, 2)^\top$, the value of the test statistic is zero and the null hypothesis $H_0 : \mu_1 = (\mu_2 + \mu_3)/2$ (or equivalently $H_0 : \mathcal{A}\mu = 0$) can not be rejected.

EXERCISE 7.11. *Let $X \sim N_2(\mu, \Sigma)$ where Σ is known to be $\Sigma = \begin{pmatrix} 2 & -1 \\ -1 & 2 \end{pmatrix}$. We have an iid sample of size $n = 6$ providing $\overline{x}^\top = (1 \; \frac{1}{2})$. Solve the following test problems ($\alpha = 0.05$):*

a) $H_0: \mu = (2, \frac{2}{3})^\top$ $\quad H_1: \mu \neq (2, \frac{2}{3})^\top$
b) $H_0: \mu_1 + \mu_2 = \frac{7}{2}$ $\quad H_1: \mu_1 + \mu_2 \neq \frac{7}{2}$
c) $H_0: \mu_1 - \mu_2 = \frac{1}{2}$ $\quad H_1: \mu_1 - \mu_2 \neq \frac{1}{2}$
d) $H_0: \mu_1 = 2$ $\quad\quad H_1: \mu_1 \neq 2$

For each case, calculate the rejection region and comment on their size and location.

a) For $X \sim N_p(\mu, \Sigma)$, the test statistic $(X - \mu_0)^\top \Sigma^{-1}(X - \mu_0)$ has under the null hypothesis $H_0 : \mu = \mu_0$ exactly a χ_p^2 distribution.

The test is based on the known distribution of the sample mean, i.e.,

$$\overline{x} \sim N_2 \left(\begin{pmatrix} 1 \\ \frac{1}{2} \end{pmatrix}, \frac{1}{6} \begin{pmatrix} 2 & -1 \\ -1 & 2 \end{pmatrix} \right)$$

Since in our case the variance matrix $\Sigma > 0$ is known, we can calculate its inverse

$$\left(\frac{1}{6}\Sigma \right)^{-1} = \begin{pmatrix} 4 & 2 \\ 2 & 4 \end{pmatrix}$$

and obtain the value of the test statistic 4.78 which is smaller than the critical value $\chi_2^2(0.05) = 5.99$. Hence, the null hypothesis can not be rejected at level $\alpha = 0.05$.

Here, the rejection region are all values greater than the critical value 5.99, i.e., the interval $(5.99, +\infty)$.

b) We could use the test statistic (7.2) with $\mathcal{A} = (1, 1)$ but it is more straightforward to use the univariate normal distribution of the random variable

$$(1, 1)\overline{x} \sim N(\mu_1 + \mu_2, 2/6).$$

The test statistic $(1, 1)\overline{x}$ has, under the null hypothesis, a normal distribution $N(7/2, 2/6)$. We reject the null hypothesis since the value of the test statistic, $\left(\frac{3}{2} - \frac{7}{2}\right) \sqrt{6/2} = 3.4641$ is smaller than the critical value of the standard normal distribution $\Phi^{-1}(0.025) = -1.96$.

The rejection region is the union of the intervals $(-\infty, -1.96) \cup (1.96, +\infty)$. We reject the null hypothesis if the observed and the hypothesized mean value are far away from each other.

c) Since $\overline{x}_1 - \overline{x}_2 = \frac{1}{2}$, the value of the test statistic (7.2) is equal to zero and we can not reject H_0 at any level $\alpha \in (0, 1)$.

d) Again, we could use formula (7.2) with $\mathcal{A} = (1, 0)$. However, we can also realize that the test concerns only the first component of the observed random vector and, since the test statistic $|1 - 2| \sqrt{6/2} = 1.7321$ is now lying between the critical values $\Phi^{-1}(0.025) = 1.96$ and $\Phi^{-1}(0.975) = 1.96$, we do not reject the null hypothesis at level $\alpha = 0.05$.

The rejection region is $(-\infty, -1.96) \cup (1.96, +\infty)$.

EXERCISE 7.12. *Repeat the Exercise 7.11 with Σ unknown and the empirical covariance matrix $S = \begin{pmatrix} 2 & -1 \\ -1 & 2 \end{pmatrix}$. Compare the results.*

a) Tests concerning the mean vector of a multivariate normal distribution can be based on the test statistic

$$(n - 1)(\overline{x} - \mu_0)^\top S^{-1}(\overline{x} - \mu_0) \sim T^2(p, n - 1),$$

or equivalently

$$\left(\frac{n - p}{p}\right)(\overline{x} - \mu_0)^\top S^{-1}(\overline{x} - \mu_0) \sim F_{p, n-p}. \tag{7.9}$$

In this case an exact rejection region may be defined as

$$\left(\frac{n - p}{p}\right)(\overline{x} - \mu_0)^\top S^{-1}(\overline{x} - \mu_0) > F_{1-\alpha; p, n-p}.$$

Alternatively, we could apply Theorem 7.1 which leads to the approximate (asymptotically valid) rejection region:

$$n \log\{1 + (\overline{x} - \mu_0)^\top S^{-1} (\overline{x} - \mu_0)\} > \chi^2_{1-\alpha;p}.$$

However, it is preferable to use the exact approach.

It is interesting to see that the test statistic is quite similar to the test statistic calculated in Exercise 7.11. The only differences are different norming constant $(n - p)/p$ instead of n and different critical values. Comparing the formula, the value of the test statistic can be calculated as

$$\frac{n - p}{pn} 4.78 = \frac{4}{24} 4.78$$

which is obviously smaller than the corresponding critical value $F_{0.95;2,4} = 6.9443$. As in Exercise 7.11, we do not reject the null hypothesis.

b) The standard univariate t-test, allowing for unknown variance matrix, is actually using the same test statistic as given in Exercise 7.11. The only difference is the critical value $t_{0.975;5} = 2.5759$. The test statistic, -3.4641, is smaller than the critical value -2.5759 and, exactly as in Exercise 7.11, we reject the null hypothesis.

Notice that the rejection region, $(-\infty, -2.5759) \cup (2.5759, +\infty)$ is smaller than in Exercise 7.11 and we can say that it is more difficult to reject the null hypothesis if the variance is not known.

c) Since $\overline{x}_1 - \overline{x}_2 = \frac{1}{2}$, the value of the test statistic will be again equal to zero and we can not reject H_0 at any level $\alpha \in (0,1)$. This decision is identical to our conclusion in Exercise 7.11.

d) The test statistic of the univariate t-test $(1 - 2) \sqrt{6/2} = 1.7321$ is lying between the corresponding critical values $t_{0.025;5} = -2.5759$ and $t_{0.975;5} = 2.5759$ which implies that we do not reject the null hypothesis at level $\alpha = 0.05$.

EXERCISE 7.13. *Test the hypothesis of the equality of the covariance matrices on two simulated 4-dimensional samples of sizes $n_1 = 30$ and $n_2 = 20$.*

Let $X_{ih} \sim N_p(\mu_h, \Sigma_h)$, $i = 1, \ldots, n_h$, $h = 1,2$, be independent random vectors. The test problem of testing the equality of the covariance matrices can be written as

$$H_0 : \Sigma_1 = \Sigma_2 \text{ versus } H_1 : \text{ no constraints}.$$

Both subsamples provide $\mathcal{S}_h$, an estimator of Σ_h, with the Wishart distribution $n_h \mathcal{S}_h \sim W_p(\Sigma_h, n_h - 1)$. Under the null hypothesis $H_0 : \Sigma_1 = \Sigma_2$, we have for the common covariance matrix that $\sum_{h=1}^2 n_h \mathcal{S}_h \sim W_p(\Sigma, n - 2)$, where $n = \sum_{h=1}^2 n_h$.

Let $S = \frac{n_1 S_1 + n_2 S_2}{n}$ be the weighted average of S_1 and S_2. The likelihood ratio test leads to the test statistic

$$-2 \log \lambda = n \log |S| - \sum_{h=1}^{2} n_h \log |S_h| \qquad (7.10)$$

which under H_0 is approximately distributed as a χ_m^2 with $m = \frac{1}{2}(2-1)p(p+1)$ degrees of freedom.

We test the equality of the covariance matrices for the three data sets given in Härdle & Simar (2003, Example 7.14) who simulated two independent normal distributed samples with $p = 4$ dimensions and the sample sizes of $n_1 = 30$ and $n_2 = 20$ leading to the asymptotic distribution of the test statistics (7.10) with $m = \frac{1}{2}(2-1)4(4+1) = 10$ degrees of freedom.

a) With a common covariance matrix in both populations $\Sigma_1 = \Sigma_2 = I_4$, we obtain the following empirical covariance matrices:

$$S_1 = \begin{pmatrix} 0.812 & -0.229 & -0.034 & 0.073 \\ -0.229 & 1.001 & 0.010 & -0.059 \\ -0.034 & 0.010 & 1.078 & -0.098 \\ 0.073 & -0.059 & -0.098 & 0.823 \end{pmatrix}$$

and

$$S_2 = \begin{pmatrix} 0.559 & -0.057 & -0.271 & 0.306 \\ -0.057 & 1.237 & 0.181 & 0.021 \\ -0.271 & 0.181 & 1.159 & -0.130 \\ 0.306 & 0.021 & -0.130 & 0.683 \end{pmatrix}$$

The determinants are $|S| = 0.594$, $|S_1| = 0.668$ and $|S_2| = 0.356$ leading to the likelihood ratio test statistic:

$$-2 \log \lambda = 50 \log(0.594) - 30 \log(0.668) - 20 \log(0.356) = 6.755$$

The value of the test statistic is smaller than the critical value $\chi_{0.95;10}^2 = 18.307$ and, hence, we do not reject the null hypothesis.

b) The second simulated samples have covariance matrices $\Sigma_1 = \Sigma_2 = 16 I_4$. Now, the standard deviation is 4 times larger than in the previous case. The sample covariance matrices from the second simulation are:

$$S_1 = \begin{pmatrix} 21.907 & 1.415 & -2.050 & 2.379 \\ 1.415 & 11.853 & 2.104 & -1.864 \\ -2.050 & 2.104 & 17.230 & 0.905 \\ 2.379 & -1.864 & 0.905 & 9.037 \end{pmatrix},$$

$$S_2 = \begin{pmatrix} 20.349 & -9.463 & 0.958 & -6.507 \\ -9.463 & 15.502 & -3.383 & -2.551 \\ 0.958 & -3.383 & 14.470 & -0.323 \\ -6.507 & -2.551 & -0.323 & 10.311 \end{pmatrix}$$

and the value of the test statistic is:

$$-2\log\lambda = 50\log(40066) - 30\log(35507) - 20\log(16233) = 21.694.$$

Since the value of the test statistic is larger than the critical value of the asymptotic distribution, $\chi^2_{0.95;10} = 18.307$, we reject the null hypothesis.

c) The covariance matrix in the third case is similar to the second case $\Sigma_1 = \Sigma_2 = 16I_4$ but, additionally, the covariance between the first and the fourth variable is $\sigma_{14} = \sigma_{41} = -3.999$. The corresponding correlation coefficient is $r_{41} = -0.9997$.

The sample covariance matrices from the third simulation are:

$$S_1 = \begin{pmatrix} 14.649 & -0.024 & 1.248 & -3.961 \\ -0.024 & 15.825 & 0.746 & 4.301 \\ 1.248 & 0.746 & 9.446 & 1.241 \\ -3.961 & 4.301 & 1.241 & 20.002 \end{pmatrix}$$

and

$$S_2 = \begin{pmatrix} 14.035 & -2.372 & 5.596 & -1.601 \\ -2.372 & 9.173 & -2.027 & -2.954 \\ 5.596 & -2.027 & 9.021 & -1.301 \\ -1.601 & -2.954 & -1.301 & 9.593 \end{pmatrix}.$$

The value of the test statistic is:

$$-2\log\lambda = 50\log(24511) - 30\log(37880) - 20\log(6602.3) = 13.175$$

The value of the likelihood ratio test statistic is now smaller than the critical value, $\chi^2_{0.95;10} = 18.307$, and we do not reject the null hypothesis. ◻ MVAtestcov

EXERCISE 7.14. *Test the equality of the covariance matrices from the two groups in the WAIS data set (Morrison 1990). The data set is given in Table A.21.*

The data set can be summarized by calculating the vectors of means,

$$\bar{x}_1 = (12.57, 9.57, 11.49, 7.97)^\top \quad \bar{x}_2 = (8.75, 5.33, 8.50, 4.75)^\top,$$

and the empirical covariance matrices

$$S_1 = \begin{pmatrix} 11.164 & 8.840 & 6.210 & 2.020 \\ 8.840 & 11.759 & 5.778 & 0.529 \\ 6.210 & 5.778 & 10.790 & 1.743 \\ 2.020 & 0.529 & 1.743 & 3.594 \end{pmatrix}$$

$$S_2 = \begin{pmatrix} 9.688 & 9.583 & 8.875 & 7.021 \\ 9.583 & 16.722 & 11.083 & 8.167 \\ 8.875 & 11.083 & 12.083 & 4.875 \\ 7.021 & 8.167 & 4.875 & 11.688 \end{pmatrix}$$

in both groups.

Let us assume that the first set of $n_1 = 37$ observations comes from 4-dimensional normal distribution $N_4(\mu_1, \Sigma_1)$ and the second set of the remaining $n_2 = 12$ observations corresponds to $N_4(\mu_2, \Sigma_2)$.

For testing the equality of the two covariance matrices, $\Sigma_1 = \Sigma_2$, we use the test described in Exercise 7.13. Formally, the null and the alternative hypotheses are:

$$H_0 : \Sigma_1 = \Sigma_2 \quad \text{versus} \quad H_1 : \Sigma_1 \neq \Sigma_2.$$

In order to calculate the likelihood ratio test statistic (7.6), we have to define the matrix $S = (n_1 S_1 + n_2 S_2)/n$, i.e., the weighted average of the observed matrices. We get

$$S = \frac{n_1 S_1 + n_2 S_2}{n} = \frac{37 S_1 + 12 S_2}{49} = \begin{pmatrix} 10.803 & 9.022 & 6.863 & 3.245 \\ 9.022 & 12.974 & 7.077 & 2.399 \\ 6.863 & 7.077 & 11.107 & 2.510 \\ 3.245 & 2.399 & 2.510 & 5.576 \end{pmatrix}$$

and we easily obtain the test statistic:

$$-2 \log \lambda = n \log |S| - \sum_{h=1}^{2} n_h \log |S_h|$$
$$= 49 \log |S| - (37 \log |S_1| + 12 \log |S_2|) = 20.7.$$

This value of the test statistics leads to the rejection of the null hypothesis of the equality of the two covariance matrices since it is larger than the critical value $\chi^2_{0.95;10} = 18.307$, where the degrees of freedom were determined for $k = 2$ groups as $m = \frac{1}{2}(k-1)p(p+1) = \frac{1}{2}(2-1)4(4+1) = 10$.

Q SMStestcovwais

EXERCISE 7.15. *Consider two independent iid samples, each of size 10, from two bivariate normal populations. The results are summarized below:*

$$\bar{x}_1 = (3,1)^\top; \quad \bar{x}_2 = (1,1)^\top$$

$$S_1 = \begin{pmatrix} 4 & -1 \\ -1 & 2 \end{pmatrix}; \quad S_2 = \begin{pmatrix} 2 & -2 \\ -2 & 4 \end{pmatrix}.$$

Provide a solution to the following tests:

a) $H_0: \mu_1 = \mu_2 \quad H_1: \mu_1 \neq \mu_2$
b) $H_0: \mu_{11} = \mu_{21} \quad H_1: \mu_{11} \neq \mu_{21}$
c) $H_0: \mu_{12} = \mu_{22} \quad H_1: \mu_{12} \neq \mu_{22}$

Compare the solutions and comment.

a) Let us start by verifying the assumption of equality of the two covariance matrices, i.e., the hypothesis:

$$H_0 : \Sigma_1 = \Sigma_2 \quad \text{versus} \quad H_1 : \Sigma_1 \neq \Sigma_2.$$

This hypothesis can be tested using the approach described in Exercise 7.13 where we used the test statistic (for $k = 2$ groups):

$$-2 \log \lambda = n \log |S| - \sum_{h=1}^{2} n_h \log |S_h|$$

which is under the null hypothesis $H_0 : \Sigma_1 = \Sigma_2$ approximately χ^2_m distributed, where $m = \frac{1}{2}(k-1)p(p+1) = \frac{1}{2}(2-1)2(2+1) = 3$.

We calculate the average of the observed variance matrices

$$S = \begin{pmatrix} 3 & -1.5 \\ -1.5 & 3 \end{pmatrix}$$

and we get the value of the test statistic

$$-2 \log \lambda = 20 \log |S| - (10 \log |S_1| + 10 \log |S_2|) = 4.8688$$

which is smaller than the critical value $\chi^2_{0.95;3} = 7.815$. Hence, the value of the test statistic is not significant, we do not reject the null hypothesis, and the assumption of the equality of the variance matrices can be used in testing the equality of the mean vectors.

Now, we can test the equality of the mean vectors:

$$H_0 : \mu_1 = \mu_2 \quad \text{versus} \quad H_1 : \mu_1 \neq \mu_2.$$

The rejection region is given by

$$\frac{n_1 n_2 (n_1 + n_2 - p - 1)}{p(n_1 + n_2)^p} (\bar{x}_1 - \bar{x}_2)^\top S^{-1} (\bar{x}_1 - \bar{x}_2) \geq F_{1-\alpha;p,n_1+n_2-p-1}.$$

For $\alpha = 0.05$ we get the test statistic $3.7778 \geq F_{0.95;2,17} = 3.5915$. Hence, the null hypothesis $H_0 : \mu_1 = \mu_2$ is rejected and we can say that the mean vectors of the two populations are significantly different.

b) For the comparison of the two mean vectors first components we calculate the 95% simultaneous confidence interval for the difference. We test the hypothesis

$$H_0 : \mu_{11} = \mu_{21} \quad \text{versus} \quad H_1 : \mu_{11} \neq \mu_{21}.$$

This test problem is only one-dimensional and it can be solved by calculating the common two-sample t-test. The test statistic

$$\frac{\overline{x}_{11} - \overline{x}_{21}}{\sqrt{\frac{4}{n_1} + \frac{2}{n_2}}} = \frac{2}{\sqrt{\frac{6}{10}}} = 2.5820$$

is greater than the corresponding critical value $t_{0.95;18} = 2.1011$ and hence we reject the null hypothesis.

c) The comparison of the second component of the mean vectors can be also based on the two-sample t-test. In this case, it is obvious that the value of the test statistic is equal to zero (since $\overline{x}_{12} = \overline{x}_{22} = 1$) and the null hypothesis can not be rejected.

In part a) we have rejected the null hypothesis that the two mean vectors are equal. From the componentwise test performed in b) and c), we observe that the reason for rejecting the equality of the two two-dimensional mean vectors was due mainly to differences in the first component.

EXERCISE 7.16. *Assume that $X \sim N_p(\mu, \Sigma)$ where Σ is unknown.*

a) *Derive the log-likelihood ratio test for testing the independence of the p components, that is $H_0 : \Sigma$ is a diagonal matrix.*

b) *Assume that Σ is a diagonal matrix (all the variables are independent). Can an asymptotic test for $H_0 : \mu = \mu_o$ against $H_1 : \mu \neq \mu_o$ be derived? How would this compare to p independent univariate t-tests on each μ_j?*

c) *Provide an easy derivation of an asymptotic test for testing the equality of the p means. Compare this to the simple ANOVA procedure.*

In order to derive the likelihood ratio test statistic, we have to calculate ℓ_0^* and ℓ_1^*, the maxima of the log-likelihood under the null and alternative hypothesis. Using the results derived in Exercise 6.7, we can write

$$\ell_0^* = \ell\{\overline{x}, \text{diag}(\mathcal{S})\} = -\frac{n}{2} \log |2\pi \, \text{diag}(\mathcal{S})| - \frac{n}{2} \text{tr} \left(\text{diag}(\mathcal{S})^{-1} \mathcal{S} \right)$$

and, from the solution of Exercise 7.7, we know that

$$\ell_1^* = \ell(\overline{x}, \mathcal{S}) = -\frac{n}{2} \log |2\pi \mathcal{S}| - \frac{n}{2} \text{tr} \left(\mathcal{S}^{-1} \mathcal{S} \right) = -\frac{n}{2} \log |2\pi \mathcal{S}| - \frac{n}{2} p,$$

where $\text{diag}(\mathcal{S}) = \text{diag}(s_{11}, \ldots, s_{pp})$ and $d = (\overline{x} - \mu)$. Then

$$-2\log\lambda = 2\left(\ell_1^* - \ell_0^*\right)$$
$$= -n\left\{\log|2\pi\mathcal{S}| - \log|2\pi\operatorname{diag}(\mathcal{S})| + p - \operatorname{tr}\left(\operatorname{diag}(\mathcal{S})^{-1}\mathcal{S}\right)\right\}$$
$$= -n\left\{\log\left|\operatorname{diag}(\mathcal{S})^{-1}\right||\mathcal{S}| + p - \operatorname{tr}\left(\operatorname{diag}(\mathcal{S})^{-1}\mathcal{S}\right)\right\}$$
$$= -n\left\{\log\left|\operatorname{diag}(\mathcal{S})^{-1/2}\mathcal{S}\operatorname{diag}(\mathcal{S})^{-1/2}\right| + p\right.$$
$$\left. - \operatorname{tr}\left(\operatorname{diag}(\mathcal{S})^{-1/2}\mathcal{S}\operatorname{diag}(\mathcal{S})^{-1/2}\right)\right\}$$
$$= -n\left(\log|\mathcal{R}| + p - \operatorname{tr}\mathcal{R}\right)$$
$$= -n\log|\mathcal{R}|,$$

where $\mathcal{R} = \operatorname{diag}(\mathcal{S})^{-1/2}\mathcal{S}\operatorname{diag}(\mathcal{S})^{-1/2}$ is the empirical correlation matrix.

According to Theorem 7.1, the test of the null hypothesis can be based on the fact that the likelihood ratio test statistics $-n\log|\mathcal{R}|$ has asymptotically $\chi^2_{p(p-1)/2}$ distribution, where the number of degrees of freedom is the difference in the number of parameters under the alternative and null hypothesis: $p(p - 1)/2 = \dim(\Omega_1) - \dim(\Omega_0) = p(p+1)/2 - p$.

b) Again, using Theorem 7.1, the test can be derived by calculating the likelihood ratio test statistics $-2\log\lambda = -2\left(\ell_1^* - \ell_0^*\right)$ comparing the maximum of the log-likelihood under the null and alternative hypothesis.

Under the null hypothesis $H_0 : \mu = \mu_0$, we maximize the log-likelihood $\ell(\mathcal{X};\mu_0,\Sigma)$ under the assumption that the variance Σ is diagonal, i.e., the unknown parameters are the diagonal elements of Σ, $\sigma = \operatorname{diag}(\Sigma)$. Similarly, as in Exercise 6.7, the log-likelihood $\ell(\mathcal{X};\mu_0,\sigma)$ is

$$-\frac{n}{2}\log(2\pi) - \frac{n}{2}\sum_{j=1}^{p}\log\sigma_{jj} - \frac{1}{2}\sum_{i=1}^{n}(x_i - \mu_0)^{\top}\operatorname{diag}(\sigma^{-1})(x_i - \mu_0).$$

Setting the partial derivative of the log-likelihood w.r.t. the vector of unknown parameters $\sigma = \operatorname{diag}(\Sigma)$ equal to zero,

$$\frac{\partial}{\partial\sigma}\ell(\mathcal{X};\mu_0,\sigma) = -\frac{n}{2}\sigma^{-1} - \frac{1}{2}\frac{\partial}{\partial\sigma}\sum_{i=1}^{n}\operatorname{tr}\{(x_i - \mu_0)^{\top}\operatorname{diag}(\sigma^{-1})(x_i - \mu_0)\}$$

$$= -\frac{n}{2}\sigma^{-1} + \frac{1}{2}\sum_{i=1}^{n}\operatorname{diag}\{(x_i - \mu_0)(x_i - \mu_0)^{\top}\}\sigma^{-2}.$$

we obtain the MLE

$$0 = -\frac{n}{2}\widehat{\sigma}^{-1} + \frac{1}{2}\sum_{i=1}^{n}\operatorname{diag}\{(x_i - \mu_0)(x_i - \mu_0)^{\top}\}\widehat{\sigma}^{-2}$$

$$\widehat{\sigma} = \operatorname{diag}\left\{\frac{1}{n}\sum_{i=1}^{n}(x_i - \mu_0)(x_i - \mu_0)^{\top}\right\} = \operatorname{diag}(\mathcal{S} + dd^{\top}),$$

where $\mathcal{S}$ is the empirical covariance matrix and $d = (\bar{x} - \mu_0)$ as in Exercise 7.6. Thus,

$$\ell_0^* = \ell(\mathcal{X}; \mu_0, \text{diag}(\mathcal{S} + dd^\top)).$$

The maximum of the log-likelihood under the alternative hypothesis has already been derived in Exercise 6.7,

$$\ell_1^* = \ell(\mathcal{X}; \bar{x}, \text{diag}\, S)$$

and we can calculate the likelihood ratio test statistic similarly as in Exercise 7.7:

$$
\begin{aligned}
-2 \log \lambda &= 2\,(\ell_1^* - \ell_0^*) \\
&= 2\left\{ \ell(\mathcal{X}; \bar{x}, \text{diag}\, S) - \ell(\mathcal{X}; \mu_0, \text{diag}(\mathcal{S} + dd^\top)) \right\} \\
&= -n \log |\text{diag}(\mathcal{S})| + n \log |\text{diag}(\mathcal{S} + dd^\top)| \\
&\quad - \text{tr}\left\{ \text{diag}(\mathcal{S}^{-1}) \sum_{i=1}^n (x_i - \bar{x})(x_i - \bar{x})^\top \right\} \\
&\quad + \text{tr}\left[\{\text{diag}(\mathcal{S} + dd^\top)\}^{-1} \sum_{i=1}^n (x_i - \mu_0)(x_i - \bar{\mu}_0)^\top \right] \\
&= n \log \frac{|\text{diag}(\mathcal{S} + dd^\top)|}{|\text{diag}(\mathcal{S})|} = n \log \prod_{j=1}^p \frac{\sum_{i=1}^n (x_{ij} - \mu_{0j})^2}{\sum_{i=1}^n (x_{ij} - \bar{x}_j)^2} \\
&= n \log \prod_{j=1}^p \frac{\sum_{i=1}^n (x_{ij} - \bar{x}_j + \bar{x}_j - \mu_{0j})^2}{\sum_{i=1}^n (x_{ij} - \bar{x}_j)^2} \\
&= n \sum_{j=1}^p \log \frac{n s_{jj} + n(\bar{x}_j - \mu_{0j})^2}{n s_{jj}} \\
&= n \sum_{j=1}^p \log \left\{ 1 + \frac{(\bar{x}_j - \mu_{0j})^2}{s_{jj}} \right\}.
\end{aligned}
$$

The derived test statistics has asymptotically a χ^2 distribution with p degrees of freedom.

Using a first order Taylor expansion of $\log(1 + x) \approx x$, the test statistics $-2 \log \lambda$ may be approximated by the expression

$$\sum_{j=1}^p n \frac{(\bar{x}_j - \mu_{0j})^2}{s_{jj}} = \sum_{j=1}^p \left(\frac{\bar{x}_j - \mu_{0j}}{s_{jj}/\sqrt{n}} \right)^2,$$

i.e., a sum of squared univariate one-sample t-test statistics calculated for each dimension separately. Hence, the multivariate test is, in this case, approximately equivalent to a combination of the p univariate t-tests.

c) The hypothesis of the equality of the p means can be equivalently written as $H_0 : \mathcal{C}\mu = 0$, where $\mathcal{C}$ is a contrast matrix

$$\mathcal{C}((p-1) \times p) = \begin{pmatrix} 1 & -1 & 0 & \cdots & 0 \\ 0 & 1 & -1 & \cdots & 0 \\ \vdots & & \ddots & \ddots & \vdots \\ 0 & \cdots & 0 & 1 & -1 \end{pmatrix}$$

and a test could be based on the statistic

$$(n-1)\bar{x}^\top \mathcal{C}^\top (\mathcal{C} \operatorname{diag} S\mathcal{C}^\top)^{-1} \mathcal{C}\bar{x} \sim T^2(p, n-1)$$

or, equivalently,

$$\frac{n-p+1}{p-1}\bar{x}^\top \mathcal{C}^\top (\mathcal{C}S\mathcal{C}^\top)^{-1}\mathcal{C}\bar{x} \sim F_{p-1,n-p+1}. \tag{7.11}$$

The analysis of variance (ANOVA) technique is, in this case, based on the test statistic

$$\frac{\{SS(\text{reduced}) - SS(\text{full})\}/\{df(r) - df(f)\}}{SS(\text{full})/df(f)} \sim F_{df(r)-df(f),df(f)},$$

i.e.,

$$\frac{\left\{n\sum_{j=1}^{p}\left(\bar{x}_j - \bar{\bar{x}}\right)^2\right\}/(p-1)}{\sum_{j=1}^{p}\sum_{i=1}^{n}\left(x_{ij} - \bar{\bar{x}}\right)^2/(np-1)} \sim F_{p-1,np-1}, \tag{7.12}$$

where $\bar{\bar{x}} = \frac{1}{p}(\bar{x}_1 + \cdots + \bar{x}_p)$.

A comparison of the test statistics and their asymptotic distributions in (7.11) and (7.12) reveals that the tests behave differently. The main difference is that the analysis of variance (7.12) assumes that the the variances $\sigma_{11}, \ldots, \sigma_{pp}$ are equal. Thus, (7.11) is, in principle, a modification of ANOVA for heteroscedastic (unequal variances within groups) observations.

EXERCISE 7.17. *The yields of wheat have been measured in 30 parcels that have been randomly attributed to 3 lots prepared by one of 3 different fertilizers A, B, and C. The data set is given in Table A.7.*

Using Exercise 7.16,

a) *test the independence between the 3 variables.*

b) *test whether $\mu = (2, 6, 4)^\top$ and compare this to the 3 univariate t-tests.*

c) *test whether $\mu_1 = \mu_2 = \mu_3$ using simple ANOVA and the χ^2 approximation.*

a) We assume that the observations, $x_1, \ldots, x_{30}$, have 3-dimensional normal distribution $N_3(\mu, \Sigma)$ where Σ is unknown. The null and alternative hypothesis are:

$$H_0 : \Sigma \text{ is diagonal} \quad \text{vs.} \quad H_1 : \text{no constraints}$$

The corresponding likelihood ratio test statistic, derived in Exercise 7.16,

$$-n \log |\mathcal{R}| = -n \log \begin{vmatrix} 1.000 & -0.400 & 0.152 \\ -0.400 & 1.000 & -0.027 \\ 0.152 & -0.027 & 1.000 \end{vmatrix} = -n \log 0.819 = 1.987$$

is smaller than the corresponding critical value of the χ_6^2 distribution $\chi_{6;0.95}^2 = 12.592$ at level $\alpha = 0.05$. Hence, we do not reject the hypothesis that the variance matrix is diagonal.

b) The corresponding test statistic:

$$(n-1)(\bar{x} - \mu_0)^\top S^{-1}(\bar{x} - \mu_0) \sim T_{p;n-1}^2$$

follows under H_0 a Hotelling T^2-distribution, with $p = 3$ and $n-1 = 9$ degrees of freedom. From the data set, we calculate the mean vector

$$\bar{x} = \begin{pmatrix} 3.2 \\ 6.7 \\ 2.2 \end{pmatrix}$$

and the inverse of the variance matrix S

$$S^{-1} = \begin{pmatrix} 0.776 & 0 & 0 \\ 0 & 0.407 & 0 \\ 0 & 0 & 0.937 \end{pmatrix}.$$

The test statistic is

$$9(1.2, 0.7, -1.8) \begin{pmatrix} 0.776 & 0 & 0 \\ 0 & 0.407 & 0 \\ 0 & 0 & 0.937 \end{pmatrix} \begin{pmatrix} 1.2 \\ 0.7 \\ -1.8 \end{pmatrix} = 39.188$$

The critical value of the Hotelling $T_{3,9}^2$ distribution is

$$T_{0.95;3,9}^2 = \frac{3 \cdot 9}{9 - 3 + 1} F_{0.95;3,9-3+1} = 16.76$$

and it follows that we reject the null hypothesis $H_0 : \mu = (2, 6, 4)^\top$ since the test statistic is larger than the critical value.

The three univariate tests for the single means are:

$$H_0 : \mu_1 = 2 \text{ vs. } H_1 : \text{no constraints,}$$
$$H_0 : \mu_2 = 6 \text{ vs. } H_1 : \text{no constraints,}$$
$$H_0 : \mu_3 = 4 \text{ vs. } H_1 : \text{no constraints.}$$

The test statistics

$$T_i = \sqrt{n}\frac{\overline{x} - \mu_i}{s_{ii}}, \quad \text{for } i = 1, 2, 3,$$

follow a Student t-distribution with $n - 1 = 9$ degrees of freedom.

In our case, we obtain

$$T_1 = 3.342, \; T_2 = 1.413, \text{ and } T_3 = -5.511.$$

The null hypothesis is rejected if the absolute value of the test statistic is larger than the critical value $t_{0.975;9} = 2.263$. The null hypothesis is rejected for $\mu_1 = 2$ and $\mu_3 = 4$.

In practice, it is not a good idea to perform a series of univariate tests instead of one overall multivariate. It is easy to see that the probability of finding false positive result (rejecting valid null hypothesis) increases with the number of performed univariate tests.

c) The ANOVA hypothesis is:

$$H_0 : \mu_1 = \mu_2 = \mu_3 \text{ vs. } H_1 : \text{no constraints.}$$

The sums of squares for the ANOVA procedure are $SS(\text{full}) = \sum_{l=1}^{3}\sum_{k=1}^{10}(x_{kl} - \overline{x}_l)^2 = 43.30$ and $SS(\text{reduced}) = \sum_{l=1}^{3}\sum_{k=1}^{10}(x_{kl} - \overline{x})^2 = 154.97$. The test statistic

$$F = \frac{\{SS(\text{reduced}) - SS(\text{full})\}/(df(r) - df(f))}{SS(\text{full})/df(f)} = 34.816$$

follows a F-distribution with $df(f) = n - 3 = 27$ and $df(r) = n - 1 = 29$ degrees of freedom.

Since the test statistic $34.816 > F_{0.95;2,27} = 3.354$, we reject the null hypothesis of equality of the three means.

Without assuming the equality of the variances (homoscedasticity) the hypothesis can be written as

$$H_0 : \mathcal{C}\mu = 0_2 \quad \text{versus} \quad H_1 : \text{no constraints}$$

under the assumption that the variance matrix Σ is diagonal, where

$$\mathcal{C} = \begin{pmatrix} 1 & -1 & 0 \\ 1 & 0 & -1 \end{pmatrix}.$$

The t-test statistic:

$$-2\log\lambda = n\log\left\{1 + (C\bar{x} - a)^\top (CSC^\top)^{-1}(C\bar{x})\right\}$$

follows under the null hypothesis H_0 asymptotically a χ^2-distribution. From the observed data set, we obtain

$$C\bar{x} = \begin{pmatrix} -3.5 \\ 1 \end{pmatrix} \quad \text{and} \quad (ASA^\top)^{-1} = \begin{pmatrix} 0.329 & -0.180 \\ -0.180 & 0.523 \end{pmatrix}.$$

The test statistic is

$$-2\log\lambda = 10\log\left\{1 + (-3.5, 1)\begin{pmatrix} 0.329 & -0.180 \\ -0.180 & 0.523 \end{pmatrix}\begin{pmatrix} -3.5 \\ 1 \end{pmatrix}\right\} = 19.19$$

and we reject the null hypothesis at level $\alpha = 0.05$ since the test statistic is larger than the corresponding critical value $\chi^2_{0.95;2} = 5.99$.

EXERCISE 7.18. *Test the first sample ($n_1 = 30$) simulated in parts b) and c) of Exercise 7.13 to see if its covariance matrix is equal to $\Sigma_0 = 4\mathcal{I}_4$ (the sample covariance matrix to be tested is given by S_1).*

a) We have a random sample from a 4-dimensional normal distribution with a sample size of 30 and the empirical covariance matrix:

$$S_1 = \begin{pmatrix} 21.907 & 1.415 & -2.050 & 2.379 \\ 1.415 & 11.853 & 2.104 & -1.864 \\ -2.050 & 2.104 & 17.230 & 0.905 \\ 2.379 & -1.864 & 0.905 & 9.037 \end{pmatrix}$$

The test of the hypothesis

$$H_0 : \Sigma = \Sigma_0 \quad \text{versus} \quad H_1 : \text{no constraints}$$

can be carried out by likelihood ratio test based on the test statistic

$$-2\log\lambda = 2(\ell_1^* - \ell_0^*)$$
$$= 2\{\ell(\mathcal{X}; \bar{x}, S) - \ell(\mathcal{X}; \bar{x}, \Sigma_0)\}$$
$$= n\,\text{tr}\left(\Sigma_0^{-1}S\right) - n\log\left|\Sigma_0^{-1}S\right| - np$$

which has, under the null hypothesis, asymptotically χ^2_m distribution with $m = p(p-1)/2$ degrees of freedom.

Plugging in the observed covariance matrix, we get $-2\log\lambda = 264.8 > \chi^2_{0.95;10} = 18.31$ and we reject the null hypothesis $H_0 : \Sigma = \mathcal{I}_4$.

b) For the second observed covariance matrix,

$$S_1 = \begin{pmatrix} 14.649 & -0.024 & 1.248 & -3.961 \\ -0.024 & 15.825 & 0.746 & 4.301 \\ 1.248 & 0.746 & 9.446 & 1.241 \\ -3.961 & 4.301 & 1.241 & 20.002 \end{pmatrix},$$

we obtain the test statistic $-2 \log \lambda = 263.526$ and, comparing it to the same critical value $\chi^2_{0.95;10} = 18.31$, we again see that the observed covariance matrix is significantly different from $\mathcal{I}_4$.

EXERCISE 7.19. *Consider the bank data set in Table A.2. For the counterfeit bank notes, we want to know if the length of the diagonal (X_6) can be predicted by a linear model in X_1 to X_5. Estimate the linear model and test if the coefficients are significantly different from zero.*

We consider the linear regression model,

$$X_6 = (1, X_1, \ldots, X_5)\beta + \varepsilon,$$

where $\beta = (\beta_0, \ldots, \beta_5)^\top$ is the vector of the regression parameters and ε is the random error distributed as $N(0, \sigma^2)$. The parameter estimates and the related tests are summarized in the following computer output:

A N O V A	SS	df	MSS	F-test	P-value
Regression	9.920	5	1.984	8.927	0.0000
Residuals	20.890	94	0.222		
Total Variation	30.810	99	0.311		

Multiple R	= 0.56743
R^2	= 0.32197
Adjusted R^2	= 0.28591
Standard Error	= 0.47142

PARAMETERS	Beta	SE	StandB	t-test	P-value
b[0,]=	47.3454	34.9350	0.0000	1.355	0.1786
b[1,]=	0.3193	0.1483	0.2016	2.153	0.0339
b[2,]=	-0.5068	0.2483	-0.2317	-2.041	0.0440
b[3,]=	0.6337	0.2021	0.3388	3.136	0.0023
b[4,]=	0.3325	0.0596	0.6747	5.576	0.0000
b[5,]=	0.3179	0.1039	0.3624	3.060	0.0029

▫ SMSlinregbank2

The first part of the output concerns the test of the hypothesis

$$H_0 : \beta_1 = \beta_5 = 0 \quad \text{vs.} \quad H_1 : \beta_i \neq 0 \text{ for some } i = 1, \ldots, 5.$$

The value of the F-statistics is 8.927 and the small p-value ($< \alpha = 0.05$) indicates that the null hypothesis is rejected. This proves that the response variable X_6 depends on the variables $X_1, \ldots, X_5$.

The second part of the computer output contains information on the parameter estimates $\widehat{\beta}_i$, $i = 0, \ldots, 5$. The parameter $\beta_0 = $ b[0,] estimates the intercept (absolute term). The remaining parameters $\beta_i = $ b[i,], $i = 1, \ldots, 5$ measure the influence of the variables X_i on the response variable X_6, see Chapter 3 for more details. Each row contains a result of the univariate t-test of the hypothesis

$$H_0 : \beta_i = 0 \quad \text{vs.} \quad H_1 : \beta_i \neq 0.$$

From the p-values given in the last column, we can see that all regression coefficients are statistically significant on level $\alpha = 0.05$.

EXERCISE 7.20. *In the vocabulary data set (Bock 1975) given in Table A.20, predict the vocabulary score of the children in eleventh grade from the results in grades 8–10. Estimate a linear model and test its significance.*

A N O V A	SS	df	MSS	F-test	P-value
Regression	166.150	3	55.383	47.386	0.0000
Residuals	70.126	60	1.169		
Total Variation	236.276	63	3.750		

```
Multiple R       = 0.83857
R^2              = 0.70320
Adjusted R^2     = 0.68836
Standard Error   = 1.08110
```

PARAMETERS	Beta	SE	StandB	t-test	P-value
b[0,]=	1.4579	0.3014	0.0000	4.838	0.0000
b[1,]=	0.1974	0.1595	0.1925	1.238	0.2206
b[2,]=	0.2265	0.1161	0.2439	1.952	0.0557
b[3,]=	0.4042	0.1313	0.4535	3.079	0.0031

Q SMSlinregvocab

Regression analysis reveals reasonably high coefficient of determination. Hypothesis of independence (H_0 : all parameters= 0) is rejected on level $\alpha = 0.05$ since the F-statistics is statistically significant (the p-value is smaller than $\alpha = 0.05$).

The vocabulary score from tenth grade (β_3 =b[3,]) is statistically significant for the forecast of performance in eleventh grade. The other two variables,
vocabulary scores from the eighth and ninth grade are not statistically significant at level $\alpha = 0.05$. More formally, the test does not reject the hypothesis
that parameters β_2 and β_3 are equal to zero.

One might be tempted to simplify the model by excluding the insignificant
variables. However, excluding only the score in eighth grade leads to the following result which shows that the variable measuring the vocabulary score
in ninth grade has changed its significance.

A N O V A	SS	df	MSS	F-test	P-value
Regression	164.359	2	82.180	69.705	0.0000
Residuals	71.917	61	1.179		
Total Variation	236.276	63	3.750		

Multiple R	= 0.83404
R^2	= 0.69562
Adjusted R^2	= 0.68564
Standard Error	= 1.08580

PARAMETERS	Beta	SE	StandB	t-test	P-value
b[0,]=	1.2210	0.2338	0.0000	5.222	0.0000
b[1,]=	0.2866	0.1059	0.3086	2.707	0.0088
b[2,]=	0.5077	0.1016	0.5696	4.997	0.0000

Q SMSlinregvocab

Hence, the final model explains the vocabulary score in grade eleven using
vocabulary scores in the previous two grades.

EXERCISE 7.21. *Assume that we have observations from two p-dimensional
normal populations, $x_{i1} \sim N_p(\mu_1, \Sigma)$, $i = 1, \ldots, n_1$, and $x_{i2} \sim N_p(\mu_2, \Sigma)$,
$i = 1, \ldots, n_2$. The mean vectors μ_1 and μ_2 are called profiles. An example of
two such 5-dimensional profiles is given in Figure 7.1. Propose tests of the
following hypotheses:*

1. Are the profiles parallel?

2. If the profiles are parallel, are they at the same level?

3. If the profiles are parallel, are they also horizontal?

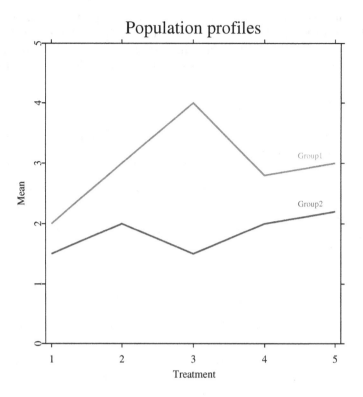

Fig. 7.1. Example of population profiles ✎ SMSprofil

The above questions are easily translated into linear constraints on the means and a test statistic can be obtained accordingly.

a) Let $\mathcal{C}$ be a $(p-1) \times p$ contrast matrix defined as

$$\mathcal{C} = \begin{pmatrix} 1 & -1 & 0 & \cdots & 0 \\ 0 & 1 & -1 & \cdots & 0 \\ 0 & \cdots & 0 & 1 & -1 \end{pmatrix}.$$

The hypothesis of parallel profiles is equivalent to

$$H_0^{(1)} : \mathcal{C}\mu_1 - \mathcal{C}\mu_2 = \mathcal{C}(\mu_1 - \mu_2) = 0_{p-1}.$$

The test of parallel profiles can be based on:

$$\mathcal{C}(\overline{x}_1 - \overline{x}_2) \sim N_{p-1}\left(\mathcal{C}(\mu_1 - \mu_2), \frac{n_1 + n_2}{n_1 n_2}\mathcal{C}\Sigma\mathcal{C}^\top\right).$$

Next, for the pooled covariance matrix $\mathcal{S} = (n_1\mathcal{S}_1 + n_2\mathcal{S}_2)/(n_1 + n_2)$ we have the Wishart distribution:

$$n_1 S_1 + n_2 S_2 \sim W_p\left(\Sigma, n_1 + n_2 - 2\right)$$
$$\mathcal{C}\left(n_1 S_1 + n_2 S_2\right)\mathcal{C}^\top \sim W_{p-1}\left(\mathcal{C}\Sigma\mathcal{C}^\top, n_1 + n_2 - 2\right).$$

Under the null hypothesis, we know that $\mathcal{C}\left(\mu_1 - \mu_2\right) = 0_{p-1}$ and it follows that the statistic

$$(n_1 + n_2 - 2)\left\{\mathcal{C}\left(\overline{x}_1 - \overline{x}_2\right)\right\}^\top \left\{\frac{n_1 + n_2}{n_1 n_2}\mathcal{C}\left(n_1 S_1 + n_2 S_2\right)\mathcal{C}^\top\right\}^{-1}\mathcal{C}\left(\overline{x}_1 - \overline{x}_2\right)$$

$$= (n_1 + n_2 - 2)\left\{\mathcal{C}\left(\overline{x}_1 - \overline{x}_2\right)\right\}^\top \left\{\frac{n_1 + n_2}{n_1 n_2}\left(n_1 + n_2\right)\mathcal{C}S\mathcal{C}^\top\right\}^{-1}\mathcal{C}\left(\overline{x}_1 - \overline{x}_2\right)$$

$$= \frac{(n_1 + n_2 - 2)n_1 n_2}{(n_1 + n_2)^2}\left\{\mathcal{C}\left(\overline{x}_1 - \overline{x}_2\right)\right\}^\top \left\{\mathcal{C}S\mathcal{C}\right\}^{-1}\mathcal{C}\left(\overline{x}_1 - \overline{x}_2\right)$$

has the Hotelling T^2 distribution $T^2\left(p - 1, n_1 + n_2 - 2\right)$ and the null hypothesis of parallel profiles is rejected if

$$\frac{n_1 n_2(n_1 + n_2 - p)}{(n_1 + n_2)^2(p - 1)}\left\{\mathcal{C}(\overline{x}_1 - \overline{x}_2)\right\}^\top \left(\mathcal{C}S\mathcal{C}^\top\right)^{-1}\mathcal{C}(\overline{x}_1 - \overline{x}_2) > F_{1-\alpha;p-1,n_1+n_2-p}.$$

$$\tag{7.13}$$

b) Assuming that the two profiles are parallel, the null hypothesis of the equality of the two levels can be formally written as

$$H_0^{(2)} : 1_p^\top\left(\mu_1 - \mu_2\right) = 0.$$

For $1_p^\top\left(\overline{x}_1 - \overline{x}_2\right)$, as a linear function of normally distributed random vectors, we have

$$1_p^\top\left(\overline{x}_1 - \overline{x}_2\right) \sim N_1\left(1_p^\top\left(\mu_1 - \mu_2\right), \frac{n_1 + n_2}{n_1 n_2}1_p^\top\Sigma 1_p\right).$$

Since

$$1_p^\top\left(n_1 S_1 + n_2 S_2\right)1_p \sim W_1\left(1_p^\top\{\Sigma 1_p, n_1 + n_2 - 2\right),$$

we have that

$$(n_1 + n_2)1_p^\top S 1_p \sim W_1(1_p^\top\Sigma 1_p, n_1 + n_2 - 2),$$

where S is the pooled empirical variance matrix. The test of equality can be based on the test statistic:

$$(n_1 + n_2 - 2)\left\{1_p^\top\left(\overline{x}_1 - \overline{x}_2\right)\right\}^\top \left\{\frac{n_1 + n_2}{n_1 n_2}\mathcal{C}\left(n_1 S_1 + n_2 S_2\right)\mathcal{C}^\top\right\}^{-1}1_p^\top\left(\overline{x}_1 - \overline{x}_2\right)$$

$$= \frac{n_1 n_2(n_1 + n_2 - 2)}{(n_1 + n_2)^2}\frac{\left\{1_p^\top\left(\overline{x}_1 - \overline{x}_2\right)\right\}^2}{1_p^\top S 1_p} \sim T^2(1, n_1 + n_2 - 2)$$

which leads directly the rejection region:

$$\frac{n_1 n_2(n_1 + n_2 - 2)}{(n_1 + n_2)^2}\frac{\left\{1_p^\top\left(\overline{x}_1 - \overline{x}_2\right)\right\}^2}{1_p^\top S 1_p} > F_{1-\alpha;1,n_1+n_2-2}.$$

$$\tag{7.14}$$

c) If it is accepted that the profiles are parallel, then we can exploit the information contained in both groups to test if the two profiles also have zero slope, i.e., the profiles are horizontal. The null hypothesis may be written as:

$$H_0^{(3)} : \mathcal{C}(\mu_1 + \mu_2) = 0.$$

The average profile $\bar{x} = (n_1\bar{x}_1 + n_2\bar{x}_2)/(n_1 + n_2)$ has a p-dimensional normal distribution:

$$\bar{x} \sim N_p \left(\frac{n_1\mu_1 + n_2\mu_2}{n_1 + n_2}, \frac{1}{n_1 + n_2}\Sigma \right).$$

Now the horizontal, $H_0^{(3)} : \mathcal{C}(\mu_1 + \mu_2) = 0_{p-1}$, and parallel, $H_0^{(1)} : \mathcal{C}(\mu_1 - \mu_2) = 0_{p-1}$, profiles imply that

$$\mathcal{C}\left(\frac{n_1\mu_1 + n_2\mu_2}{n_1 + n_2} \right) = \frac{\mathcal{C}}{n_1 + n_2}(n_1\mu_1 + n_2\mu_2)$$

$$= \frac{\mathcal{C}}{2(n_1 + n_2)}\{(n_1 + n_2)(\mu_1 + \mu_2) + (n_1 - n_2)(\mu_1 - \mu_2)\}$$

$$= 0_{p-1}.$$

So, under parallel and horizontal profiles we have

$$\mathcal{C}\bar{x} \sim N_{p-1}\left(0_{p-1}, \frac{1}{n_1 + n_2}\mathcal{C}\Sigma\mathcal{C}^\top \right).$$

and

$$\mathcal{C}(n_1 + n_2)\mathcal{S}\mathcal{C}^\top = \mathcal{C}\left(n_1\mathcal{S}_1 + n_2\mathcal{S}_2\right)\mathcal{C}^\top \sim W_{p-1}\left(\mathcal{C}\Sigma\mathcal{C}^\top, n_1 + n_2 - 2\right).$$

Again, we get under the null hypothesis that

$$(n_1 + n_2 - 2)(\mathcal{C}\bar{x})^\top(\mathcal{C}\mathcal{S}\mathcal{C}^\top)^{-1}\mathcal{C}\bar{x} \sim T^2(p - 1, n_1 + n_2 - 2)$$

which leads to the rejection region:

$$\frac{n_1 + n_2 - p}{p - 1}(\mathcal{C}\bar{x})^\top(\mathcal{C}\mathcal{S}\mathcal{C}^\top)^{-1}\mathcal{C}\bar{x} > F_{1-\alpha;p-1,n_1+n_2-p}. \qquad (7.15)$$

EXERCISE 7.22. *In Olkin & Veath (1980), the evolution of citrate concentrations in plasma is observed at three different times of day for two groups of patients who follow different diet. (The patients were randomly attributed to each group under a balanced design $n_1 = n_2 = 5$). The data set is given in Table A.14.*

Test if the profiles of the groups are parallel, if they are at the same level and if they are horizontal.

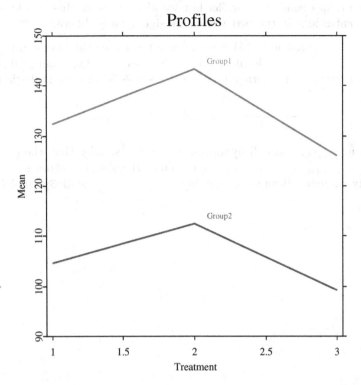

Fig. 7.2. Groups profiles of the evolution of citrate concentrations in plasma observed at 3 different times of day. ⊙ SMSprofplasma

The observed profiles are plotted in Figure 7.2. We apply the test statistics derived in Exercise 7.21 to test the statistical significance of the difference between the observed profiles.

a) The test for parallel profiles (7.13) leads to the test statistic:

$$\frac{n_1 n_2 (n_1 + n_2 - p)}{(n_1 + n_2)^2 (p-1)} (\mathcal{C}\overline{x})^\top (\mathcal{C}\mathcal{S}\mathcal{C}^\top)^{-1} \mathcal{C}\overline{x} = 0.95$$

and we do not reject the null hypothesis since $0.95 < F_{1-\alpha;p-1,n_1+n_2-p} = F_{0.95;2,7} = 4.74$.

b) Let us now use (7.14) to test the equality of the profiles: The test statistic

$$\frac{n_1 n_2 (n_1 + n_1 - 2)\{1_p^\top (\overline{x}_1 - \overline{x}_2)\}^2}{(n_1 + n_2)^2 1_p^\top \mathcal{S} 1_p} = 10.9$$

is larger than the critical value $F_{1-\alpha;1,n_1+n_2-2} = F_{0.95;1,8} = 5.32$ and the hypothesis of equal profiles is rejected.

Hence, the profiles could be parallel but we already know that the level of citrate concentrations in the two groups are significantly different.

c) Using the test statistic (7.15), we can assert whether the horizontality of the observed profiles, i.e., decide whether the concentrations change during the day. Considering the average profile $\bar{x} = \frac{n_1\bar{x}_1 + n_2\bar{x}_2}{n_1+n_2}$, the test statistic is

$$\frac{n_1 + n_2 - p}{p - 1}(\mathcal{C}\bar{x})^{\top}(\mathcal{C}\mathcal{S}\mathcal{C}^{\top})^{-1}\mathcal{C}\bar{x} = 0.30$$

and we do not reject the null hypothesis since it is smaller than the critical value $F_{1-\alpha;p-1,n_1+n_2-p} = F_{0.95;2,7} = 4.74$. Thus, the observed profiles are not significantly changing throughout the day.   ⊡ SMSprofplasma

Part III

Multivariate Techniques

8

Decomposition of Data Matrices by Factors

It is of the highest importance in the art of detection to be able to recognize, out of a number of facts, which are incidental and which vital. Otherwise your energy and attention must be dissipated instead of being concentrated.
Sherlock Holmes in "The Reigate Puzzle"

In this chapter, we take a descriptive perspective and show how using a geometrical approach can be a good way to reduce the dimension of a data matrix. We derive the interesting projections with respect to a least-squares criterion. The results will be low-dimensional graphical pictures of the data matrix. This involves the decomposition of the data matrix into factors. These factors will be sorted in decreasing order of importance. The approach is very general and is the core idea of many multivariate techniques. We deliberately use the word "factor" here as a tool or transformation for structural interpretation in an exploratory analysis.

In practical studies, we apply this factorial technique to the Swiss bank notes, the time budget data, and the French food data. We will see that these transformations provide easier interpretations in lower-dimensional spaces. An important measure of resolution of information in a low-dimensional projected space is the notion of inertia. We will calculate this inertia for several practical examples.

Representation of the p-Dimensional Data Cloud

On one hand, the data set $\mathcal{X}$ can be understood as a cloud of n points in $\mathbb{R}^p$. The best representation of the p-dimensional data set in $q < p$ dimensions can be found by searching for directions $u_j \in \mathbb{R}^p$, $j = 1, \ldots, q$, minimizing the distance

$$\sum_{i=1}^{n} \|x_i - p_{x_i}\|^2, \tag{8.1}$$

where

$$p_{j,x_i} = x_i^\top \frac{u_j}{\|u_j\|} = x_i^\top u_j \tag{8.2}$$

are projections of observations x_i into the jth direction u_j. The best subspace is generated by $u_1, u_2, \ldots, u_q$, the orthonormal eigenvectors of $\mathcal{X}^\top \mathcal{X}$ associated with the corresponding eigenvalues $\lambda_1 \geq \lambda_2 \geq \ldots \geq \lambda_q$.

The coordinates of the n individuals on the kth factorial axis, u_k, are given by the kth factorial variable $z_k = \mathcal{X} u_k$ for $k = 1, \ldots, q$. Each factorial variable $z_k = (z_{1k}, z_{2k}, \ldots, z_{nk})^\top$ is a linear combination of the original variables whose coefficients are given by the elements of the corresponding eigenvector u_k, i.e., $z_{ik} = x_i^\top u_k$.

In general, the scalar product $y^\top y$ is called the inertia of $y \in \mathbb{R}^n$ w.r.t. the origin. Note that $\lambda_k = (\mathcal{X} u_k)^\top (\mathcal{X} u_k) = z_k^\top z_k$. Thus, λ_k is the inertia of the jth factorial variable w.r.t. the origin.

Representation of the n-Dimensional Data Cloud

On the other side, we can interpret the data set $\mathcal{X}$ as a cloud of p variables observed in n-dimensional space $\mathbb{R}^n$.

The best q-dimensional subspace is generated by the orthonormal eigenvectors $v_1, v_2, \ldots, v_q$ of $\mathcal{X} \mathcal{X}^\top$ associated with the eigenvalues $\mu_1 \geq \mu_2 \geq \ldots \geq \mu_q$.

The coordinates of the p variables on the kth factorial axis are given by the factorial variables $w_k = \mathcal{X}^\top v_k$, $k = 1, \ldots, q$. Each factorial variable $w_k = (w_{k1}, w_{k2}, \ldots, w_{kp})^\top$ is a linear combination of the original n-dimensional vectors $x_{[i]}$ whose coefficients are given by the kth eigenvector, i.e., $w_{ki} = x_{[i]}^\top v_k$.

Duality Relations

Both views at the data set are closely related. The precise description of this relationship is given in the following theorem.

THEOREM 8.1. *Let r be the rank of $\mathcal{X}$. For $k \leq r$, the eigenvalues λ_k of $\mathcal{X}^\top \mathcal{X}$ and $\mathcal{X} \mathcal{X}^\top$ are the same and the eigenvectors (u_k and v_k, respectively) are related by*

$$u_k = \frac{1}{\sqrt{\lambda_k}} \mathcal{X}^\top v_k \quad and \quad v_k = \frac{1}{\sqrt{\lambda_k}} \mathcal{X} u_k.$$

Note that u_k and v_k provide the singular value decomposition (SVD) of $\mathcal{X}$. Letting $\mathcal{U} = (u_1, u_2, \ldots, u_r)$, $\mathcal{V} = (v_1, v_2, \ldots, v_r)$, and $\Lambda = \text{diag}(\lambda_1, \ldots, \lambda_r)$, we have

$$\mathcal{X} = \mathcal{V}\Lambda^{1/2}\mathcal{U}^\top.$$

EXERCISE 8.1. *Prove Theorem 8.1.*

Consider the eigenvector equations in the n-dimensional space, $(\mathcal{X}\mathcal{X}^\top)v_k = \mu_k v_k$, for $k \leq r$, where $r = \text{rank}(\mathcal{X}\mathcal{X}^\top) = \text{rank}(\mathcal{X}) \leq \min(p, n)$. Multiplying by $\mathcal{X}^\top$, we have

$$(\mathcal{X}^\top \mathcal{X})(\mathcal{X}^\top v_k) = \mu_k(\mathcal{X}^\top v_k)$$

so that each eigenvector v_k of $\mathcal{X}\mathcal{X}^\top$ corresponds to an eigenvector $(\mathcal{X}^\top v_k)$ of $\mathcal{X}^\top \mathcal{X}$ associated with the same eigenvalue μ_k. This means that every non-zero eigenvalue of $\mathcal{X}\mathcal{X}^\top$ is also an eigenvalue of $\mathcal{X}^\top \mathcal{X}$. The corresponding eigenvectors are related by $u_k = c_k\mathcal{X}^\top v_k$, where c_k is some constant.

Now consider the eigenvector equations in the p-dimensional space, $(\mathcal{X}^\top \mathcal{X})u_k = \lambda_k u_k$, for $k \leq r$. Multiplying by $\mathcal{X}$, we have

$$(\mathcal{X}\mathcal{X}^\top)(\mathcal{X}u_k) = \lambda_k(\mathcal{X}u_k),$$

i.e., each eigenvector u_k of $\mathcal{X}^\top \mathcal{X}$ corresponds to an eigenvector $\mathcal{X}u_k$ of $\mathcal{X}\mathcal{X}^\top$ associated with the same eigenvalue $\lambda_k = \mu_k$. Therefore, every non-zero eigenvalue of $(\mathcal{X}^\top \mathcal{X})$ is an eigenvalue of $\mathcal{X}\mathcal{X}^\top$. The corresponding eigenvectors are related by $v_k = d_k\mathcal{X}u_k$, where d_k is some constant.

Now, since $u_k^\top u_k = v_k^\top v_k = 1$ we have

$$1 = u_k^\top u_k = v_k^\top \mathcal{X}c_k^2\mathcal{X}^\top v_k = c_k^2 v_k^\top \mathcal{X}\mathcal{X}^\top v_k = c_k^2 v_k^\top \lambda_k v_k = c_k^2 \lambda_k$$
$$1 = v_k^\top v_k = u_k^\top \mathcal{X}^\top d_k^2\mathcal{X}u_k = d_k^2 u_k^\top \mathcal{X}^\top \mathcal{X}u_k = d_k^2 u_k^\top \lambda_k u_k = d_k^2 \lambda_k$$

and it follows that

$$c_k = d_k = \frac{1}{\sqrt{\lambda_k}}.$$

EXERCISE 8.2. *Describe the relation between the projections of the individuals and the variables on the factorial axes.*

Note that the projection of the p variables on the kth factorial axis v_k is given by

$$w_k = \mathcal{X}^\top v_k = \frac{1}{\sqrt{\lambda_k}}\mathcal{X}^\top \mathcal{X}u_k = \sqrt{\lambda_k}\, u_k.$$

Therefore, the projections on the factorial axis v_k are rescaled eigenvectors of $\mathcal{X}^\top \mathcal{X}$. Consequently, the eigenvectors v_k do not have to be explicitly recomputed to get the projections w_k.

Similarly, we have also for the projections z_k of the n observations on the kth factorial axis u_k that

$$z_k = \mathcal{X}u_k = \frac{1}{\sqrt{\lambda_k}}\mathcal{X}\mathcal{X}^\top v_k = \sqrt{\lambda_k}\, v_k.$$

EXERCISE 8.3. *Let u_k, $k = 1,\ldots,r$ be the first r eigenvectors of $\mathcal{X}^\top\mathcal{X}$. Define $z_k = \mathcal{X}u_k$ and prove that $n^{-1}\mathcal{Z}^\top\mathcal{Z}$ is the covariance of the centered data matrix, where $\mathcal{Z}$ is the matrix formed by the columns z_k, $k = 1,\ldots,r$.*

Let us write the spectral decomposition of the matrix $\mathcal{X}^\top\mathcal{X}$ as $\mathcal{X}^\top\mathcal{X} = \mathcal{U}\Lambda\mathcal{U}^\top$. Then, we have $\mathcal{Z} = \mathcal{X}\mathcal{U}$ and we obtain:

$$n^{-1}\mathcal{Z}^\top\mathcal{Z} = n^{-1}\mathcal{U}^\top\mathcal{X}^\top\mathcal{X}\mathcal{U} = n^{-1}\mathcal{U}^\top\mathcal{U}\Lambda\mathcal{U}^\top\mathcal{U} = n^{-1}\Lambda.$$

For the mean of $\mathcal{Z}$ we have

$$\bar{z}^\top = 1_n^\top\mathcal{Z} = 1_n^\top\mathcal{X}\mathcal{U} = \bar{x}^\top\mathcal{U}$$

and it follows that performing the factorial technique on a centered data set $\mathcal{X}$ leads to a centered data set $\mathcal{Z}$. The empirical covariance matrix $\mathcal{S}_{\mathcal{Z}}$ of the centered data set $\mathcal{Z}$ can now be written as

$$\mathcal{S}_{\mathcal{Z}} = \frac{1}{n}\mathcal{Z}^\top\mathcal{Z} = \frac{1}{n}\Lambda.$$

Observe that the marginal variances of $\mathcal{Z}$ are the eigenvalues of $\mathcal{X}^\top\mathcal{X}$ and that the vectors of $\mathcal{Z}$ are orthogonal.

EXERCISE 8.4. *Apply the factorial technique to the French food data (Table A.9) and relate the results to the SVD of the same data matrix.*

The French food data set gives the food expenditures of various types of French families (manual workers = MA, employees = EM, managers = CA) with varying numbers of children (2, 3, 4 or 5 children).

We shall now represent food expenditures and households simultaneously using two factors. First, note that in this particular problem the origin has no specific meaning (it represents a "zero" consumer). So it makes sense to compare the consumption of any family to that of an "average family" rather than to the origin. Therefore, the data is first centered (the origin is translated to the center of gravity, $\bar{x}$). Furthermore, since the dispersions of the 7 variables are quite different each variable is standardized so that each has the same weight in the analysis (mean 0 and variance 1). Finally, for convenience, we divide each element in the matrix by $\sqrt{n} = \sqrt{12}$. (This will only change the scaling of the plots in the graphical representation.)

The data matrix to be analyzed is therefore:

$$X_* = \frac{1}{\sqrt{n}} \mathcal{H} \mathcal{X} \mathcal{D}^{-1/2},$$

where $\mathcal{H}$ is the centering matrix and $\mathcal{D} = \text{diag}(s_{X_i X_i})$. Note that from standardizing by $\sqrt{n}$, it follows that $X_*^\top X_* = \mathcal{R}$ where $\mathcal{R}$ is the correlation matrix of the original data.

A standard way of evaluating the quality of the factorial representations in a subspace of dimension q is given by the ratio

$$\tau_q = \frac{\lambda_1 + \lambda_2 + \ldots + \lambda_q}{\lambda_1 + \lambda_2 + \ldots + \lambda_p}. \tag{8.3}$$

The sum $\sum_{j=1}^{q} \lambda_j$ is the sum of the inertia of the first q factorial variables $z_1, z_2, \ldots, z_q$. The denominator in (8.3) is a measure of the total inertia of the p variables because

$$\sum_{j=1}^{p} \lambda_j = \text{tr}(X_*^\top X_*) = \sum_{j=1}^{p} \sum_{i=1}^{n} x_{ij}^2 = \sum_{j=1}^{p} x_{[j]}^\top x_{[j]}.$$

Therefore, the ratio τ_q (8.3) is usually interpreted as the percentage of the inertia explained by the first q factors.

Calculating the eigenvalues $\lambda = (4.33, 1.83, 0.63, 0.13, 0.06, 0.02, 0.00)^\top$ shows that the directions of the first two eigenvectors play a dominant role ($\tau_2 = 88\%$), whereas the other directions contribute less than 15% of inertia. A two-dimensional plot should therefore suffice for interpreting this data set.

The representation of the n individuals on a plane is then obtained by plotting $z_1 = X_* u_1$ versus $z_2 = X_* u_2$ ($z_3 = X_* u_3$ may eventually be added if a third dimension is helpful). Using Theorem 8.1, representations for the p variables can easily be obtained. These representations can be visualized in a scatterplot of $w_1 = \sqrt{\lambda_1}\, u_1$ against $w_2 = \sqrt{\lambda_2} u_2$.

In the first window of Figure 8.1 we see the representation of the $p = 7$ variables given by the first two factors. The plot shows the factorial variables w_1 and w_2. We see that the points for meat, poultry, vegetables and fruits are close to each other in the lower left of the graph. The expenditures for bread and milk can be found in the upper left whereas wine stands alone in the upper right. The first factor, w_1, may be interpreted as the meat/fruit factor of consumption, the second factor, w_2, as the bread/wine component.

On the right-hand side of Figure 8.1, we show the factorial variables z_1 and z_2 from the fit of the $n = 12$ household types. Note that by the duality relations of Theorem 8.1, the factorial variables z_j are linear combinations of the factors w_k from the left window. The points displayed in the consumer window (graph

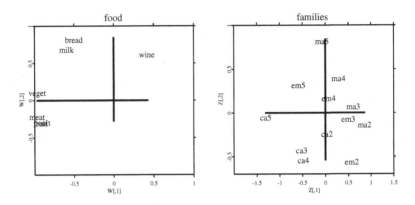

Fig. 8.1. Representation of food expenditures and family types in two dimensions.
Q SMSdecofood

on the right) are plotted relative to an average consumer represented by the origin. The manager families are located in the lower left corner of the graph whereas the manual workers and employees tend to be in the upper right. The factorial variables for CA5 (managers with five children) lie close to the meat/fruit factor. Relative to the average consumer this household type is a large consumer of meat/poultry and fruits/vegetables.

The SVD of the centered and standardized French food data set, $\mathcal{X}_*$, is given as

$$\mathcal{X}_* = \Gamma \Lambda \Delta^\top,$$

where

$$
\Gamma = \begin{pmatrix}
-0.41 & 0.08 & -0.15 & 0.30 & 0.28 & -0.14 & -0.24 \\
-0.27 & 0.40 & 0.50 & -0.14 & -0.12 & -0.15 & 0.24 \\
-0.02 & 0.16 & -0.54 & 0.17 & -0.56 & 0.34 & -0.13 \\
-0.30 & -0.07 & 0.04 & 0.09 & 0.01 & 0.15 & 0.62 \\
-0.25 & 0.04 & 0.19 & 0.13 & -0.22 & -0.12 & -0.31 \\
0.25 & 0.30 & -0.38 & -0.36 & -0.10 & -0.59 & 0.14 \\
-0.14 & -0.31 & 0.10 & -0.22 & 0.12 & -0.15 & -0.52 \\
-0.04 & -0.14 & -0.10 & 0.24 & 0.20 & 0.24 & 0.14 \\
0.23 & 0.39 & -0.04 & -0.34 & 0.53 & 0.48 & -0.13 \\
0.03 & -0.62 & -0.22 & -0.21 & 0.15 & -0.14 & 0.24 \\
0.28 & -0.25 & 0.38 & -0.27 & -0.41 & 0.28 & -0.06 \\
0.63 & 0.02 & 0.22 & 0.61 & 0.10 & -0.21 & 0.01
\end{pmatrix}
\begin{matrix}
\text{MA2} \\
\text{EM2} \\
\text{CA2} \\
\text{MA3} \\
\text{EM3} \\
\text{CA3} \\
\text{MA4} \\
\text{EM4} \\
\text{CA4} \\
\text{MA5} \\
\text{EM5} \\
\text{CA5}
\end{matrix}
\; ,
$$

$$\Delta = \begin{pmatrix} 0.24 & -0.62 & 0.01 & -0.54 & 0.04 & -0.51 & -0.02 \\ 0.47 & -0.10 & 0.06 & -0.02 & -0.81 & 0.30 & 0.16 \\ 0.45 & 0.21 & -0.15 & 0.55 & -0.07 & -0.63 & -0.20 \\ 0.46 & 0.14 & -0.21 & -0.05 & 0.41 & 0.09 & 0.74 \\ 0.44 & 0.20 & -0.36 & -0.32 & 0.22 & 0.35 & -0.60 \\ 0.28 & -0.52 & 0.44 & 0.45 & 0.34 & 0.33 & -0.15 \\ -0.21 & -0.48 & -0.78 & 0.31 & -0.07 & 0.14 & 0.04 \end{pmatrix} \begin{matrix} \text{bread} \\ \text{vegetables} \\ \text{fruits} \\ \text{meat} \\ \text{poultry} \\ \text{milk} \\ \text{wine} \end{matrix} \quad,$$

$$\Lambda = \mathrm{diag}\{(2.08, 1.35, 0.79, 0.36, 0.24, 0.14, 0.03)^\top\}.$$

It is easy to see that the singular values are equal to the square roots of the eigenvalues of the correlation matrix $\mathcal{R}$ of the original data.

The coordinates of the representation of the n points and p variables given in Figure 8.1 are given by the first two columns of Γ and Δ multiplied by the corresponding singular values. The only difference might be an opposite sign—since multiplication of any eigenvector by -1 leads to an equivalent SVD.

EXERCISE 8.5. *Recall the factorial analysis of the French food data of Exercise 8.4 and compute* $\tau_3, \tau_4, \ldots$.

The eigenvalues of the correlation matrix, corresponding to the centered and standardized data matrix $\mathcal{X}_*$ are calculated in Exercise 8.4:

$$\lambda = (4.33, 1.83, 0.63, 0.13, 0.06, 0.02, 0.00)^\top.$$

It follows that

$$\tau_3 = \frac{4.33 + 1.83 + 0.63}{4.33 + 1.83 + 0.63 + 0.13 + 0.06 + 0.02 + 0.00} = 0.970$$

$$\tau_4 = 0.989$$
$$\tau_5 = 0.997$$
$$\tau_6 = \tau_7 = 1.000.$$

As we have seen in Exercise 8.4, each τ_q can be interpreted as the percentage of the inertia explained by the first q factors. We see that 97% of the inertia is explained by the first three factors. Recalling that $\tau_1 = 0.619$ and $\tau_2 = 0.880$, we see that the third factor explains 9% of the inertia. The fourth and fifth factor explain together less than 3% of the inertia.

EXERCISE 8.6. *How do the eigenvalues and eigenvectors in Exercise 8.4 change if we take the prices in USD instead of in EUR? Does it make a difference if some of the prices are in EUR and others in USD?*

The eigenvalues and eigenvectors in Exercise 8.4 do not change because they are calculated from the correlation matrix which does not change for different units of measurement.

If some prices are quoted in EUR and some in USD, the standardization of the prices performed in Exercise 8.4 leads to the same result: the eigenvalues and eigenvectors are unaffected by such a scale change.

To make an example, assume that the prices in the United States are: $\mathcal{X}_{US} = 1.2\mathcal{X}_{EUR}$. Then, the SVD of $\mathcal{X}_{EUR} = \Gamma\Lambda\Delta^{\top}$ leads to a SVD of $\mathcal{X}_{US} = \Gamma(1.2\Lambda)\Delta^{\top}$, i.e., the matrix $\mathcal{X}_{US}$ has the same eigenvectors as $\mathcal{X}_{EUR}$. The singular values of $\mathcal{X}_{US}$ are equal to the singular values of $\mathcal{X}_{EUR}$ are multiplied by the exchange rate 1.2. The eigenvalues of $\mathcal{X}_{US}^{\top}\mathcal{X}_{US}$ are equal to the eigenvalues of $\mathcal{X}_{EUR}^{\top}\mathcal{X}_{EUR}$ multiplied by the constant $1.2^2 = 1.44$.

Hence, reporting all the prices in different currency affects only the eigenvalues. The proportions of explained inertia, defined as the ratio of the eigenvalues remain the same. The projections on the factorial axes (which are proportional to the square root of the eigenvalue) will be multiplied by the constant $\sqrt{1.44}$, i.e., by the exchange rate 1.2.

EXERCISE 8.7. *Apply the factorial techniques to the Swiss bank notes (Table A.2). Give an interpretation of the factorial variables.*

We follow the same steps as in Exercise 8.4. Centering the data matrix bases the decomposition of the data matrix on differences from "average banknote". Standardizing the data set makes the measurements of different lengths comparable, i.e., the importance of the different measurements does not depend on the scale.

The vector of the eigenvalues is:

$$\lambda = (2.95, 1.28, 0.87, 0.45, 0.27, 0.19)^{\top}$$

and it leads immediately the following proportions of explained inertia:

$$\tau = (0.49, 0.70, 0.85, 0.92, 0.97, 1.00)^{\top}.$$

The choice of the number of factorial variables can be based on various criteria. A reasonable approach is to choose the factorial variables that explain "larger than average" percentage of inertia. In this case, this rule leads to $q = 2$. However, in this example the third factorial variable is still rather important with 15% of the explained inertia and we choose $q = 3$ in order to demonstrate the factorial analysis in three dimensions.

The three factorial variables are presented using the multivariate tools described in Chapter 1. In Figures 8.2 and 8.3, we plot the projections onto the factorial axes in a scatterplot matrix. In Figure 8.4, we plot the projections onto the 3 factorial axes in a 3D-scatterplot, running the program ⌑ SMSdecobank allows interactive rotation of the graphic. The genuine and forged bank notes are denoted by letters 'G' and 'F', respectively.

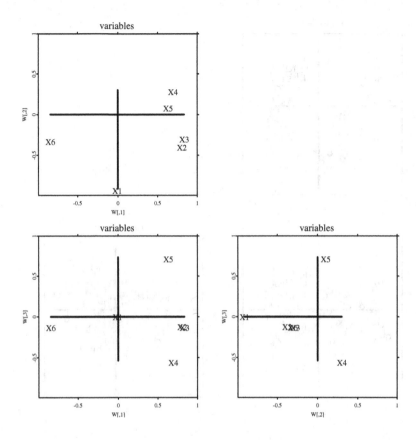

Fig. 8.2. Representation of the variables for Swiss bank notes in three dimensions.
 Ⓠ SMSdecobank

The data set contains six variables of various distances: height and length of the bank note, the length of the diagonal and also some measurements concerning the position of the central picture on the bank note. For detailed description see Table A.2.

In Figure 8.2, we observe the projections of the variables. The first factorial variable, w_1, measures the contrast between X_6 (length of the diagonal) and X_2–X_5 (distances related to the height of the bank notes). The second factorial variable consists mainly of X_1, the length of the bank note. The third factorial variable could be interpreted as a contrast between X_4 (distance of inner frame to the lower border) and X_5 (distance of inner frame to the upper border). A possible explanation of the third factor could be that it measures the position of the central picture on the bank note. Note that these three

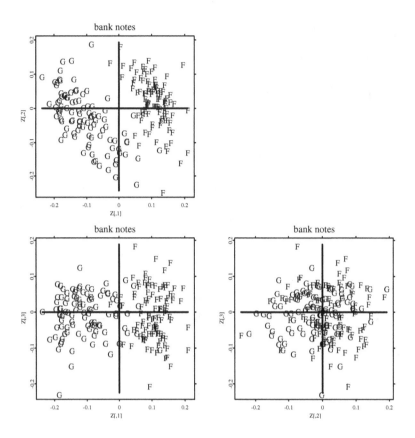

Fig. 8.3. Representation of the individuals for Swiss bank notes in a scatterplot matrix. Q SMSdecobank

factorial variables explain almost 98% of the total inertia of the (centered and standardized) data set.

In Figures 8.3 and 8.4, we show the projections of the individuals. It seems that in both graphics, the separation of the forged and of the genuine bank notes is quite good. However, the separation would be more apparent if we would rotate the three-dimensional graphics displayed in Figure 8.4.

The factorial analysis of the Swiss bank notes provides interesting insights into the structure of the data set. The 3-dimensional representation of the data set keeps 97.8% of the inertia of the complete 6-dimensional data set that would be very difficult to visualize.

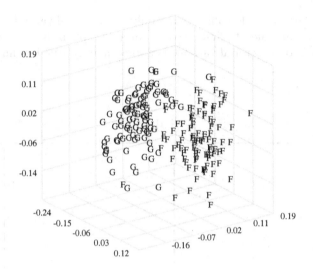

Fig. 8.4. Representation of swiss bank notes in 3-dimensional scatterplot.
◫ SMSdecobank

EXERCISE 8.8. *Apply the factorial techniques to the time budget data (Table A.15) which gives the amount of time a person spent on ten activities over 100 days in 1976 (Volle 1985).*

The following analysis is based on the centered data set as it seems to be more natural to consider the differences from the average time spent on various activities. However, the times spent on different activities are left on the original scale: here, the scale is the same for all variables and this approach guarantees that the analysis will concentrate on the activities that really occupy the largest share of the time.

The vector of the eigenvalues is:

$$\lambda = (87046, 7085, 2623.7, 1503.4, 315.9, 156.6, 71.5, 42.6, 25.8, 0.0)^\top .$$

The last eigenvalue has to be equal to zero since the time spend on all activities has to sum to 24 hours/day and the data matrix thus cannot have full rank.

The proportions of the explained inertia:

$$\tau = (0.8804, 0.9521, 0.9786, 0.9938, 0.9970, 0.9986, 0.9993, 0.9997, 1.00, 1.00)^\top$$

suggest that here it would suffice to use only one factorial variable. Notice the large difference in scale: the first factorial variable explains 88% of total inertia whereas the second factorial variable is approximately 10× less important.

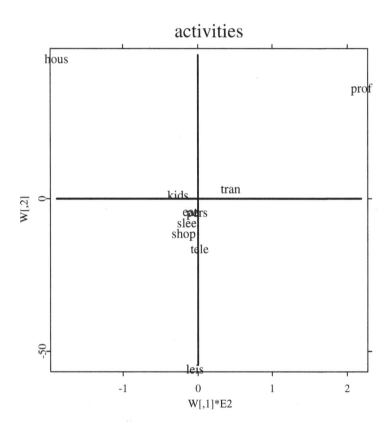

Fig. 8.5. Representation of the variables for the time budget data. ⓠ SMSdecotime

In Figures 8.5 and 8.6, we present the two-dimensional projections, see the description in Table A.15 for the names of the various activities used in Figure 8.5.

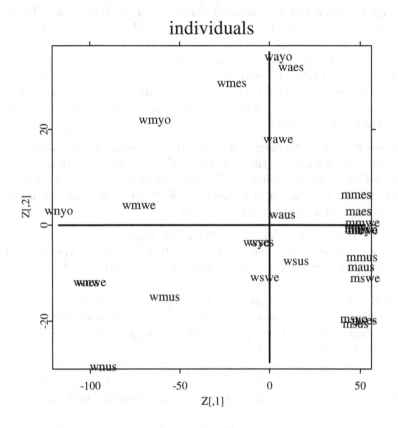

Fig. 8.6. Representation of the individuals for the time budget data.
Q SMSdecotime

The factorial representation of the variables in Figure 8.5 shows that the first factor, explaining 88% of the inertia, is just the contrast between the household and professional activities. The second factorial variable, explaining 7% of the inertia, is the contrast between leisure activities on one side and household and professional activities on the other side. The fact that the other activities lie close to the origin is partially caused by the fact that these activities either do not vary a lot among the observed individuals or they take less time. Notice that "kids" are lying in the direction of household activities and that "transportation" lies in the direction of professional activities.

The four letter codes in Figure 8.6 indicate the sex (m: man, w: woman), activity (a: active, n: nonactive, m: married, s: single) and country (us: U.S., we: West, yo: Yugoslavia, es: East). For example, "mmus" denotes married man in United States. The projections of the individuals in Figure 8.6 allow us to judge the effects of the factorial variables on the individuals. We see that all men are lying close to each other on the right hand side of the plot: comparison with Figure 8.5 suggests that men at that year 1976 are more involved in professional than in household activities. On the left hand side of Figure 8.6 you will find married and nonactive women whereas single and active women are located in the central region. It seems that married women are involved mainly in household activities while single women balance between the household and professional activities.

The second direction distinguishes between "professional and household" and "leisure" activities. In the direction of "professional activities" you will find active women (without U.S.) and married women from Eastern countries and Yugoslavia. In the direction of "leisure" we can see mainly singles and married and nonactive women in U.S.

The factorial analysis provides again interesting insights into the structure of the data set. For the time budget data, two factorial variables explain 95% of the inertia of the (centered but not standardized) data set.

EXERCISE 8.9. *Assume that you wish to analyze a data matrix consisting of p orthogonal, standardized, and centered columns. What is the percentage of the inertia explained by the first factor? What is the percentage of the inertia explained by the first q factors?*

If the columns of the matrix $\mathcal{X}$ are orthogonal, centered and standardized, then $\mathcal{X}^\top \mathcal{X} = n \operatorname{diag}(1_p)$ and the spectral decomposition can be written as

$$\mathcal{X}^\top \mathcal{X} = \mathcal{I}_p n \operatorname{diag}(1_p) \mathcal{I}_p.$$

Hence, all eigenvalues, $\lambda_1, \ldots, \lambda_p$, of $\mathcal{X}^\top \mathcal{X}$ are equal to n. The total inertia of such data set is equal to np and the proportion of inertia explained by the first factorial variable is obviously

$$\tau_1 = \frac{\lambda_1}{\lambda_1 + \cdots + \lambda_p} = \frac{n}{np} = \frac{1}{p}.$$

The proportion of inertia explained by the first q factors is

$$\tau_q = \frac{\lambda_1 + \cdots + \lambda_q}{\lambda_1 + \cdots + \lambda_p} = \frac{nq}{np} = \frac{q}{p}.$$

EXERCISE 8.10. *Reconsider the setup of the Exercise 8.9. What does the eigenvector, corresponding to the first factor, look like.*

Let us return to the spectral decomposition of the matrix $\mathcal{X}^\top \mathcal{X}$ derived in Exercise 8.9:

$$\mathcal{X}^\top \mathcal{X} = n\mathcal{I}_p = \Gamma \Lambda \Gamma^\top = \mathcal{I}_p n \operatorname{diag}(1_p)\mathcal{I}_p.$$

Since all eigenvalues are equal to n, we have that $\Lambda = n \operatorname{diag}(1_p)$ and it follows that the matrix Γ has to satisfy the equation

$$n\mathcal{I}_p = \mathcal{X}^\top \mathcal{X} = \Gamma n \operatorname{diag}(1_p)\Gamma^\top = n\Gamma\Gamma^\top,$$

i.e., Γ can be chosen as any matrix satisfying the condition $\Gamma^\top \Gamma = \Gamma\Gamma^\top = \mathcal{I}_p$. Hence, the first eigenvector γ_1 can be any vector with norm $\|\gamma_1\| = 1$. A reasonable choice would be $\gamma_1 = (1, 0, 0, \dots, 0)^\top$.

EXERCISE 8.11. *Suppose that the data matrix consists of two columns, $x_{[1]}$ and $x_{[2]}$, and that $x_{[2]} = 2x_{[1]}$. What do the eigenvalues and eigenvectors of the empirical correlation matrix $\mathcal{R}$ look like? How many eigenvalues are nonzero?*

The correlation matrix is

$$\mathcal{R} = \begin{pmatrix} 1 & 1 \\ 1 & 1 \end{pmatrix}.$$

It has rank 1, one eigenvalue must therefore be zero. The eigenvalues can be found by solving the equation

$$0 = |\mathcal{R} - \lambda\mathcal{I}_2| = \left| \begin{pmatrix} 1-\lambda & 1 \\ 1 & 1-\lambda \end{pmatrix} \right| = (1-\lambda)^2 - 1 = \lambda^2 - 2\lambda = \lambda(\lambda - 2),$$

i.e., the eigenvalues are $\lambda_1 = 2$ and $\lambda_2 = 0$.

The corresponding eigenvectors can be found by solving the systems of equations

$$\begin{pmatrix} 1 & 1 \\ 1 & 1 \end{pmatrix} \begin{pmatrix} \gamma_{1i} \\ \gamma_{2i} \end{pmatrix} = \begin{pmatrix} \gamma_{1i} \\ \gamma_{2i} \end{pmatrix} \lambda_i,$$

for $i = 1, 2$. For the first eigenvalue, $\lambda_1 = 2$, we obtain that

$$\gamma_{11} + \gamma_{21} = 2\gamma_{11} = 2\gamma_{21}.$$

Since the length of the eigenvector, $(\gamma_{12}^2 + \gamma_{21}^2)^{1/2}$, has to be equal to 1, we obtain $\gamma_{11} = \gamma_{21}$ and $|\gamma_{11}| = 1/\sqrt{2}$.

For the second eigenvalue, $\lambda_2 = 0$, we have

$$\gamma_{12} + \gamma_{22} = 0$$

which leads to conditions $\gamma_{12} = -\gamma_{22}$ and $|\gamma_{12}| = 1/\sqrt{2}$.

Notice that the sign of the eigenvectors is not determined uniquely.

From the derived eigenvalues and eigenvectors, we have the spectral decomposition of the correlation matrix

$$\mathcal{R} = \begin{pmatrix} 1 & 1 \\ 1 & 1 \end{pmatrix} = \Gamma \Lambda \Gamma^\top = \begin{pmatrix} 1/\sqrt{2} & 1/\sqrt{2} \\ 1/\sqrt{2} & -1/\sqrt{2} \end{pmatrix} \begin{pmatrix} 2 & 0 \\ 0 & 0 \end{pmatrix} \begin{pmatrix} 1/\sqrt{2} & 1/\sqrt{2} \\ 1/\sqrt{2} & -1/\sqrt{2} \end{pmatrix}.$$

EXERCISE 8.12. *What percentage of inertia is explained by the first factor in Exercise 8.11?*

In Exercise 8.11, the eigenvalues of the correlation matrix are $\lambda_1 = 2$ and $\lambda_2 = 0$. Hence, the percentage of inertia explained by the first factor is

$$\tau_1 = \frac{2}{2} = 100\%$$

and 1-dimensional representation explains all inertia contained in the data set.

9

Principal Component Analysis

I tried one or two explanations, but, indeed, I was completely puzzled myself. Our friend's title, his fortune, his age, his character, and his appearance are all in his favour, and I know nothing against him, unless it be the dark fate which runs in his family.
"The Hound of the Baskervilles"

This chapter addresses the issue of reducing the dimensionality of a multivariate random variable by using linear combinations (the principal components). The identified principal components are ordered in decreasing order of importance. When applied in practice to a data matrix, the principal components will turn out to be the factors of a transformed data matrix (the data will be centered and eventually standardized).

For a random vector X with $E(X) = \mu$ and $\text{Var}(X) = \Sigma = \Gamma \Lambda \Gamma^\top$, the principal component (PC) transformation is defined as

$$Y = \Gamma^\top (X - \mu). \tag{9.1}$$

It will be demonstrated in Exercise 9.1 that the components of the random vector Y have zero correlation. Furthermore, it can be shown that they are also standardized linear combinations with the largest variance and that the sum of their variances, $\sum \text{Var} Y_i$, is equal to the sum of the variances of $X_1, \ldots, X_p$.

In practice, the PC transformation is calculated using the estimators $\bar{x}$ and S instead of μ and Σ. If $S = \mathcal{G}\mathcal{L}\mathcal{G}^\top$ is the spectral decomposition of the empirical covariance matrix S, the principal components are obtained by

$$\mathcal{Y} = (\mathcal{X} - 1_n \bar{x}^\top)\mathcal{G}. \tag{9.2}$$

Theorem 9.1 describes the relationship between the eigenvalues of Σ and the eigenvalues of the empirical variance matrix S.

THEOREM 9.1. *Let $\Sigma > 0$ with distinct eigenvalues and let $\mathcal{U} \sim m^{-1} W_p(\Sigma, m)$ with spectral decompositions $\Sigma = \Gamma \Lambda \Gamma^\top$ and $\mathcal{U} = \mathcal{G}\mathcal{L}\mathcal{G}^\top$. Then*

$$\sqrt{m}(\ell - \lambda) \xrightarrow{\mathcal{L}} N_p(0, 2\Lambda^2),$$

where $\ell = (\ell_1, \ldots, \ell_p)^\top$ and $\lambda = (\lambda_1, \ldots, \lambda_p)^\top$ are the diagonals of $\mathcal{L}$ and Λ.

The proof and the asymptotic distribution of $\mathcal{G}$ can be found, e.g., in Härdle & Simar (2003, theorem 9.4).

The resulting PCA (principal component analysis) or NPCA (normalized PCA) is presented in a variety of examples, including U.S. crime and health data. A PCA is also performed for an OECD data set on variables of political nature (life expectancy, literacy, etc.).

EXERCISE 9.1. *Calculate the expected value and the variance of the PC transformation Y defined in (9.1). Interpret the results.*

For the expected value, EY, we have

$$EY = E\Gamma^\top(X - \mu) = \Gamma^\top E(X - \mu) = \Gamma^\top(EX - \mu) = 0_p.$$

The variance matrix, $\text{Var}(Y)$, can be calculated as

$$\text{Var}(Y) = \text{Var}\{\Gamma^\top(X - \mu)\} = \Gamma^\top \Sigma \Gamma = \Gamma^\top \Gamma \Lambda \Gamma^\top \Gamma = \Lambda.$$

Hence, the random vector Y is centered (its expected value is equal to zero) and its variance matrix is diagonal.

The eigenvalues $\lambda_1, \ldots, \lambda_p$ are variances of the principal components $Y_1, \ldots, Y_p$. Notice that

$$\sum_{i=1}^{p} Var(X_i) = \text{tr}\,\Sigma = \text{tr}\{\Gamma \Lambda \Gamma^\top\} = \text{tr}\{\Gamma^\top \Gamma \Lambda\} = \text{tr}\,\Lambda = \sum_{i=1}^{p} \lambda_i = \sum_{i=1}^{p} Var(Y_i).$$

Hence, the variances of X_i are decomposed into the variances of Y_i which are given by the eigenvalues of Σ. The sum of variances of the first q principal components, $\sum_{i=1}^{q} \lambda_i$, thus measures the variation of the random vector X explained by $Y_1, \ldots, Y_q$. The proportion of the explained variance,

$$\psi_q = \frac{\lambda_1 + \cdots + \lambda_q}{\lambda_1 + \cdots + \lambda_p},$$

will be important for the interpretation of results of the practical analyses presented in the following exercises.

EXERCISE 9.2. *Calculate the correlation between X and its PC transformation Y.*

The covariance between the PC vector Y and the original vector X is:

$$\text{Cov}(X,Y) = \text{Cov}\{X, \Gamma^\top(X-\mu)\} = \text{Cov}(X,Y)\Gamma = \Sigma\Gamma = \Gamma\Lambda\Gamma^\top\Gamma = \Gamma\Lambda.$$

The correlation, $\rho_{X_iY_j}$, between variable X_i and the PC Y_j is

$$\rho_{X_iY_j} = \frac{\gamma_{ij}\lambda_j}{(\sigma_{X_iX_i}\lambda_j)^{1/2}} = \gamma_{ij}\left(\frac{\lambda_j}{\sigma_{X_iX_i}}\right)^{1/2}.$$

The correlations describe the relations between the PCs and the original variables. Note that $\sum_{j=1}^p \lambda_j\gamma_{ij}^2 = \gamma_i^\top \Lambda\gamma_i$ is the (i,i)-element of the matrix $\Gamma\Lambda\Gamma^\top = \Sigma$, so that

$$\sum_{j=1}^p \rho_{X_iY_j}^2 = \frac{\sum_{j=1}^p \lambda_j\gamma_{ij}^2}{\sigma_{X_iX_i}} = \frac{\sigma_{X_iX_i}}{\sigma_{X_iX_i}} = 1.$$

Hence, the correlation $\rho_{X_iY_j}^2$ may be seen as the proportion of variance of the ith variable X_i explained by the jth principal component Y_j.

Notice that the percentage of variance of X_i explained by the first q PCs $Y_1,\ldots,Y_q$ is $\sum_{j=1}^q \rho_{X_iY_j}^2 < 1$. The distance of the point with coordinates $(\rho_{X_iY_1},\ldots,\rho_{X_iY_q})$ from the surface of the unit ball in q-dimensional space can be used as a measure of the explained variance of X_i.

EXERCISE 9.3. *Apply the PCA to the car marks data in Table A.5. Interpret the first two PCs. Would it be necessary to look at the third PC?*

The eigenvalues of the covariance matrix,

$$\lambda = (5.56, 1.15, 0.37, 0.10, 0.08, 0.05, 0.04, 0.02)^\top,$$

lead to the following proportions of the explained variance:

$$\psi = (0.76, 0.91, 0.96, 0.98, 0.99, 0.99, 1.00, 1.00)^\top.$$

Observing that the first two principal components explain more than 90% of the variability of the data set, it does not seem necessary to include also the third PC which explains only 5% of the variability. A graphical display of the eigenvalues, the screeplot, is plotted in the lower right part in Figure 9.1.

The first two eigenvectors of the covariance matrix are

$$\gamma_1 = (-0.22, 0.31, 0.44, -0.48, 0.33, 0.39, 0.42, -0.01)^\top$$

and

$$\gamma_2 = (0.54, 0.28, 0.22, 0.30, -0.14, -0.16, 0.46, 0.49)^\top.$$

Hence, the first two principal components are defined as:

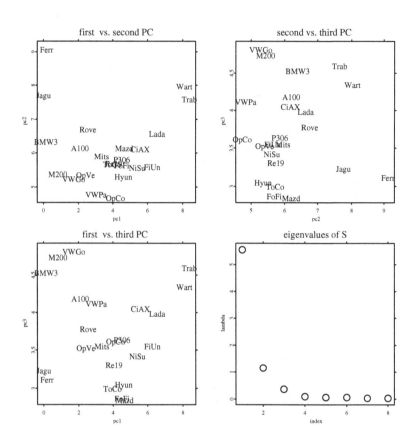

Fig. 9.1. Scatterplots of the first three principal components and a screeplot of the eigenvalues, car marks data set. Q SMSpcacarm

$$Y_1 = -0.22 \times \text{econ} + 0.31 \times \text{serv} + 0.44 \times \text{value} - 0.48 \times \text{price} + 0.33 \times \text{desi}$$
$$+ 0.39 \times \text{sport} + 0.42 \times \text{safe} - 0.01 \times \text{easy},$$
$$Y_2 = 0.54 \times \text{econ} + 0.28 \times \text{serv} + 0.22 \times \text{value} + 0.30 \times \text{price} - 0.14 \times \text{desi}$$
$$- 0.16 \times \text{sport} + 0.46 \times \text{safe} + 0.49 \times \text{easy}.$$

Using the coefficients of the PCs for interpretation might be misleading especially when the variables are observed on different scales. It is advisable to base the interpretations on the correlations of PCs with the original variables which are plotted in Figure 9.2.

For the car marks data set both the coefficients of the PCs and their correlations with the original variables in Figure 9.2 suggest that the first principal components distinguishes the expensive and design cars from the cheap and

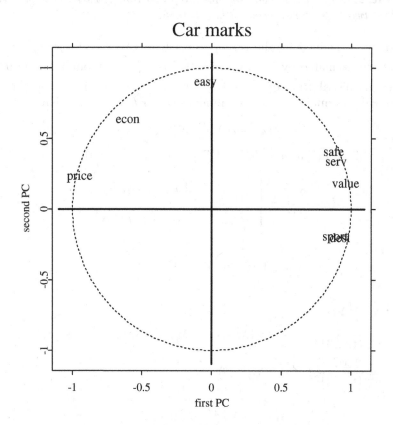

Fig. 9.2. Correlations of the first two principal components with the original variables in the car marks data set. ⌕ SMSpcacarm

less sporty vehicles. This interpretation is confirmed by the plot of the first principal component, Y_1, on Figure 9.1. On the right hand side, we observe the not so cool brands such as Wartburg, Trabant, Lada or Fiat, whereas on the left hand side, we see Jaguar, Ferrari, BMW, and Mercedes-Benz.

The second PC distinguishes economic cars that are easy to handle, such as Volkswagen and Opel, from the cars that consume a lot of gas and their handling is more problematic such as Ferrari, Wartburg, Jaguar, and Trabant.

Figure 9.2 shows that all of the original variables are very well explained by the first two PCs since all points can be found very close to the unit circle, see the explanation in Exercise 9.2.

EXERCISE 9.4. *Test the hypothesis that the proportion of variance explained by the first two PCs in Exercise 9.3 is $\psi = 0.85$.*

The variance explained by the first q PCs, $\psi_q = (\lambda_1 + \cdots + \lambda_q)/\sum_{j=1}^{p} \lambda_j$, is in practice estimated by $\widehat{\psi}_q = (\ell_1 + \cdots + \ell_q)/\sum_{j=1}^{p} \ell_j$. From Theorem 9.1 we know the distribution of $\sqrt{n-1}(\ell - \lambda)$ and, since $\widehat{\psi}_q$ is a function of asymptotically normally distributed random vector ℓ, we obtain that

$$\sqrt{n-1}(\widehat{\psi}_q - \psi_q) \xrightarrow{\mathcal{L}} N(0, \mathcal{D}^{\top} \mathcal{V} \mathcal{D})$$

where $\mathcal{V} = 2\Lambda^2$ from Theorem 9.1 and $\mathcal{D} = (d_1, \ldots, d_p)^{\top}$ with

$$d_j = \frac{\partial \psi_q}{\partial \lambda_j} = \begin{cases} \dfrac{1 - \psi_q}{\mathrm{tr}(\Sigma)} & \text{if } 1 \leq j \leq q, \\[2mm] \dfrac{-\psi_q}{\mathrm{tr}(\Sigma)} & \text{if } q + 1 \leq j \leq p. \end{cases}$$

It follows that

$$\sqrt{n-1}(\widehat{\psi}_q - \psi_q) \xrightarrow{\mathcal{L}} N(0, \omega^2),$$

where

$$\begin{aligned} \omega^2 &= \mathcal{D}^{\top} \mathcal{V} \mathcal{D} \\ &= \frac{2}{\{\mathrm{tr}(\Sigma)\}^2} \left\{ (1 - \psi)^2 (\lambda_1^2 + \cdots + \lambda_q^2) + \psi^2 (\lambda_{q+1}^2 + \cdots + \lambda_p^2) \right\} \\ &= \frac{2\,\mathrm{tr}(\Sigma^2)}{\{\mathrm{tr}(\Sigma)\}^2} (\psi^2 - 2\beta\psi_q + \beta) \end{aligned}$$

and

$$\beta = \frac{\lambda_1^2 + \cdots + \lambda_q^2}{\lambda_1^2 + \cdots + \lambda_p^2} = \frac{\lambda_1^2 + \cdots + \lambda_q^2}{\mathrm{tr}(\Sigma^2)}.$$

In practice, we work with an estimate $\widehat{\omega}^2$ based on the spectral decomposition of the empirical covariance matrix.

In Exercise 9.3 we have calculated the eigenvalues:

$$\lambda = (5.56, 1.15, 0.37, 0.10, 0.08, 0.05, 0.04, 0.02)^{\top}$$

and the proportions of the explained variance:

$$\psi = (0.76, 0.91, 0.96, 0.98, 0.99, 0.99, 1.00, 1.00)^{\top}.$$

It follows that, for $q = 2$, we obtain $\hat{\beta} = 0.99524$ and $\widehat{\omega}^2 = 0.0140$. Under the null hypothesis, $H_0 : \psi_2 = 0.85$, the test statistic $\sqrt{n-1}(\widehat{\psi}_2 - 0.85)/\omega$ has asymptotically standard normal distribution. In our case the value of the test statistic, 2.4401, is in absolute value larger than the critical value of the normal distribution $\Phi^{-1}(0.975) = 1.96$ and we reject the null hypothesis.

Hence, on confidence level $\alpha = 0.95$, we have proved that the proportion of variance explained by the first two principal components is larger than 85%.

EXERCISE 9.5. *Take the athletic records for 55 countries given in Table A.1 and apply the NPCA. Interpret your results.*

The athletic records data set contains national records in 8 disciplines (100m, 200m, 400m, 800m, 1500m, 5km, 10km, and marathon) for $n = 55$ countries. Clearly, the times and hence also the differences between countries will be much larger for longer tracks. Hence, before running the PC analysis, the dataset is normalized by dividing each variable by its estimated standard deviation. The resulting analysis will be called Normalized PCA (NPCA).

In principle, the same results can be obtained by calculating the spectral decomposition of the empirical correlation matrix of the original data set. One only has to be very careful and keep in mind that the derived coefficients of the PCs apply to the normalized variables. Combining these coefficients with the original variables would lead to misleading results.

The eigenvalues and the proportions of explained variance are

$$\lambda = (6.04, 0.99, 0.60, 0.13, 0.10, 0.07, 0.05, 0.02)^\top$$

and

$$\psi = (0.75, 0.88, 0.95, 0.97, 0.98, 0.99, 1.00, 1.00)^\top.$$

Notice that the sum of all eigenvalues is equal to 8. This follows from the fact that the variances of the standardized variables are equal to 1 and from the relationship $\sum_{i=1}^{p} \lambda_i = \operatorname{tr} \mathcal{S} = \sum_{i=1}^{p} 1 = p = 8$.

Considering the above eigenvalues and proportions of explained variance, it would be reasonable to investigate only 1 principal component, see also the screeplot in Figure 9.3. A commonly accepted rule says that it suffices to keep only PCs that explain larger than the average number of the total variance. For NPCA, it is easy to see that larger than average proportion of variance is explained by PCs with corresponding eigenvalue larger than 1.

However, the second eigenvalue $\lambda_2 = 0.99$ is so close to 1 that we have decided to discuss also the second PC. The coefficients of the linear combinations are given by the eigenvectors

$$\gamma_1 = (0.32, 0.16, 0.37, 0.38, 0.39, 0.39, 0.39, 0.37)^\top$$

and

$$\gamma_2 = (0.39, 0.85, 0.03, -0.04, -0.13, -0.16, -0.17, -0.22)^\top.$$

In this exercise, it is very important to keep in mind the meaning of the measurements. Larger values correspond here to longer, i.e., worse times. The first PC is positively related to all original variables and it can be interpreted as the arithmetic average of the records with slightly smaller weight of the record on 200m track, see also the correlations in Figure 9.4. In Figure 9.3,

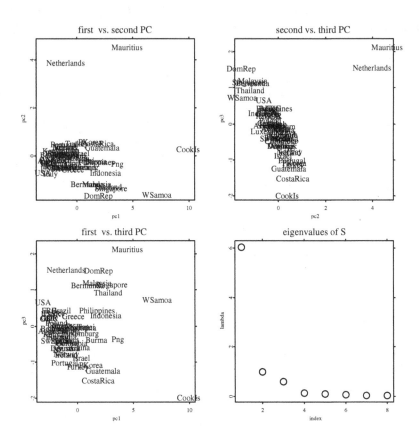

Fig. 9.3. Scatterplots of the first three principal components and a screeplot of the eigenvalues, athletic records data set. ⊙ SMSnpcathletic

we can see that large values of this "average time" component are achieved in Cook Islands, West Samoa, and Mauritius. On contrary, fastest times are achieved in USA.

The second principal component is strongly positively related to 200m and important positive component is also the 100m record whereas longer tracks show mostly negative relationship. The second principal components separates Mauritius and Netherlands which shows poor records in 200m.

In Figure 9.4, we see that two principal components explain very well all original variables. Using only one PC would lead to much worse explanation of the 200m records.

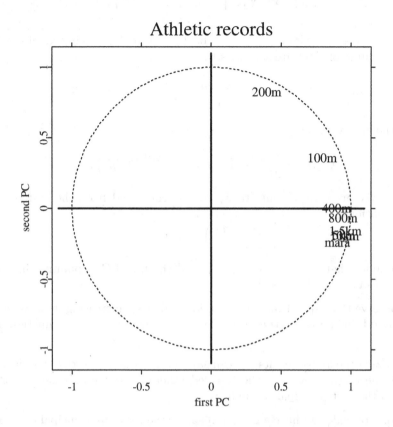

Fig. 9.4. Correlations of the first two principal components with the original variables in the athletic records data set. ⬚ SMSnpcathletic

EXERCISE 9.6. *Apply a PCA to* $\Sigma = \begin{pmatrix} 1 & \rho \\ \rho & 1 \end{pmatrix}$, *where* $0 < \rho < 1$. *Now change the scale of* X_1, *i.e., consider the covariance of* cX_1 *and* X_2, *where* $c > 1$. *How do the PC directions change with the screeplot?*

The spectral decomposition of matrix Σ has already been investigated in Exercise 2.7. Recall that we have

$$\Sigma = \Gamma \Lambda \Gamma^{\top} = \frac{1}{\sqrt{2}} \begin{pmatrix} 1 & 1 \\ 1 & -1 \end{pmatrix} \begin{pmatrix} 1+\rho & 0 \\ 0 & 1-\rho \end{pmatrix} \frac{1}{\sqrt{2}} \begin{pmatrix} 1 & 1 \\ 1 & -1 \end{pmatrix}.$$

Since $\rho > 0$, the PCs are $Y_1 = (X_1 + X_2)/\sqrt{2}$ and $Y_1 = (X_1 - X_2)/\sqrt{2}$.

Multiplying X_1 by constant $c > 0$ leads the covariance matrix:

$$\text{Var}\{(cX_1, X_2)^\top\} = \Sigma(c) = \begin{pmatrix} c^2 & c\rho \\ c\rho & 1 \end{pmatrix}.$$

The spectral decomposition of $\Sigma(c)$ can be derived similarly as in Exercise 2.7. The eigenvalues of $\Sigma(c)$ are solutions to:

$$\begin{vmatrix} c^2 - \lambda & c\rho \\ c\rho & 1 - \lambda \end{vmatrix} = 0.$$

Hence the eigenvalues are

$$\lambda_{1,2}(c) = \frac{1}{2}\left(c^2 + 1 \pm \sqrt{(c^2 - 1)^2 + 4c^2\rho^2}\right).$$

The eigenvector corresponding to λ_1 can be computed from the system of linear equations:

$$\begin{pmatrix} c^2 & c\rho \\ c\rho & 1 \end{pmatrix}\begin{pmatrix} x_1 \\ x_2 \end{pmatrix} = \lambda_1 \begin{pmatrix} x_1 \\ x_2 \end{pmatrix}$$

which implies that $x_1 = x_2(\lambda_1 - 1)/c\rho$ and the first PC is pointing in the direction $(cX_1)(\lambda_1 - 1)/c\rho + X_2$.

Next, observe that $\lambda_1 > 1$ and the function $\lambda_1(c)/c$ is increasing in c. Hence, $x_1 > x_2$ and, furthermore, the ratio of x_1 and x_2 is an increasing function of c.

Summarizing the above results, we can say that as c increases, the first eigenvalue λ_1 becomes larger and the rescaled random variable cX_1 gains more weight in the first principal component.

The choice of scale can have a great impact on the resulting principal components. If the scales differ, it is recommended to perform the Normalized PCA (NPCA), i.e., to standardize each variable by its standard deviation.

EXERCISE 9.7. *Suppose that we have standardized some data using the Mahalanobis transformation. Would it be reasonable to apply a PCA?*

Standardizing any given data set $\mathcal{X}$ by the Mahalanobis transformation leads to a data set $\mathcal{Z} = \mathcal{X}\mathcal{S}^{-1/2}$ with the covariance matrix

$$\mathcal{S}_{\mathcal{Z}} = \mathcal{S}^{-1/2}\mathcal{S}\mathcal{S}^{-1/2} = \mathcal{I}_p.$$

It immediately follows that all eigenvalues of $\mathcal{S}_{\mathcal{Z}}$ are equal to 1 and that the principal components of $\mathcal{Z}$ have exactly the same variances as the original variables. Hence, such analysis would be entirely useless.

Principal components analysis of $\mathcal{Z}$ leads always to this same uninteresting result.

EXERCISE 9.8. *Apply a NPCA to the U.S. crime data set in Table A.18. Interpret the results. Would it be necessary to look at the third PC? Can you see any difference between the four regions?*

The U.S. crime data set consists of the reported number of crimes in the 50 U.S. states in 1985. The crimes were classified according to 7 categories: murder, rape, robbery, assault, burglary, larceny, and auto theft. The dataset also contains identification of the region: Northeast, Midwest, South, and West.

The Normalized PCA means that, before running the analysis, all observed variables are put on the same scale.

The eigenvalues of the correlation matrix are:

$$\lambda = (4.08, 1.43, 0.63, 0.34, 0.25, 0.14, 0.13)^{\top}$$

and we obtain the proportions of explained variance:

$$\psi = (0.58, 0.79, 0.88, 0.93, 0.96, 0.98, 1.00)^{\top}.$$

The data set is well described by the first two NPCs, each of the first two NPCs describes larger than average amount of variance. The first two NPCs describe together 79% of the total variability, see also the screeplot in Figure 9.5.

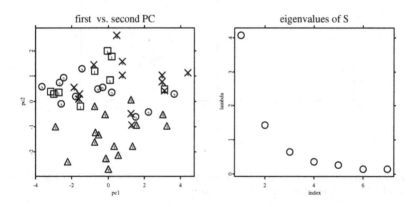

Fig. 9.5. Scatterplot of the first two principal components and a screeplot of the eigenvalues, U.S. crime data set. ⚙ SMSnpcacrime

The first two eigenvectors are:

$$\gamma_1 = (0.28, 0.42, 0.39, 0.39, 0.44, 0.36, 0.35)^{\top},$$
$$\gamma_2 = (-0.64, -0.12, 0.05, -0.46, 0.26, 0.40, 0.37)^{\top}.$$

The first principal component combines the numbers of all crimes with approximately constant (0.28–0.44) weights and we can interpret it as the overall crime rate, see also the correlations in Figure 9.6. The second principal component is negatively correlated with 1st and 4th variable (murder and assault) and positively correlated with the 5th till 7th variable (burglary, larceny, auto theft). The second NPC can be interpreted as "type of crime" component.

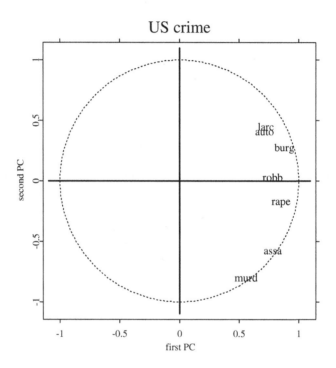

Fig. 9.6. Correlations of the first two principal components with the original variables in the U.S. crime data set. Q SMSnpcacrime

In Figure 9.5, we denote each of the four regions by a different plotting symbol. It looks as if the symbol changes in the direction of the second, type of crime, principal component. In the upper part of the graph, we see mainly circles, squares, and crosses corresponding to the regions 1, 2, and 4. In the lower part, we observe mainly triangles corresponding to the third South region. Hence, it seems that in region 3 occur more murders and assaults and less burglaries, larcenies and auto thefts than in the rest of USA.

EXERCISE 9.9. *Repeat Exercise 9.8 using the U.S. health data set in Table A.19.*

The U.S. health data set consists of reported number of deaths in the 50 U.S. states classified according to 7 categories: accident, cardiovascular, cancer, pulmonary, pneumonia flu, diabetes, and liver.

Here, we have decided to run the usual PC analysis. Normalizing the data set would mean that, in certain sense, all causes of death would have the same importance. Without normalization, we can expect that the variables responsible for the largest number of deaths will play the most prominent role in our analysis, see also Exercise 9.6 for theoretical justification.

The eigenvalues of the covariance matrix are:

$$\lambda = (8069.40, 189.22, 76.03, 25.21, 10.45, 5.76, 3.47)^\top$$

and the huge first eigenvalue stresses the importance of the first principal component. Calculating the proportions of the explained variance,

$$\psi = (0.96, 0.99, 0.99, 1.00, 1.00, 1.00, 1.00)^\top,$$

we see that the first PC explains 96% of the total variability. The screeplot is plotted in Figure 9.7.

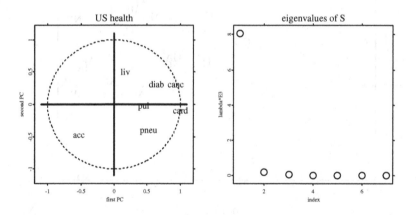

Fig. 9.7. Correlations of the first two principal components with the original variables and the screeplot for the U.S. health data set. ✎ SMSpcahealth

The first (most important) eigenvectors is:

$$\gamma_1 = (-0.06, 0.94, 0.34, 0.03, 0.02, 0.03, 0.01)^\top$$

and we see that the first PC reflects the most common causes of death: cardiovascular diseases and, with smaller weight, cancer. The second eigenvector,

$$\gamma_2 = (-0.34, -0.34, 0.86, 0.01, -0.11, 0.09, 0.11)^\top,$$

is strongly positively correlated with cancer and less strongly negatively correlated with cardiovascular and pulmonary diseases, see also Figure 9.7. The first principal component explains satisfactorily only variables cardiovascular and cancer.

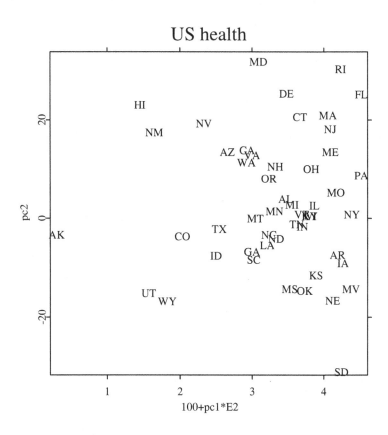

Fig. 9.8. Scatterplot of the first two principal components for U.S. health data set. Q SMSpcahealth

In Figure 9.8, we show the values of the first two PCs for the 50 observed U.S. states. Keeping in mind the meaning of the principal components, we should see the states with large number of deaths due to cardiovascular diseases and cancer on the right hand side (Florida, New York, Pennsylvania). From the point of view of the first PC, the best quality of life can be found in Arkansas, Hawaii, New Mexico, Wyoming, and Colorado. The much less important second PC suggests that cancer is more common cause of death in Maryland than in South Dakota.

EXERCISE 9.10. *Do a NPCA on the Geopol data set, Table A.10, which compares 41 countries with respect to different aspects of their development. Why or why not would a PCA be reasonable here?*

The Geopol data set contains a comparison of 41 countries according to 10 political and economic parameters. We will perform the analysis without the first variable, size of population. The variables to be analyzed, X_2–X_9 are: gross internal product per habitant (giph), rate of increase of the population (ripo), rate of urban population (rupo), rate of illiteracy (rlpo), rate of students (rspo), expected lifetime (eltp), rate of nutritional needs realized (rnnr), number of newspaper and magazines per 1000 habitants (nunh), and number of televisions per 1000 inhabitants (nuth).

Clearly, these variables are measured on very different scales and, in order to produce trustworthy results, the data set has to be normalized. In this exercise, we have to perform NPCA.

The eigenvalues of the correlation matrix are:

$$\lambda = (5.94, 0.87, 0.70, 0.54, 0.43, 0.18, 0.15, 0.12, 0.08)^\top$$

and we obtain the percentages of explained variance:

$$\psi = (0.66, 0.76, 0.83, 0.89, 0.94, 0.96, 0.98, 0.99, 1.00)^\top.$$

The screeplot is plotted in Figure 9.9. It would suffice to keep only one NPC, but we decide to keep the first three principal components although Y_2 and Y_3 contribute only little to the total variability.

The coefficients of the first three normalized principal components are given by the first three eigenvectors:

$$\gamma_1 = (0.34, -0.34, 0.29, -0.36, 0.30, 0.37, 0.28, 0.33, 0.37)^\top,$$
$$\gamma_2 = (0.41, 0.38, 0.23, 0.20, 0.16, -0.20, -0.61, 0.36, 0.19)^\top,$$
$$\gamma_3 = (-0.18, 0.37, 0.34, -0.02, 0.66, -0.05, 0.14, -0.49, 0.06)^\top.$$

The correlations of $Y_1, \ldots, Y_3$ with the original variables are plotted in Figure 9.10.

From the correlations plotted in Figure 9.10, we can interpret the first PC as the overall quality of life component: notice that it is positively related to the

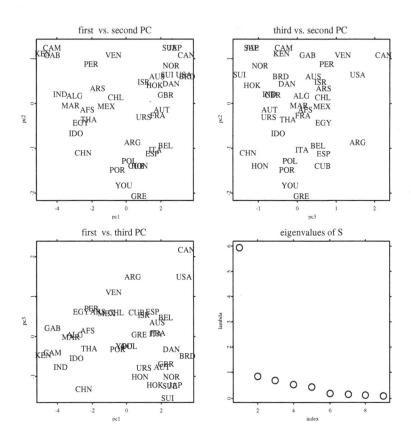

Fig. 9.9. Scatterplots of the first three principal components and a screeplot of the eigenvalues, Geopol data set. ⊂ SMSnpcageopol

all variables apart of rate of increase of the population and rate of illiteracy. In Figure 9.9, we can see that large values of this component are achieved in the former West Germany (BRD), Canada, and USA. Smallest values of this component are observed in Kenya, Cameroon, Gabon, and India.

The second PC seems to point mainly in the direction opposite to the rnnr (rate of nutritional needs realized). The third PC is positively correlated to the rate of students and negatively correlated to the number of newspapers. From Figure 9.9, we can see that already one PC is enough to explain substantial part of the variability of all variables.

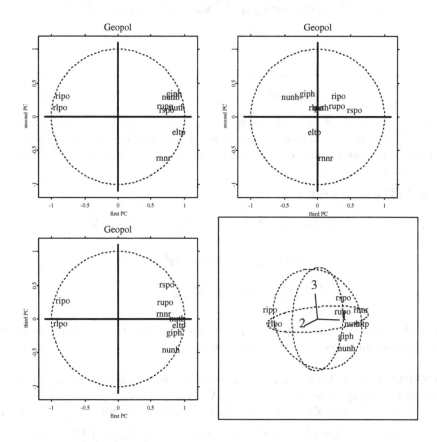

Fig. 9.10. Correlations of the first three principal components with the original variables in the Geopol data set. ◘ SMSnpcageopol

EXERCISE 9.11. *Let U be an uniform random variable on $[0,1]$. Let $a = (a_1, a_2, a_3)^\top \in \mathbb{R}^3$ be a vector of constants. Suppose that $X = (X_1, X_2, X_3)^\top = aU$. What do you expect the NPCs of X to be?*

Let us assume that $a_i \neq 0$, $i = 1, 2, 3$. Next, normalizing the random vector X by subtracting its expected value and by dividing it by its standard deviation leads to the normalized random vector

$$Z = \left\{ \text{diag}\left(a^2\sigma_U^2\right)\right\}^{-1/2}(X - EX) = \left\{ \text{diag}\left(a^2\sigma_U^2\right)\right\}^{-1/2}a(U - E\overset{.}{U})$$

with the variance matrix

$$\text{Var}(Z) = \left\{ \text{diag} \left(a^2 \sigma_U^2 \right) \right\}^{-1/2} \text{Var}(X) \left\{ \text{diag} \left(a^2 \sigma_U^2 \right) \right\}^{-1/2}$$

$$= \left\{ \text{diag} \left(a^2 \sigma_U^2 \right) \right\}^{-1/2} a \sigma_U^2 a^\top \left\{ \text{diag} \left(a^2 \sigma_U^2 \right) \right\}^{-1/2}$$

$$= \left(\frac{a_i a_j}{\text{abs}\, a_i \,\text{abs}\, a_j} \right)_{i,j=1,2,3}$$

$$= \left\{ \text{sign}(a_i a_j) \right\}_{i,j=1,2,3} \, .$$

Clearly, the rank of the variance matrix $\text{Var}(Z)$ is equal to 1 and it follows that it has only one nonzero eigenvalue. Hence, the spectral decomposition of $\text{Var}(Z)$ leads to only one principal component explaining 100% of total variability of $\mathcal{Z}$.

The NPC can be written as

$$Y_1 = \frac{1}{\sqrt{3}} \left\{ \text{sign}(a_1) Z_1 + \text{sign}(a_2) Z_2 + \text{sign}(a_3) Z_3 \right\}$$

$$= \frac{1}{\sqrt{3}} \left\{ \text{sign}(a_1) a_1 U + \text{sign}(a_2) a_2 U + \text{sign}(a_3) a_3 U \right\}$$

$$= U \frac{\text{abs}(a_1) + \text{abs}(a_2) + \text{abs}(a_3)}{\sqrt{3}},$$

i.e., the normalized principal components analysis of $X = aU$ leads us back to the one-dimensional random variable U.

EXERCISE 9.12. *Let U_1 and U_2 be two independent uniform random variables on $[0,1]$. Suppose that $X = (X_1, X_2, X_3, X_4)^\top$ where $X_1 = U_1$, $X_2 = U_2$, $X_3 = U_1 + U_2$ and $X_4 = U_1 - U_2$. Compute the correlation matrix P of X. How many PCs are of interest? Show that $\gamma_1 = \left(\frac{1}{\sqrt{2}}, \frac{1}{\sqrt{2}}, 1, 0 \right)^\top$ and $\gamma_2 = \left(\frac{1}{\sqrt{2}}, \frac{-1}{\sqrt{2}}, 0, 1 \right)^\top$ are eigenvectors of P corresponding to the non trivial λ's. Interpret the first two NPCs obtained.*

For random variables U_1 and $U_2 \sim U[0,1]$, we have $EU_1 = 1/2$ and $Var\, U_1 = Var\, U_2 = 1/12$. It follows that also $Var\, X_1 = Var\, X_2 = 1/12$.

For the variance of $X_3 = U_1 + U_2$ and $X_4 = U_1 - U_2$, we obtain

$$Var(X_3) = Var(X_4) = Var(U_1) + Var(U_2) = \frac{1}{6}$$

since U_1 and U_2 are independent. The covariances can be calculated as

$$\text{Cov}(X_1, X_3) = \text{Cov}(U_1, U_1 + U_2) = Var(U_1) + \text{Cov}(U_1, U_2) = \frac{1}{12}$$

and

$$\text{Cov}(X_3, X_4) = \text{Cov}(U_1 + U_2, U_1 - U_2) = Var(U_1) - Var(U_2) = 0.$$

The remaining elements of the variance matrix can be calculated in the same way leading to

$$
\text{Var}(X) = \frac{1}{12}
\begin{pmatrix}
1 & 0 & 1 & 1 \\
0 & 1 & 1 & -1 \\
1 & 1 & 2 & 0 \\
1 & -1 & 0 & 2
\end{pmatrix}.
$$

Dividing each row and each column by the square root of the corresponding diagonal element gives the correlation matrix

$$
P =
\begin{pmatrix}
1 & 0 & \frac{1}{\sqrt{2}} & \frac{1}{\sqrt{2}} \\
0 & 1 & \frac{1}{\sqrt{2}} & -\frac{1}{\sqrt{2}} \\
\frac{1}{\sqrt{2}} & \frac{1}{\sqrt{2}} & 1 & 0 \\
\frac{1}{\sqrt{2}} & -\frac{1}{\sqrt{2}} & 0 & 1
\end{pmatrix}.
$$

Now it is easy to verify that γ_1 and γ_2 are indeed eigenvectors of the correlation matrix P since

$$
P\gamma_1 =
\begin{pmatrix}
1 & 0 & \frac{1}{\sqrt{2}} & \frac{1}{\sqrt{2}} \\
0 & 1 & \frac{1}{\sqrt{2}} & -\frac{1}{\sqrt{2}} \\
\frac{1}{\sqrt{2}} & \frac{1}{\sqrt{2}} & 1 & 0 \\
\frac{1}{\sqrt{2}} & -\frac{1}{\sqrt{2}} & 0 & 1
\end{pmatrix}
\cdot
\begin{pmatrix}
\frac{1}{\sqrt{2}} \\
\frac{1}{\sqrt{2}} \\
1 \\
0
\end{pmatrix}
=
\begin{pmatrix}
\sqrt{2} \\
\sqrt{2} \\
2 \\
0
\end{pmatrix}
= 2\gamma_1.
$$

and, similarly, $P\gamma_2 = 2\gamma_2$. This, by the way, implies that also $P(\gamma_2 + \gamma_1) = 2(\gamma_1 + \gamma_2)$ and hence, any linear combination of γ_1 and γ_2 is also an eigenvector of P with the same eigenvalue.

Thus, we have the eigenvalues $\lambda_1 = \lambda_2 = 2$. The remaining two eigenvalues, λ_3 and λ_4 are equal to 0 because the rank of the correlation matrix is equal to 2.

The first two NPCs are not determined uniquely. Choosing the coefficients as γ_1 and γ_2 and keeping in mind that these coefficients correspond to the normalized variables we have:

$$
Y_1 = \frac{1}{\sqrt{2}}X_1 + \frac{1}{\sqrt{2}}X_2 + \frac{X_3}{\sqrt{2}} = \sqrt{2}(U_1 + U_2)
$$

$$
Y_2 = \frac{1}{\sqrt{2}}X_1 - \frac{1}{\sqrt{2}}X_2 + \frac{X_4}{\sqrt{2}} = \sqrt{2}(U_1 - U_2).
$$

The NPCs, Y_1 and Y_2, can be now interpreted respectively as the sum and the difference of U_1 and U_2.

EXERCISE 9.13. *Simulate a sample of size $n = 50$ for the r.v. X in Exercise 9.12 and analyze the results of a NPCA.*

Performing the NPCA for the simulated data set, we obtain the eigenvalues:

$$\widehat{\lambda} = (2.11, 1.89, 0.00, 0.00)^{\top}$$

and the proportions of the explained variance:

$$\widehat{\psi} = (0.53, 1.00, 1.00, 1.00)^{\top}.$$

These numbers correspond well to the theoretical values $\lambda_1 = \lambda_2 = 2$ derived in Exercise 9.12. The remaining two eigenvalues are equal to zero because of the linear dependencies in the data set. The screeplot is plotted in Figure 9.11 and we see that the first two NPCs explain each approximately 50% of the variability whereas the other two NPCs do not explain anything.

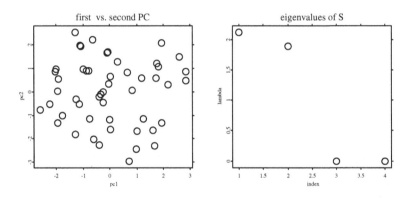

Fig. 9.11. Scatterplots of the first two principal components and a screeplot of the eigenvalues, simulated data set. ❑ SMSnpcasimu

The first two eigenvectors are

$$\widehat{\gamma}_1 = (0.32, -0.64, -0.26, 0.65)^{\top}$$

and

$$\widehat{\gamma}_2 = (0.65, 0.28, 0.67, 0.23)^{\top}$$

and the resulting values for the 50 NPCs are plotted in Figure 9.11. Rewriting the resulting NPCs in terms of the original variables and rounding the coefficients leads that the first NPC points approximately in the direction $U_1 - 2U_2$ and the second NPC in the direction $2U_1 + U_2$. This result differs from the eigenvectors γ_1 and γ_2 calculated in Exercise 9.12 because γ_1 and γ_2 are not uniquely defined.

In Figure 9.12, we plot the correlation of the NPCs with the normalized variables $X_1, \ldots, X_4$. The correlations correspond to the coefficients of the NPCs.

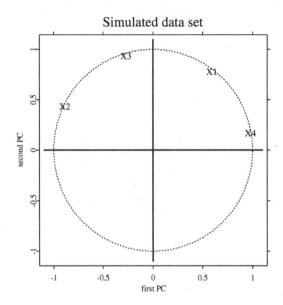

Fig. 9.12. Correlations of the first two principal components with the original variables in the simulated data set. ⊙ SMSnpcasimu

All of the original variables are perfectly explained by two NPCs because all four points are lying on the unit circle.

The simulated data set changes with every simulation. One can observe that the eigenvalues $\widehat{\lambda}$ do not vary a lot for different runs of the simulation. However, the eigenvectors can vary a lot due to the fact that they are not defined uniquely.

Factor Analysis

A certain selection and discretion must be used in producing a realistic effect.

Sherlock Holmes in "A Case of Identity"

In factor analysis, we address the same problem of reducing the dimension of a multivariate random variable, but we want to fix, from the start, the number of factors. Each factor will then be interpreted as a latent characteristic of the individuals revealed by the original variables.

From a statistical point of view, the essential purpose of factor analysis is to describe, if possible, the covariance relationships among many variables in terms of a few underlying, but unobservable, random quantities called factors.

The ultimate goal is to find underlying reasons that explain the data variation. In achieving this goal we need to check the relation of the factors and original variables and give them an interpretation in the framework of how the data were generated.

Factor Analysis Model

The factor analysis model used in practice is:

$$X = \mathcal{Q}F + U + \mu, \qquad (10.1)$$

where $\mathcal{Q}$ is a $(p \times k)$ matrix of the (nonrandom) loadings of the common factors $F(k \times 1)$ and U is a $(p \times 1)$ matrix of the (random) specific factors. It is assumed that the common factors F are uncorrelated random variables and that the specific factors are uncorrelated and have zero covariance with the common factors. More precisely, it is assumed that: $EF = 0$, $\text{Var}(F) = \mathcal{I}_k$, $EU = 0$, $\text{Cov}(U_i, U_j) = 0$, $i \neq j$, and $\text{Cov}(F, U) = 0$.

The random vectors F and U are unobservable. Define $\text{Var}(U) = \Psi = \text{diag}(\psi_{11}, \ldots, \psi_{pp})$; then the variance matrix of X can be written as $\text{Var}(X) = \Sigma = \mathcal{Q}\mathcal{Q}^\top + \Psi$, and we have for the ith component of the random vector X that $\sigma_{X_j X_j} = \text{Var}(X_j) = \sum_{\ell=1}^{k} q_{j\ell}^2 + \psi_{jj}$. The quantity $h_j^2 = \sum_{\ell=1}^{k} q_{j\ell}^2$ is called the communality and ψ_{jj} the specific variance. The objective of factor analysis is to find a small number, k, of common factors leading to large communalities and small specific variances.

Estimation of the Factor Model

In practice, we have to find estimates $\widehat{\mathcal{Q}}$ of the loadings $\mathcal{Q}$ and estimates $\widehat{\Psi}$ of the specific variances Ψ such that $\mathcal{S} = \widehat{\mathcal{Q}}\widehat{\mathcal{Q}}^\top + \widehat{\Psi}$, where $\mathcal{S}$ denotes the empirical covariance of $\mathcal{X}$. The most commonly used methods are the following:

The maximum likelihood method is based on the assumption of normality. The equations resulting from the maximization of the log-likelihood under the assumption $\Sigma = \mathcal{Q}\mathcal{Q}^\top + \Psi$ are complicated and have to be solved by iterative numerical algorithms.

The method of principal factors starts with a preliminary estimate of $\widehat{h}_j^2$ and the specific variances $\widehat{\psi}_{jj} = 1 - \widehat{h}_j$. In the next step, the matrix of loadings is estimated from the spectral decomposition of the reduced covariance matrix $\mathcal{S} - \widehat{\Psi}$. This procedure can be iterated until convergence is reached.

The principal component method starts by obtaining estimated loadings $\widehat{\mathcal{Q}}$ from a spectral decomposition of the matrix $\mathcal{S}$. The specific variances are then estimated by the diagonal elements of the matrix $\mathcal{S} - \widehat{\mathcal{Q}}\widehat{\mathcal{Q}}^\top$.

Rotation

Suppose that $\mathcal{G}$ is an orthogonal matrix. Then X in (10.1) can also be written as $X = (\mathcal{Q}\mathcal{G})(\mathcal{G}^\top F) + U + \mu$. This implies that the factors are not defined uniquely because equivalent models with factors $\mathcal{G}^\top F$ and loadings $\mathcal{Q}\mathcal{G}$ are valid for an arbitrary orthogonal matrix $\mathcal{G}$. In practice, the choice of an appropriate rotation $\mathcal{G}$ of the loadings $\mathcal{Q}$ results in a matrix of loadings $\mathcal{Q}^* = \mathcal{Q}\mathcal{G}$ that are easier to interpret.

A well-known algorithm for choosing a reasonable rotation of the factor loadings is given by the varimax rotation method proposed by Kaiser (1985). The idea of this popular method is to find the angles that maximize the sum of the variances of the squared loadings q_{ij}^* within each column of $\mathcal{Q}^*$. The varimax criterion attempts to split the variables automatically into disjoint sets, each associated with one factor.

Strategy for Factor Analysis

1. Perform a principal component factor analysis, look for suspicious observations, try varimax rotation.

2. Perform maximum likelihood factor analysis, including varimax rotation.

3. Compare the factor analyses: do the loadings group in the same manner?

4. Repeat the previous steps for other numbers of common factors.

After the estimation and interpretation of factor loadings and communalities,
estimate the factor values. The estimated values of the factors are called the
factor scores and may be useful in the interpretation as well as in the diagnostic
analysis. To be more precise, the factor scores are estimates of the unobserved
k-dimensional random vectors F for each individual x_i, $i = 1, \ldots, n$. Johnson
& Wichern (1998) describe three methods that in practice yield very similar
results. The regression method (see Exercise 10.6) is also described in Härdle
& Simar (2003, section 10.3).

EXERCISE 10.1. *Compute the orthogonal factor model for*

$$\Sigma = \begin{pmatrix} 1.0 & 0.9 & 0.7 \\ 0.9 & 1.0 & 0.4 \\ 0.7 & 0.4 & 1.0 \end{pmatrix}.$$

We have to find loadings $\mathcal{Q}$ and specific variances Ψ satisfying the decomposition $\Sigma = \mathcal{Q}\mathcal{Q}^\top + \Psi$. The problem is difficult to solve due to the non-uniqueness
of the solutions. An acceptable technique is to impose some additional con-
straints such as: $\mathcal{Q}^\top \Psi^{-1} \mathcal{Q}$ is diagonal.

The factor analysis without any constraints has $pk + k$ unknown parameters of
the matrix $\mathcal{Q}$ and the diagonal of Ψ. The diagonality of $\mathcal{Q}^\top \Psi^{-1} \mathcal{Q}$ introduces
$\frac{1}{2}\{k(k-1)\}$ constraints. Therefore, the degrees of freedom of a model with k
factors is $d = \frac{1}{2}(p-k)^2 - \frac{1}{2}(p+k)$.

If $d < 0$, then there are infinitely many solutions. If $d = 0$ the there is an
unique solution to the problem (except for rotation). In practice we usually
have that $d > 0$ and an exact solution does not exist. Evaluating the degrees
of freedom, d, is particularly important, because it already gives an idea of
the upper bound on the number of factors we can hope to identify in a factor
model.

If $p = 3$, we can identify at most $k = 1$ factor. This factor is then given
uniquely since $d = \frac{1}{2}(3-1)^2 - \frac{1}{2}(3+1) = 0$. Implementing a simple iterative
procedure, i.e., the principal factor method described in the introduction, we
arrive to the following exact solution:

$$\Sigma = \begin{pmatrix} 1.0 & 0.9 & 0.7 \\ 0.9 & 1.0 & 0.4 \\ 0.7 & 0.4 & 1.0 \end{pmatrix}$$

$$= \begin{pmatrix} 1.2549 \\ 0.7172 \\ 0.5578 \end{pmatrix} (1.2549, 0.7172, 0.5578) + \begin{pmatrix} -0.5748 & 0.0000 & 0.0000 \\ 0.0000 & 0.4857 & 0.0000 \\ 0.0000 & 0.0000 & 0.6889 \end{pmatrix}.$$

The obvious disadvantage of this unique solution is that it cannot be interpreted as a factor analysis model since the specific variance ψ_{11} cannot be negative.

Hence, the ability to find a unique solution of the orthogonal factor model does not have to lead to the desired result.   Q SMSfactsigma

EXERCISE 10.2. *Using the bank data set in Table A.2, how many factors can you find with the method of principal factors?*

The number of variables is $p = 6$. For $k = 3$ factors, the orthogonal factor model would have

$$d = \frac{1}{2}(p - k)^2 - \frac{1}{2}(p + k) = 4.5 - 4.5 = 0$$

degrees of freedom, see Exercise 10.1. It follows that for 3 factors, we would have an exact solution. Unfortunately, as we have seen in Exercise 10.1, the unique exact solution does not have to be interpretable. In this situation, it is advisable to work with at most $k = 2$ factors.

The empirical correlation analysis calculated from the given 6-dimensional data set is:

$$\mathcal{R} = \begin{pmatrix} 1.0000 & 0.2313 & 0.1518 & -0.1898 & -0.0613 & 0.1943 \\ 0.2313 & 1.0000 & 0.7433 & 0.4138 & 0.3623 & -0.5032 \\ 0.1518 & 0.7433 & 1.0000 & 0.4868 & 0.4007 & -0.5165 \\ -0.1898 & 0.4138 & 0.4868 & 1.0000 & 0.1419 & -0.6230 \\ -0.0613 & 0.3623 & 0.4007 & 0.1419 & 1.0000 & -0.5940 \\ 0.1943 & -0.5032 & -0.5165 & -0.6230 & -0.5940 & 1.0000 \end{pmatrix}.$$

The communalities h_j^2, $j = 1, \ldots, 6$, measure the part of variance of each variable that can be assigned to the common factors. One possibility to define a reasonable starting estimates is to set $\widehat{h}_j^2 = \max_{i \neq j, i=1,\ldots,6} |r_{X_j X_i}|$. For the Swiss bank notes, we obtain

$$\widehat{h}^2 = (\widehat{h}_1^2, \ldots, \widehat{h}_6^2)^\top = (0.2313, 0.7433, 0.7433, 0.6230, 0.5940, 0.6230)^\top.$$

The estimates of the specific variances ψ_{jj}, $j = 1, \ldots, 6$ are

$$\widehat{\psi} = (\widehat{\psi}_{11}, \ldots, \widehat{\psi}_{66})^\top = (0.7687, 0.2567, 0.2567, 0.3770, 0.4060, 0.3770)^\top$$

and the reduced correlation matrix $\mathcal{R} - \widehat{\Psi}$ is

$$
\mathcal{R} - \mathrm{diag}(\widehat{\psi}) = \begin{pmatrix}
0.2313 & 0.2313 & 0.1518 & -0.1898 & -0.0613 & 0.1943 \\
0.2313 & 0.7433 & 0.7433 & 0.4138 & 0.3623 & -0.5032 \\
0.1518 & 0.7433 & 0.7433 & 0.4868 & 0.4007 & -0.5165 \\
-0.1898 & 0.4138 & 0.4868 & 0.6230 & 0.1419 & -0.6230 \\
-0.0613 & 0.3623 & 0.4007 & 0.1419 & 0.5940 & -0.5940 \\
0.1943 & -0.5032 & -0.5165 & -0.6230 & -0.5940 & 0.6230
\end{pmatrix}.
$$

The vector of the eigenvalues of the reduced correlation matrix is:

$$
\lambda = (2.6214, 0.7232, 0.4765, 0.0054, -0.0845, -0.1841)^{\top}.
$$

At this step, some of the eigenvalues can be negative. The possibility that the reduced correlation matrix does not have to be positive definite has to be taken into account in the computer implementation of the factor analysis.

The matrix of eigenvectors of the reduced correlation matrix is:

$$
\Gamma = \begin{pmatrix}
-0.0011 & -0.6225 & 0.0488 & -0.1397 & 0.7663 & 0.0582 \\
0.4832 & -0.4510 & -0.0727 & -0.5783 & -0.4575 & -0.1185 \\
0.5019 & -0.3314 & -0.1077 & 0.7670 & -0.1328 & 0.1438 \\
0.3974 & 0.3489 & -0.6039 & -0.0434 & 0.3510 & -0.4802 \\
0.3543 & 0.1661 & 0.7768 & 0.0604 & 0.1328 & -0.4714 \\
-0.4807 & -0.3872 & -0.1125 & 0.2285 & -0.2123 & -0.7135
\end{pmatrix}.
$$

With $k = 2$ factors, we obtain the factor loadings

$$
\widehat{Q} = \begin{pmatrix}
-0.0011 & -0.6225 \\
0.4832 & -0.4510 \\
0.5019 & -0.3314 \\
0.3974 & 0.3489 \\
0.3543 & 0.1661 \\
-0.4807 & -0.3872
\end{pmatrix}
\begin{pmatrix}
\sqrt{2.6214} & 0 \\
0 & \sqrt{0.7232}
\end{pmatrix}
= \begin{pmatrix}
-0.0018 & -0.5294 \\
0.7824 & -0.3835 \\
0.8127 & -0.2819 \\
0.6435 & 0.2967 \\
0.5736 & 0.1412 \\
-0.7783 & -0.3293
\end{pmatrix}.
$$

If the variables are normalized, i.e., if the analysis is based on the correlation matrix, the factor loadings Q are the correlations between the original variables and the unobserved factors.

The final estimates of the two factor model, given in Table 10.1, were obtained by several iterations of the described algorithm. It is interesting to notice that the final estimates are rather different from the starting values.

The next step in the analysis is a rotation of the two factor loadings leading to better interpretable results. In Figure 10.1 you can see both the original factor loadings as given in Table 10.1 and the same factor loadings rotated by the angle $5\pi/12$ counterclockwise. The rotation, i.e., multiplication of the factor loadings by the rotation matrix

	Estimated factor loadings		Communalities	Specific variances
	$\hat{q}_1$	$\hat{q}_2$	$\hat{h}_j^2$	$\hat{\psi}_{jj} = 1 - \hat{h}_j^2$
1 length	−0.0046	−0.5427	0.2946	0.7054
2 height measured left	0.7888	−0.4107	0.7910	0.2090
3 height measured right	0.7996	−0.2982	0.7283	0.2717
4 lower frame distance	0.5929	0.1953	0.3896	0.6104
5 upper frame distance	0.5109	0.1068	0.2724	0.7276
6 length of the diagonal	−0.8784	−0.4436	0.9683	0.0317

Table 10.1. Estimated factor loadings, communalities, and specific variances, PFM, Swiss bank notes data set. ⊙ SMSfactbank

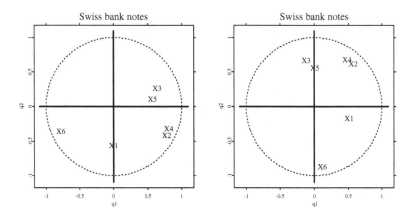

Fig. 10.1. Rotation of the factor loadings in the Swiss bank notes data set. The original and rotated factor loadings are on the left and right hand side, respectively. ⊙ SMSfactbank

$$\mathcal{G}(\theta) = \begin{pmatrix} \cos\theta & \sin\theta \\ -\sin\theta & \cos\theta \end{pmatrix},$$

where $\theta = 5\pi/12$ changes only the factor loadings and their interpretation. In Figure 10.1, we suggest rotation leading to one factor positively correlated to X_1, X_2, and X_4 whereas the second factor is strongly positively related to X_2, X_3, X_4, and X_5 and strongly negatively related to X_6.

Further insight into the factors might be achieved by estimating their values for our observations. This part of the factor analysis will be demonstrated in detail in Exercise 10.6.

EXERCISE 10.3. *An example of an orthogonal matrix in two-dimensions is the so-called rotation matrix*

$$\mathcal{G}(\theta) = \begin{pmatrix} \cos\theta & \sin\theta \\ -\sin\theta & \cos\theta \end{pmatrix},$$

representing a clockwise rotation of the coordinate axes by the angle θ. Generalize the two-dimensional rotation matrix $\mathcal{G}(\theta)$ to 3-dimensional space.

The two-dimensional rotation matrix $\mathcal{G}(\theta)$ rotates two-dimensional coordinates counterclockwise by angle θ with respect to the origin $(0,0)^{\top}$, see Figure 10.1 for an illustration.

In 3-dimensional space, we can fix three angles, θ_1, θ_2, and θ_3 specifying three two-dimensional rotations. In the first step, we can rotate the given three-dimensional points in the first two coordinates and keep the third coordinate fixed, this can be achieved by the rotation matrix:

$$\mathcal{G}_{12}(\theta_3) = \begin{pmatrix} \cos\theta_3 & \sin\theta_3 & 0 \\ -\sin\theta_3 & \cos\theta_3 & 0 \\ 0 & 0 & 1 \end{pmatrix}.$$

Rotating the points only in the first coordinates can be described as a rotation of the thee-dimensional cloud of points around the third axis by angle θ_3.

The rotation in the first and third coordinate (around the second axis) is achieved by:

$$\mathcal{G}_{13}(\theta_2) = \begin{pmatrix} \cos\theta_2 & 0 & \sin\theta_2 \\ 0 & 1 & 0 \\ -\sin\theta_2 & 0 & \cos\theta_2 \end{pmatrix}$$

and for the rotation in the second and third coordinate (around the first axis), we have:

$$\mathcal{G}_{23}(\theta_1) = \begin{pmatrix} 1 & 0 & 0 \\ 0 & \cos\theta_1 & \sin\theta_1 \\ 0 & -\sin\theta_1 & \cos\theta_1 \end{pmatrix}.$$

Arbitrary rotation in three-dimensional space can now be written as a combination of the two-dimensional rotations $\mathcal{G}_{23}(\theta_1)$, $\mathcal{G}_{13}(\theta_2)$, and $\mathcal{G}_{12}(\theta_3)$. We define the general three-dimensional rotation matrix:

$$\mathcal{G}_{123}(\theta_1, \theta_2, \theta_3) = \mathcal{G}_{23}(\theta_1)\mathcal{G}_{13}(\theta_2)\mathcal{G}_{12}(\theta_3).$$

Similarly, the two-dimensional rotation matrices can be used to define a rotation in n-dimensional space.

EXERCISE 10.4. *Perform a factor analysis on the type of families in the French food data set A.9. Rotate the resulting factors in a way which provides a reasonable interpretation. Compare your result to the varimax method.*

The French food data set contains average expenditures on seven types of food for different types of families (manual workers, employees, managers) in France. The abbreviations MA, EM, and CA denote respectively manual workers, employees, and managers. The number denotes the number of children. In this exercise, we consider the dataset as consisting of 7 measurement of the 12 type of family variables.

A first look at the data set reveals that the structure of expenditures strongly depends on the type of food. Hence, before running the factor analysis, we put all measurements on the same scale by standardizing the expenditures for each type of food separately.

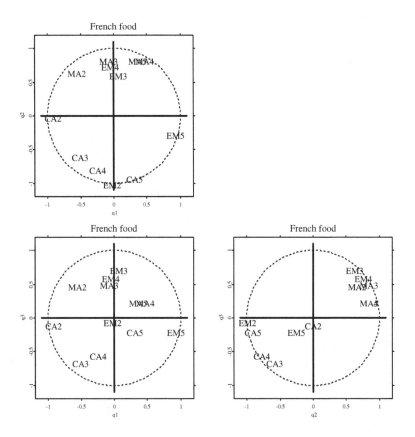

Fig. 10.2. Factor loadings for the French food data set after manual rotation of the factor loading obtained by PFM method. ⬛ SMSfactfood

We choose $k = 3$ factors. The corresponding factor loadings were estimated by the principal factors method. In order to obtain more interpretable results, we have rotated the factor loadings in Figure 10.2. After the manual rotation of the factor loadings, the first factor seems to be related to the number of children. The second and the third factor are related to the type of family. The main disadvantage of this approach are that a manual rotation of the factor loadings is rather time consuming and that the final result might be strongly influenced by prior beliefs of the data analyst.

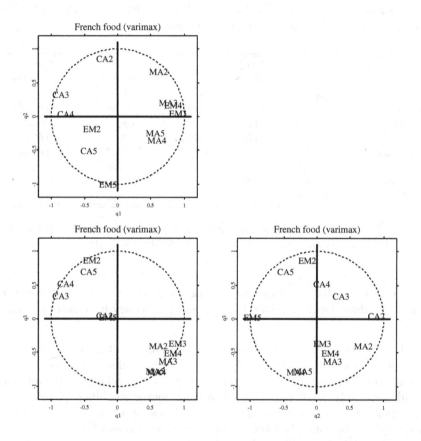

Fig. 10.3. Varimax rotation for French food data set. ▢ SMSfactfood

Hence, in practice, we recommend to use the varimax rotation which in this case leads to very similar result, see Figure 10.3. A comparison of Figures 10.2 and 10.3 shows that the varimax methods find automatically a rotation which is very similar to the result obtain by manual rotation of factor loadings. The main difference seems to be the order and the signs of the factors.

EXERCISE 10.5. *Perform a factor analysis on the variables X_4 to X_{10} in the U.S. health data set in Table A.19. Would it make sense to use all of the variables for the factor analysis?*

From the discussion of the degrees of freedom of the factor analysis model in Exercises 10.1 and 10.2 it follows that we can estimate at most $k = 3$ factors in this 7-dimensional data set. The results of the factor analysis are given in Table 10.2 and Figure 10.4. The factor analysis model was estimated by the maximum likelihood method with varimax rotation.

	Estimated factor loadings			Communalities	Specific variances
	$\hat{q}_1$	$\hat{q}_2$	$\hat{q}_3$	$\hat{h}_j^2$	$\hat{\psi}_{jj} = 1 - \hat{h}_j^2$
1 accident	−0.5628	0.0220	−0.1958	0.3556	0.6448
2 cardiovascular	0.7354	0.1782	0.5955	0.9271	0.0735
3 cancer	0.8381	−0.1166	0.5246	0.9913	0.0087
4 pulmonary	0.1709	−0.0682	0.5476	0.3337	0.6666
5 pneumonia flu	0.0098	0.4338	0.7631	0.7706	0.2252
6 diabetes	0.8046	−0.0488	0.0569	0.6531	0.3477
7 liver	0.1126	−0.8082	0.3321	0.7762	0.2173

Table 10.2. Estimated factor loadings after varimax rotation, communalities, and specific variances, MLM, U.S. health data set. ⊙ SMSfactushealth

Table 10.2 shows that the three factor model explains very well most of the original variables. Only variables accident and pulmonary have lower communalities.

The plots of the factor loadings in Figure 10.4 suggests that the first factor corresponds to causes of death related by cardiovascular problems, cancer, and diabetes. The second factor seems to be positively related to pneumonia flu and negatively related to liver. The third factor combines all causes of death apart of accidents and diabetes. The discussion of the meaning of the factors will be continued in Exercise 10.7, where we present the estimation of the corresponding factor scores for each state.

Let us now investigate the question whether the three factors derived in this exercise describe sufficiently the dependencies within the U.S. health data set. This question can be answered by formal statistical test based on the likelihood ratio approach that has been demonstrated in Chapter 7.

Assuming that $\widehat{\mathcal{Q}}$ and $\widehat{\Psi}$ are the estimates obtained by the maximum likelihood method, the likelihood ratio (LR) test statistic for the null hypothesis H_0 :

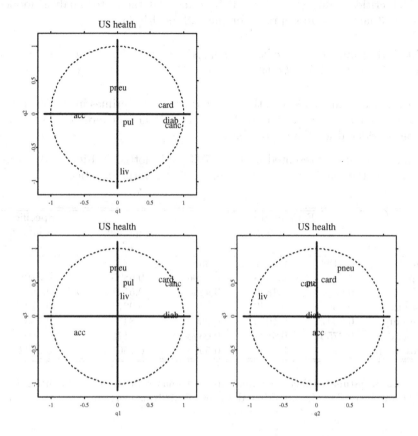

Fig. 10.4. Factor loadings for the U.S. health data set after varimax rotation.
🔾 SMSfactushealth

$\Sigma = \mathcal{Q}\mathcal{Q}^\top + \Psi$ can be derived as:

$$-2\log\left(\frac{\text{maximized likelihood under } H_0}{\text{maximized likelihood}}\right) = n\log\left(\frac{|\widehat{\mathcal{Q}}\widehat{\mathcal{Q}}^\top + \widehat{\Psi}|}{|\mathcal{S}|}\right). \quad (10.2)$$

Under the null hypothesis, the LR test statistic has asymptotically the $\chi^2_{\frac{1}{2}\{(p-k)^2-p-k\}}$ distribution. Bartlett (1954) suggested a correction which improves the above χ^2 approximation by replacing n by $n-1-(2p+4k+5)/6$ in (10.2). The LR test can be applied only if the degrees of freedom are positive, see also the discussion of the degrees of freedom in Exercise 10.1.

Let us now test the null hypothesis $H_0 : k = 3$. The value of the LR test statistic with Bartlett correction is 3.66 and we cannot reject the null hypothesis $H_0 : \Sigma = \mathcal{Q}\mathcal{Q}^\top + \Psi$ since the observed value of the test statistic is smaller than the critical value $\chi^2_{0.95;3} = 7.81$. It seems that the factor analysis model with $k = 3$ factors is appropriate for the U.S. health data set.

EXERCISE 10.6. *Perform a factor analysis on the U.S. crime data set in Table A.18 and estimate the factor scores.*

The U.S. crime data set states the reported number of crimes in the 50 states of the USA classified according to 7 categories. Hence, at most $k = 3$ factors can be considered for the factor analysis.

The factor loadings presented in Table 10.3 and plotted in Figure 10.5 were obtained by the maximum likelihood method and varimax rotation.

	Estimated factor loadings			Communalities	Specific variances
	$\hat{q}_1$	$\hat{q}_2$	$\hat{q}_3$	$\hat{h}_j^2$	$\hat{\psi}_{jj} = 1 - \hat{h}_j^2$
1 murder	0.4134	−0.7762	−0.0651	0.7777	0.2225
2 rape	0.7938	−0.2438	−0.0006	0.6895	0.3108
3 robbery	0.6148	−0.1866	0.4494	0.6147	0.3855
4 assault	0.6668	−0.6940	0.0368	0.9275	0.0723
5 burglary	0.8847	0.1073	0.2302	0.8472	0.1534
6 larceny	0.8753	0.3834	−0.0625	0.9172	0.0808
7 auto theft	0.6132	0.1435	0.5995	0.7561	0.2432

Table 10.3. Estimated factor loadings after varimax rotation, communalities, and specific variances, MLM, U.S. crime data set. ⬛ SMSfactuscrime

The LR test of the hypothesis that three factors are enough to described the dependencies within the U.S. crime data set leads p-value 0.8257 and the null hypothesis $H_0 : k = 3$ cannot be rejected.

The first factor could be described as the overall criminality factor. The second factor is positively related to larceny and negatively related to more violent crimes such as murder and assault. The third factor is related mainly to robbery and auto theft.

In order to describe the differences between different states, we have to estimate the values of the factor scores for individual observations. The idea of the commonly used regression method is based on the joint distribution of $(X - \mu)$ and F. The joint covariance matrix of $(X - \mu)$ and F is:

$$Var \begin{pmatrix} X - \mu \\ F \end{pmatrix} = \begin{pmatrix} \mathcal{Q}\mathcal{Q}^\top + \Psi & \mathcal{Q} \\ \mathcal{Q}^\top & \mathcal{I}_k \end{pmatrix} = \begin{pmatrix} \Sigma & \mathcal{Q} \\ \mathcal{Q}^\top & \mathcal{I}_k \end{pmatrix}. \tag{10.3}$$

In practice, we replace the unknown $\mathcal{Q}$, Σ and μ by corresponding estimators, leading to the estimated individual factor scores:

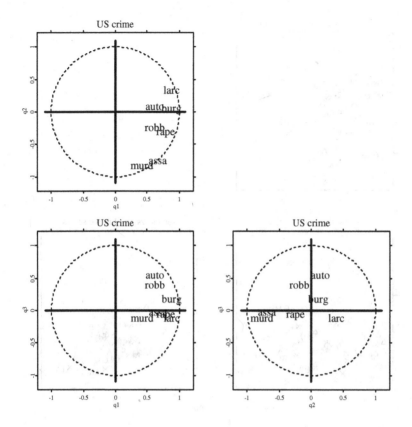

Fig. 10.5. Factor loadings for the U.S. crime data set after varimax rotation. ⬜ SMSfactuscrime

$$\widehat{f}_i = \widehat{\mathcal{Q}}^\top \mathcal{S}^{-1}(x_i - \overline{x}).$$

The same rule can be followed when using $\mathcal{R}$ instead of $\mathcal{S}$. Then (10.3) remains valid when standardized variables, i.e., $Z = \mathcal{D}_\Sigma^{-1/2}(X - \mu)$, are considered if $\mathcal{D}_\Sigma = \text{diag}(\sigma_{11}, \ldots, \sigma_{pp})$. In this case the factors are given by

$$\widehat{f}_i = \widehat{\mathcal{Q}}^\top \mathcal{R}^{-1}(z_i),$$

where $z_i = \mathcal{D}_S^{-1/2}(x_i - \overline{x})$, $\widehat{\mathcal{Q}}$ is the loading obtained with the matrix $\mathcal{R}$, and $\mathcal{D}_S = \text{diag}(s_{11}, \ldots, s_{pp})$.

The factor scores corresponding to the factor loadings given in Table 10.3 are plotted in Figure 10.6. The estimated factor scores for the first factor,

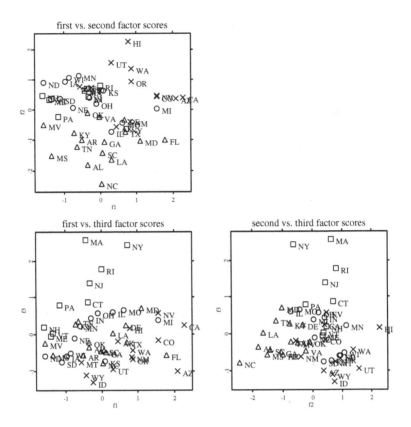

Fig. 10.6. Factor scores for the U.S. crime data set estimated by the regression method. Northeast (squares), Midwest (circles), South (triangles) and West (crosses). ℚ SMSfactuscrime

overall criminality, seem to be largest in California, Arizona, and Florida. The second factor suggests that murder and assault are common mainly in North Carolina. The third factor, auto theft and robbery, reaches the highest estimated factor scores in Massachusetts and New York.

EXERCISE 10.7. *Estimate the factor scores for the U.S. health data set analyzed in Exercise 10.5 and compare the estimated factor scores to the scores obtained for the U.S. crime data set in Exercise 10.6.*

The factor scores for the U.S. health data set, corresponding to the factor loadings obtained in Exercise 10.5, are plotted in Figure 10.7.

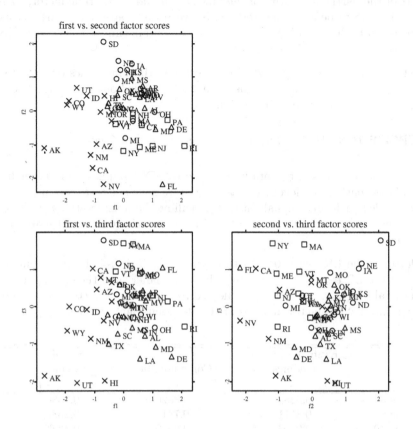

Fig. 10.7. Factor scores for the U.S. health data set estimated by the regression method. Northeast (squares), Midwest (circles), South (triangles) and West (crosses). ◻ SMSfactushealth

The first factor, corresponding to diabetes, cancer, and cardiovascular problems, leads to higher factor scores in Richmond, Delaware, and Pennsylvania. On the other side, these causes of death are less common mainly in Arkansas, Wyoming, Colorado, and Utah. This factor looks a bit like the third factor obtained for the U.S. crime data set in Exercise 10.6.

The second health factor, strongly negatively related to liver and positively related to pneumonia flu has highest values in South Dakota and Nebraska and smallest values in Nevada, Florida, and California. The third health factor has high values in South Dakota, New York, and Massachusetts and small values in Utah, Hawaii, and Arkansas.

Apart of the partial similarity of the first health and third crime factor, there does not seem to be any other relation between the health and crime factors. However, in both plots of the factor scores, we obtain similar factor scores for states coming from the same region.

The factor analysis is not designed to investigate the similarities between two sets of variables. Such comparisons ought to be carried out by the method of canonical correlations described in Chapter 14.

EXERCISE 10.8. *Analyze the vocabulary data given in Table A.20.*

The vocabulary data set contains test scores of 64 pupils from the eighth through eleventh grade levels. For each pupil we have one test score per grade which leads to a 4-dimensional data set. Recalling the considerations presented in Exercises 10.1 and 10.2, we see that in this exercise we can estimate only one factor.

Performing the LR test (10.2) of the hypothesis $H_0 : k = 1$, we obtain the value of the LR test statistic 1.6101, which is smaller than the corresponding critical value $\chi^2_{0.95;2} = 5.9915$ (p-value 0.4470). Hence, one factor seems to be appropriate for the factor analysis of this 4-dimensional data set.

	Estimated factor loadings $\hat{q}_1$	Communalities $\hat{h}_j^2$	Specific variances $\hat{\psi}_{jj} = 1 - \hat{h}_j^2$
1 Grade 8	0.9284	0.8620	0.1380
2 Grade 9	0.8611	0.7415	0.2585
3 Grade 10	0.9306	0.8659	0.1341
4 Grade 11	0.8618	0.7427	0.2573

Table 10.4. Estimated factor loadings, communalities, and specific variances, MLM, vocabulary data set. ✑ SMSfactvocab

The results obtained by maximum likelihood method are summarized in Table 10.4. The rotation on the one-dimensional factor loadings would not have any meaning. The resulting factor can be interpreted as an overall vocabulary score strongly positively related to the test score in all four grades.

The estimated one-dimensional factor scores are plotted in Figure 10.8 by means of a dot-plot. The position of each observation on the horizontal axis is given by the estimated factor score. The values on the vertical axis are randomly chosen so that the plotted numbers are readable. The best values were achieved in observations 36 and 38 whereas the 5th observation seems to be extremely bad.

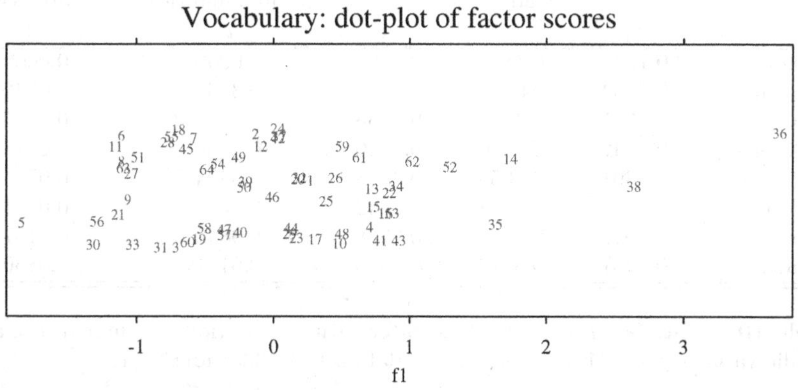

Fig. 10.8. Dot-plot of the one-dimensional factor scores for the vocabulary data set estimated by the regression method. ▢ SMSfactvocab

EXERCISE 10.9. *Analyze the athletic records data set in Table A.1. Can you recognize any patterns if you sort the countries according to the estimates of the factor scores?*

The athletic records data set provides data on athletic records in 100m up to a marathon for 55 countries.

Performing the estimation of the factor loadings by the maximum likelihood method allows us to test the hypothesis $H_0 : k = 3$ by means of the likelihood ratio test statistic (10.2). In this exercise, we obtain the test statistic 7.5207 which is smaller than the critical value $\chi^2_{0.95;7} = 14.0671$. The p-value of the test is 0.3767. The hypothesis that 3 factors are enough to describe the athletic records data set thus cannot be rejected.

The estimated factor loadings obtained by maximum likelihood method and varimax rotation are given in Table 10.5 and plotted in Figure 10.9. The communalities and specific variances show that three factors explain very well all of the original variables up to the record in 200m.

The first factor is most strongly related to times achieved in 100 and 200m, the second factor is positively related mainly to the records in longer distances. The third factor has positive relationship to the records in middle distances and 100m. It is important to keep in mind that high numbers here correspond to worse times. Hence, the athletic nations should exhibit small values of the factor scores.

| | Estimated factor loadings | | | Communalities | Specific variances |
	$\hat{q}_1$	$\hat{q}_2$	$\hat{q}_3$	$\hat{h}_j^2$	$\hat{\psi}_{jj} = 1 - \hat{h}_j^2$
1 100 m	0.7642	0.1803	0.6192	1.0000	0.0000
2 200 m	0.5734	0.0711	0.0474	0.3361	0.6642
3 400 m	0.4617	0.4869	0.6468	0.8686	0.1315
4 800 m	0.3442	0.6530	0.6060	0.9120	0.0878
5 1.5 km	0.3391	0.7655	0.4894	0.9404	0.0596
6 5 km	0.3771	0.8612	0.2842	0.9647	0.0354
7 10 km	0.4022	0.8636	0.2768	0.9842	0.0157
8 marathon	0.3231	0.8813	0.1843	0.9151	0.0850

Table 10.5. Estimated factor loadings after varimax rotation, communalities, and specific variances, MLM, athletic records data set. ꞓ SMSfacthletic

Rank	1	2	3
1	Italy	Portugal	GB
2	Colombia	NZ	Bermuda
3	USA	Ireland	DomRep
4	USSR	Netherlands	Thailand
5	Canada	Kenya	USA
6	Poland	Norway	FRG
⋮	⋮	⋮	⋮
50	Kenya	Bermuda	Colombia
51	PKorea	Malaysia	PNG
52	Netherlands	Singapore	WSamoa
53	Philippines	DomRep	Guatemala
54	Mauritius	Thailand	CostaRica
55	CookIs	WSamoa	CookIs

Table 10.6. Countries sorted according to the factor scores estimated for the athletic records data set. ꞓ SMSfacthletic

The factor scores estimated by the regression method are plotted in Figure 10.10. Furthermore, Table 10.6 lists the best and the worst countries according to each factor.

Keeping in mind the interpretation of the factors, we can say that Italy, Colombia, USA, USSR, Canada, and Poland possess the best sprinters. On long distances, the best countries are Portugal, New Zealand, Ireland, Netherlands, and Kenya. The best times on 100m, 400m, and 800m are on average achieved by Great Britain, Bermuda, Dominican Republic, Thailand, and USA.

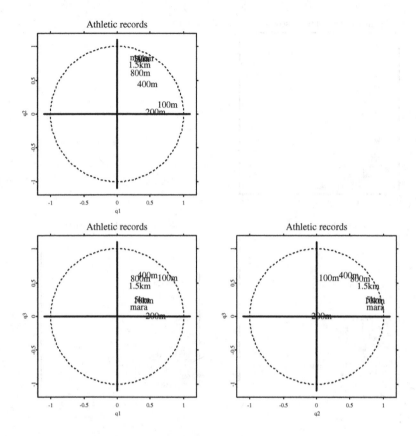

Fig. 10.9. Factor loadings for the athletic records data set after varimax rotation.

It is also interesting to notice that some of the countries which have very good factor scores for third or second factor, have, at the same time, very bad first or second factor scores. See, for example, Dominican Republic, Netherlands, and Kenya.

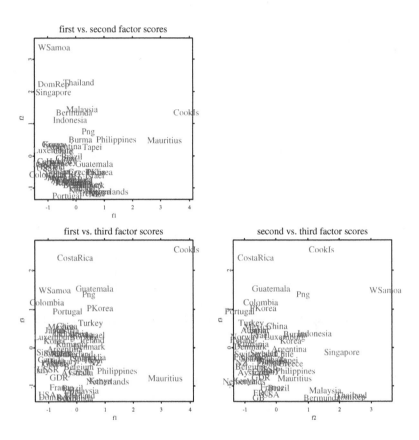

Fig. 10.10. Factor scores for the athletic records data set estimated by the regression method. Q `SMSfacthletic`

11

Cluster Analysis

> From a drop of water, a logician could infer the possibility of an Atlantic or a Niagara without having seen or heard of one or the other. So all life is a great chain, the nature of which is known whenever we are shown a single link of it.
> Sherlock Holmes in "Study in Scarlet"

When considering groups of objects in a multivariate data set, two situations can arise. Given a data set containing measurements on individuals, in some cases we want to see if some natural groups or classes of individuals exist, and in other cases, we want to classify the individuals according to a set of existing groups. Cluster analysis develops tools and methods concerning the former case, that is, given a data matrix containing multivariate measurements on a large number of individuals (or objects), the objective is to build subgroups or clusters of individuals. This is done by grouping individuals that are "similar" according to some appropriate criterion.

Cluster analysis is applied in many fields, including the natural sciences, the medical sciences, economics, and marketing. In marketing, for instance, it is useful to build and describe the different segments of a market from a survey of potential consumers. An insurance company, on the other hand, might be interested in the distinction among classes of potential customers so that it can derive optimal prices for its services. Other examples are provided in this chapter.

In this chapter we will concentrate on the so-called agglomerative hierarchical algorithms. The clustering algorithms start by calculating the distances between all pairs of observations, followed by stepwise agglomeration of close observations into groups.

Agglomerative Algorithm

1. Compute the distance matrix $\mathcal{D} = (d_{ij})_{i,j=1,...,n}$.

2. Find two observations with the smallest distance and put them into one cluster.

3. Compute the distance matrix between the $n-1$ clusters.

4. Find two clusters with the smallest intercluster distance and join them.

5. Repeat step 4 until all observations are combined in one cluster.

The properties of the clustering algorithm are driven mainly by the choice of distance.

Intercluster Distance

Assume that two observations or clusters, P and Q, are combined in a cluster denoted by $P \cup Q$. Let $d(P, Q)$ denote the distance between clusters P and Q and n_P and n_Q the number of observations belonging to clusters P and Q, respectively. Some common methods for defining the distance between the cluster $P \cup Q$ and some other cluster, say R, are:

Single linkage: $d(P \cup Q, R) = \min\{d(P, R), d(Q, R)\}$.

Complete linkage: $d(P \cup Q, R) = \max\{d(P, R), d(Q, R)\}$.

Average linkage: $d(P \cup Q, R) = \{d(P, R) + d(Q, R)\}/2$.

Average linkage (weighted):

$$d(P \cup Q, R) = \{n_P d(P, R) + n_Q d(Q, R)\}/(n_P + n_Q).$$

Median: $d^2(P \cup Q, R) = \{d^2(P, R) + d^2(Q, R)\}/2 - d^2(P, Q)/4$.

Centroid: $d^2(P \cup Q, R)$ is defined as the squared distance between R and the weighted (coordinatewise) average of P and Q; see Exercise 11.1.

Ward method: the heterogeneity of group R is measured by the inertia $I_R = \sum_{i=1}^{n_R} d^2(x_i, \overline{x}_R)$ (Ward 1963). In each step, we join the groups P and Q that give the smallest increase, $\Delta(P, Q)$, of the overall inertia; see Exercises 11.2 and 11.3.

Dendrogram

The successive joining of observations to the clusters is finally plotted in the so-called dendrogram. The construction of the dendrogram is explained in detail in Exercise 11.4.

EXERCISE 11.1. *Prove that the centroid distance $d^2(R, P \cup Q)$, defined as the (squared) distance between $R = (r_1, \ldots, r_p)^\top$ and the weighted average $\{n_P(p_1, \ldots, p_p)^\top + n_Q(q_1, \ldots, q_p)^\top\}/(n_P + n_Q)$ of P and Q, can be calculated as*

$$\frac{n_P}{n_P + n_Q} d^2(R, P) + \frac{n_Q}{n_P + n_Q} d^2(R, Q) - \frac{n_P n_Q}{(n_P + n_Q)^2} d^2(P, Q).$$

Let us calculate the Euclidean distance between the center $(r_1, \ldots, r_p)^\top$ of the cluster R and the weighted "center of gravity" of clusters P and Q:

$$d^2(P \cup Q, R)$$
$$= \sum_{i=1}^{p} \left\{ r_i - \frac{p_i n_P + q_i n_Q}{n_Q + n_P} \right\}^2$$
$$= \sum_{i=1}^{p} \left[r_i^2 - 2r_i \frac{p_i n_P + q_i n_Q}{n_Q + n_P} + \left\{ \frac{p_i n_P + q_i n_Q}{n_Q + n_P} \right\}^2 \right]$$
$$= \sum_{i=1}^{p} \left[\frac{n_P}{n_P + n_Q}(r_i - p_i)^2 + \frac{n_Q}{n_P + n_Q}(r_i - q_i)^2 - \frac{n_P n_Q}{(n_P + n_Q)^2}(q_i - p_i)^2 \right]$$
$$= \frac{n_P}{n_P + n_Q} d^2(R, P) + \frac{n_Q}{n_P + n_Q} d^2(R, Q) - \frac{n_P n_Q}{(n_P + n_Q)^2} d^2(P, Q).$$

Hence, the intercluster distance between R and $P \cup Q$ can be calculated from the distance between R, P, and Q. This property greatly simplifies the software implementation of the clustering algorithm since all calculations can be carried out using only the distance matrix between the n observations.

EXERCISE 11.2. *Derive the formula for the increase of the inertia $\Delta(P, Q)$ in the Ward method.*

In the Ward method, the heterogeneity of group R is measured by the inertia defined as:

$$I_R = \sum_{i=1}^{n_R} d^2(x_i, \overline{x}_R),$$

where $\overline{x}_R$ is the arithmetic average and n_R the number of observations within group R. If the usual Euclidean distance is used, then I_R represents the sum of the variances of the p components of x_i inside group R, see Exercise 11.3.

The Ward algorithm joins the groups P and Q that give the smallest increase, $\Delta(P, Q)$, of the inertia. The common inertia of the new group $P \cup Q$ can be written as:

$$I_{P \cup Q} = \sum_{i=1}^{n_P + n_Q} d^2(x_i - \overline{x}_{P \cup Q}) = \sum_{i=1}^{n_P + n_Q} \sum_{j=1}^{p} (x_{ij} - \overline{x}_{P \cup Q, j})^2$$

$$= \sum_{j=1}^{p} \left\{ \sum_{i=1}^{n_P} (x_{P,ij} - \overline{x}_{P \cup Q, j})^2 + \sum_{i=1}^{n_Q} (x_{Q,ij} - \overline{x}_{P \cup Q, j})^2 \right\}$$

$$= \sum_{j=1}^{p} \left\{ \sum_{i=1}^{n_P} (x_{P,ij} - \overline{x}_{P,j})^2 + n_P (\overline{x}_{P,j} - \overline{x}_{P \cup Q, j})^2 \right.$$

$$\left. + \sum_{i=1}^{n_Q} (x_{Q,ij} - \overline{x}_{Q,j})^2 + n_Q (\overline{x}_{Q,j} - \overline{x}_{P \cup Q, j})^2 \right\}$$

$$= I_P + I_Q + \sum_{j=1}^{p} \left\{ n_P (\overline{x}_{P,j} - \overline{x}_{P \cup Q, j})^2 + n_Q (\overline{x}_{Q,j} - \overline{x}_{P \cup Q, j})^2 \right\}$$

Hence, the inertia of $P \cup Q$ can be split into the sum of I_P and I_Q and a remainder term $\Delta(P, Q)$ for which we have:

$$\Delta(P, Q) = \sum_{j=1}^{p} \left\{ n_P (\overline{x}_{P,j} - \overline{x}_{P \cup Q, j})^2 + n_Q (\overline{x}_{Q,j} - \overline{x}_{P \cup Q, j})^2 \right\}$$

$$= \sum_{j=1}^{p} \left\{ n_P \left(\frac{n_Q \overline{x}_{P,j} - n_Q \overline{x}_{Q,j}}{n_P + n_Q} \right)^2 + n_Q \left(\frac{n_P \overline{x}_{P,j} - n_P \overline{x}_{Q,j}}{n_P + n_Q} \right)^2 \right\}$$

$$= \frac{n_P n_Q}{n_P + n_Q} \sum_{j=1}^{p} (\overline{x}_{P,j} - \overline{x}_{Q,j})^2 = \frac{n_P n_Q}{n_P + n_Q} d^2(P, Q).$$

The change of inertia $\Delta(P, Q)$ resulting from the joining of the groups P and Q can be considered as a distance of the clusters P and Q. In order to implement the Ward method numerically, we have to derive a formula for the intercluster distance between cluster R and the newly created cluster $P \cup Q$.

Applying the result of Exercise 11.1, we can write:

$$\Delta(R, P \cup Q) = \frac{n_R (n_P + n_Q)}{n_R + n_P + n_Q} d^2(R, P \cup Q)$$

$$= \frac{n_R (n_P + n_Q)}{n_R + n_P + n_Q} \left\{ \frac{n_P}{n_P + n_Q} d^2(R, P) + \frac{n_Q}{n_P + n_Q} d^2(R, Q) \right.$$

$$\left. - \frac{n_P n_Q}{(n_P + n_Q)^2} d^2(P, Q) \right\}$$

$$
= \frac{1}{n_R + n_P + n_Q} \left\{ n_R n_P \, d^2(R, P) + n_R n_Q \, d^2(R, Q) \right.
$$

$$
\left. - \frac{n_R n_P n_Q}{n_P + n_Q} \, d^2(P, Q) \right\}
$$

$$
= \frac{n_R + n_P}{n_R + n_P + n_Q} \, \Delta(R, P) + \frac{n_R + n_Q}{n_R + n_P + n_Q} \, \Delta(R, Q)
$$

$$
- \frac{n_R}{n_R + n_P + n_Q} \, \Delta(P, Q).
$$

The ability to express $\Delta(R, P \cup Q)$ using the distances $\Delta(R, P)$, $\Delta(R, Q)$, and $\Delta(P, Q)$ greatly simplifies the computer implementation of the Ward algorithm.

EXERCISE 11.3. *Prove that in the Ward method, the inertia $I_R = n_R \, tr(\mathcal{S}_R)$, where $\mathcal{S}_R$ denotes the empirical covariance matrix of the observations contained in group R.*

The inertia is defined as:

$$
I_R = \sum_{i=1}^{n_R} d^2(x_i, \overline{x}_R).
$$

Assuming that $d(x_i, \overline{x}_R)$ is the usual Euclidean distance between the ith observation $x_i = (x_{i1}, \ldots, x_{ip})^\top$ and the sample mean within group R, $\overline{x}_R = (x_{R1}, \ldots, x_{Rp})^\top$, we have:

$$
I_R = \sum_{i=1}^{n_R} d^2(x_i, \overline{x}_R) = \sum_{i=1}^{n_R} \sum_{j=1}^{p} (x_{ij} - \overline{x}_{Rj})^2
$$

$$
= n_R \sum_{j=1}^{p} \frac{1}{n_R} \sum_{i=1}^{n_R} (x_{ij} - \overline{x}_{Rj})^2 = n_R \sum_{j=1}^{p} s_{X_j X_j} = n_R \, \mathrm{tr} \, \mathcal{S}_R.
$$

EXERCISE 11.4. *Explain the differences between various proximity measures by means of the 8 points example given in Härdle & Simar (2003, example 11.5).*

The eight points from Example 11.5 in Härdle & Simar (2003) are plotted in Figure 11.1. Selected distances between some of the points are marked by lines. Different proximity measures assign different values to these interpoint distances. It is clear that the choice of the proximity measure can influence the behavior of the clustering algorithm.

In Figure 11.2, we plot the dendrograms obtained for the eight points example using two different simple distances. In both dendrograms, we can see how

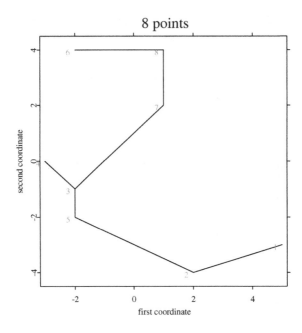

8 points

Fig. 11.1. 8 points example using single linkage. `SMSclus8pd`

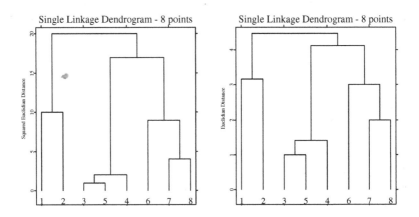

Fig. 11.2. Single linkage using squared Euclidean and Euclidean distance. `SMSclus8pd`

the n points were consecutively joined into only one cluster. The intercluster distances are given on the vertical axis. In both plots in Figure 11.2 we can see that in the first step of the algorithm, the points 3 and 5 were combined. Both the Euclidean and squared Euclidean distance between these points is equal to 1, see also Figure 11.1.

The distance in the right plot of Figure 11.2 is equal to the square root of the distance in the left plot. Thanks to the single linkage algorithm which defines the intercluster distance as the distance between closest points, we obtain exactly the same clustering in both plots. The only difference is the change of scale on the vertical axis.

The last step in cluster analysis is the choice of a number of cluster. For example, three clusters in the 8 points example can be obtained by cutting the dendrogram given in Figure 11.2 at a specified level. In this case, we would obtain clusters $\{1, 2\}$, $\{3, 4, 5\}$, and $\{6, 7, 8\}$.

EXERCISE 11.5. *Repeat the 8 point example (Exercise 11.4) using the complete linkage and the Ward algorithm. Explain the difference to single linkage.*

The dendrograms obtained by complete linkage and Ward method are plotted on the right hand side in Figures 11.3 and 11.4. The left plots contain the original points with lines describing the successive joining of the clusters.

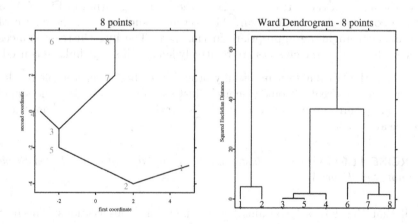

Fig. 11.3. Ward algorithm. ◻ SMSclus8p

The lines plotted in Figure 11.4 demonstrate how the intercluster distances are calculated in the complete linkage. For example, the line connecting points

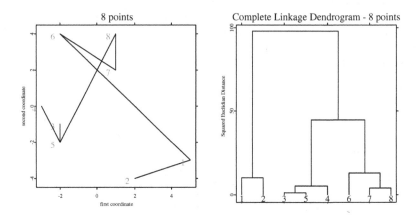

Fig. 11.4. Complete linkage. ⊙ SMSclus8p

5 and 8 gives the distance between the clusters consisting of points {3,4,5} and {6,7,8}. In the single linkage method used in Exercise 11.4, the distance between these clusters would be given by the distance of the closest points, i.e., by the distance of points 3 and 7.

Comparing the dendrograms in Figures 11.2–11.4, we see that, in this example, the three clustering algorithms arrive to the same result. The only difference lies in the scale on the vertical axis. Both the Ward algorithm in Figure 11.3 and the complete linkage in Figure 11.4 strongly suggest that the choice of three clusters might be appropriate in this case. The intercluster distances between the same three clusters are relatively smaller if single linkage is used.

In practice, the Ward algorithm usually provides the best interpretable results since it tends to create "homogeneous" clusters. On the contrary, the single linkage algorithm often finds chains of observations which do not have any other clear structure.

EXERCISE 11.6. *Perform a cluster analysis for 20 randomly selected Swiss bank notes in Table A.2.*

Recall that the data set contains 200 6-dimensional observations. The first 100 observations correspond to genuine and the other half to counterfeit bank notes. Here, we use only a subsample of size 20 so that the resulting dendrogram in Figure 11.5 is still readable. On the left plot in Figure 11.5 we plot the first two principal components for the data set. From Chapter 9 we know that this is, in some sense, the best two-dimensional representation of the data set. One can observe that the plot consists of two point clouds: on the left hand

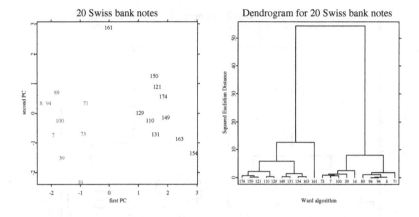

Fig. 11.5. Cluster analysis of 20 Swiss bank notes using Ward algorithm and squared Euclidean distance. ◘ SMSclusbank

side, we have the genuine bank notes with numbers smaller than 100 and, on the right hand side, we observe point cloud of the counterfeit bank notes. The observation 161 is a bit separated from both these groups.

The dendrogram, resulting from the Ward algorithm using the squared Euclidean distance, is plotted on the right hand side of Figure 11.5. If the dendrogram is cut to two clusters, we obtain exactly the genuine and counterfeit bank notes. The outlying observation 161 was correctly put into the counterfeit cluster but the dendrogram shows that the distance from the other counterfeit bank notes is largest from all (counterfeit) observations.

The dendrograms obtained by the single and complete linkage clustering algorithms are given in Exercise 11.7.

EXERCISE 11.7. *Repeat the cluster analysis of the bank notes example in Exercise 11.6 with single and complete linkage clustering algorithms.*

The dendrograms for both the single and complete linkage are plotted in Figures 11.6 and 11.7. The complete linkage plotted in Figure 11.6 provides better result since it correctly puts the observation 161 into the counterfeit group. However, comparing the complete linkage and the dendrogram obtained by the Ward algorithm in Figure 11.5, the Ward distance seems to be more appropriate in this case.

The single linkage dendrogram in Figure 11.7 shows the chain building tendency of this method. The observations are usually added one by one and the result of this method often consists of two clusters: one containing almost all

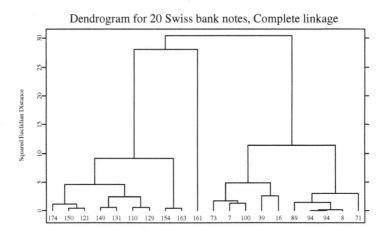

Fig. 11.6. Cluster analysis of 20 Swiss bank notes using squared Euclidean distance with complete linkage. ⓠ SMSclusbank2

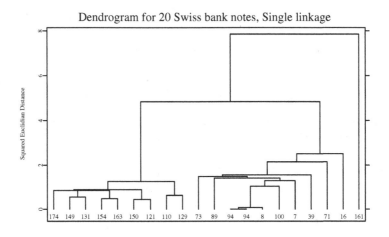

Fig. 11.7. Cluster analysis of 20 Swiss bank notes using squared Euclidean distance with single linkage. ⓠ SMSclusbank2

observations and the other one or two outliers. This is exactly what happened in Figure 11.7, where the outlying observation 161 was put into a cluster by itself.

EXERCISE 11.8. *Repeat the cluster analysis of the bank notes example in Exercise 11.6 using the L_1 distance.*

The Euclidean distance is just a special case of the L_r-norms, $r \geq 1$,

$$d_{ij} = ||x_i - x_j||_r = \left\{ \sum_{k=1}^{p} |x_{ik} - x_{jk}|^r \right\}^{1/r}, \qquad (11.1)$$

where x_{ik} denotes the value of the kth variable measured on the ith individual.

Apart of the usual Euclidean distance (L_2-norm), the L_1-norm is the most popular member of this family. The L_1 distance has very simple interpretation since from (11.1) it is easy to see that the L_1 distance is just the sum of the absolute values of the differences observed in each variable. The L_1 metric is useful whenever we want to assign less weight to the outlying observations.

In the previous exercises, it appeared that the Ward method leads to nice and interpretable results. Hence, we apply the Ward method with L_1 distance to obtain the dendrogram plotted in Figure 11.8. The same analysis with

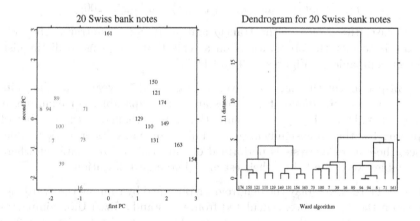

Fig. 11.8. Cluster analysis of 20 Swiss bank notes using Ward algorithm and L_1 distance. ⊠ SMSclusbank3

the squared Euclidean distance was carried out in Exercise 11.6. Instead of the squared Euclidean distance, we have now selected the L_1 distance which should assign less weight to outlying observations.

The overall shape of the dendrogram plotted in Figure 11.8 looks very similar to the dendrogram given in Figure 11.5. Again, the bank notes are clearly split into two groups. However, in Figure 11.5, the counterfeit observation 161 lies in one cluster with the genuine bank notes.

EXERCISE 11.9. *Analyze the U.S. companies data set in Table A.17 using the Ward algorithm and L_1 distance.*

The six dimensional data set contains the information on the assets, sales, market value, profits, cash flow and number of employees of 79 U.S. companies. The companies are classified according to their type: Communication, Energy, Finance, Hi-Tech, Manufacturing, Medical, Other, Retail, and Transportation.

In Figure 11.9, we plot the first two principal components for a rescaled version of the data set. The rescaling is in this case necessary since otherwise we observe most of the points concentrated in the lower left corner with the two largest companies (IBM and General Electric) dominating the plot. The transformation was used only for plotting in Figures 11.9 and 11.11 and the cluster analysis was performed using the L_1 distances calculated from the original data set.

The transformation which is used on all columns of the data set for plotting is

$$f(x) = \log[x - \min(x) + \{\max(x) - \min(x)\}/200].$$

In this case, the choice of the transformation is quite arbitrary. The only
. purpose is to plot the observations on a scale that allows us to distinguish different companies in Figures 11.9 and 11.11.

Short inspection of the data set given in Table A.17 reveals that the units of measurements for different variables are not comparable. For example, it would not make much sense to assume that a unit change in the number of employees has the same significance as a unit change in sales or market value. Hence, the cluster analysis is performed on the standardized data set where all variables were divided by their estimated standard deviation.

In Figure 11.10, we display the dendrogram obtained by running the Ward algorithm on the L_1 distances calculated from the standardized U.S. companies data set. From the graphics, it looks reasonable to split the data set into 3 or 5 clusters. In Figure 11.10, we give also the first two letter of the type of the company. It is interesting that in Figure 11.10, the same types of company are often close to each other. See, for example, the large groups of financial or energy companies. However, if we choose lower number of cluster, these groups are mixed with other types of companies.

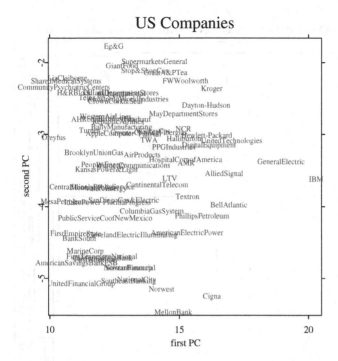

Fig. 11.9. Plot of the first two principal components for the rescaled U.S. companies data set. ◻ SMScluscomp

The resulting five clusters are plotted in Figure 11.11 where different plotting symbols were used for each cluster. The type of each company is also specified by the first two letters. Two hi-tech companies form a cluster by themselves: IBM and General Electric. In the upper part of Figure 11.11, we can observe a large group of retail companies. Unfortunately, the Ward algorithm puts this group into two different clusters. The same could be said for the group of financial companies visible in the lower left part of Figure 11.11.

The cluster analysis could be summarized in the following way: the clusters seem to split the data set mainly in the direction of the first principal component which seems to be related mainly to the size of the company. Hence, the clustering algorithm does not recover the type of company which seems to be better explained by the (less important) second principal component.

An improvement in clustering might be achieved also by transforming the data set before calculating the distance matrix used in the clustering algorithm. One possible transformation might be the logarithmic transformation

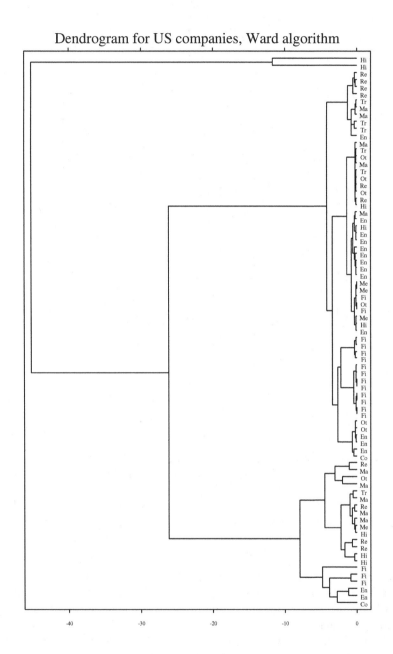

Fig. 11.10. Dendrogram for U.S. companies using Ward algorithm and L_1 distance.
Q SMScluscomp

Five Clusters for US Companies

Fig. 11.11. Plot of the first two principal components for the rescaled U.S. companies data set with five clusters denoted by different symbols. ▣ SMScluscomp

used for plotting in Figures 11.9 and 11.11 or possibly another transformation correcting for the effect of the size of the company.

EXERCISE 11.10. *Analyze the U.S. crime data set in Table A.18 with the Ward algorithm. Use the χ^2-metric measuring differences between rows of a contingency table and compare the results to the usual L_2-norm on standardized variables.*

The U.S. crime data set contains the reported number of 7 types of crimes in the 50 USA states. The entries in this data set can be interpreted as counts and the data set as a (50×7) contingency table.

In a given contingency table, the ith row can be interpreted as the conditional frequency distribution $\frac{x_{ik}}{x_{i\bullet}}$, $k = 1, \ldots, p$, where $x_{i\bullet} = \sum_{j=1}^{p} x_{ij}$. The distance between the ith and jth row can be defined as a χ^2 distance between the respective frequency distributions:

$$d^2(i,j) = \sum_{k=1}^{p} \frac{1}{\left(\frac{x_{\bullet k}}{x_{\bullet \bullet}}\right)} \left(\frac{x_{ik}}{x_{i\bullet}} - \frac{x_{jk}}{x_{j\bullet}}\right)^2,$$

see, e.g., Härdle & Simar (2003, section 11.2).

χ^2 *Distance*

The χ^2 distances between the rows (observations) in the U.S. crime data set are used to construct the distance matrix. The dendrogram plotted in Figure 11.12 was obtained by the Ward method. Each observation displayed

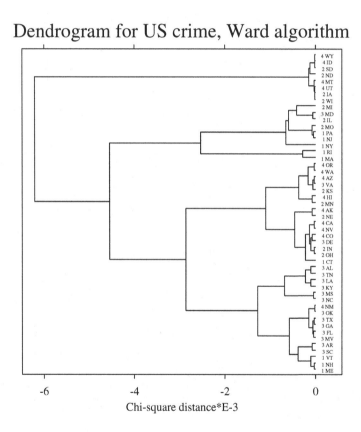

Fig. 11.12. Cluster analysis of U.S. crime data set using Ward algorithm and χ^2 distance. ⌕ SMScluscrimechi2

in the dendrogram in Figure 11.12 is marked by the abbreviation of the state and by the region number (1=Northeast, 2=Midwest, 3=South, 4=West).

The dendrogram suggests that it would be reasonable to split the data set into 5 or 7 clusters. Let us try to consider 5 clusters and let us define cluster one as ME, NH, VT, MV, NC, SC, GA, FL, KY, TN, AL, MS, AR, LA, OK, TX, and NM. Cluster 2 consists of CT, OH, IN, MN, NE, KS, DE, VA, CO, AZ, NV, WA, OR, CA, AK, and HI, cluster 3 contains MA and RI, cluster 4 NY, NJ, PA, IL, MI, MO, and MD. Cluster 5 is WI, IA, ND, SD, MT, ID, WY, UT. In Table 11.1, we give the average relative frequencies within the five clusters. The information given in Table 11.1 allows us to describe

	murder	rape	robbery	assault	burglary	larceny	auto theft
1	0.00	0.01	0.02	0.06	0.30	0.52	0.09
2	0.00	0.00	0.02	0.03	0.26	0.57	0.11
3	0.00	0.00	0.02	0.03	0.27	0.46	0.22
4	0.00	0.00	0.06	0.04	0.27	0.49	0.13
5	0.00	0.00	0.01	0.02	0.21	0.70	0.06

Table 11.1. The average relative frequencies for U.S. crimes within the 5 clusters obtained with χ^2 distance. ⊙ SMScluscrimechi2

the differences between the clusters. It seems that larceny is very "popular" mainly in cluster 5 consisting mainly of only from West and Midwest states (region code 4). Auto theft is relatively more spread out in cluster 3 consisting only from Massachusetts and Richmond. Cluster 4 (NY, NJ, ...) contains more robberies. Cluster 1, consisting mainly of southern states (region code 3), slightly overrepresents rape and burglaries.

Euclidean Distance

The results of the Ward algorithm performed on the Euclidean distances between standardized observations are summarized in Figure 11.13 and Table 11.2. Here, we have chosen to consider four clusters.

The first cluster contains the states: ME, NH, VT, PA, WI, IA, ND, SD, NE, MV, MT, ID, and WY. The second cluster is MA, RI, CT, NJ, OH, IN, MN, KS, UT, WA, OR, and HI. The third cluster consists of VA, NC, SC, GA, KY, TN, AL, MS, AR, and OK. The fourth cluster contains NY, IL, MI, MO, DE, MD, FL, LA, TX, CO, NM, AZ, NV, CA, and AK. From the regional point of view, it is interesting to notice that the third cluster contains only southern states.

Table 11.2 allows us to describe the differences between clusters. Cluster 1 contains the states with low criminality since the average of the standardized

Dendrogram for US crime, Ward algorithm

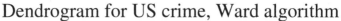

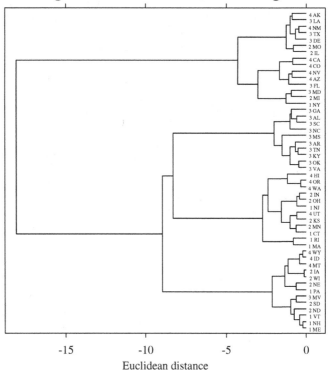

Euclidean distance

Fig. 11.13. Cluster analysis of U.S. crime data set using Ward algorithm and Euclidean distance. ◘ SMScluscrime

	murder	rape	robbery	assault	burglary	larceny	auto theft
1	−0.96	−0.91	−0.80	−1.03	−1.07	−0.70	−0.97
2	−0.72	−0.37	−0.06	−0.63	0.37	0.40	0.62
3	1.06	−0.14	−0.43	0.55	−0.40	−0.82	−0.66
4	0.70	1.18	1.03	1.03	0.91	0.83	0.78

Table 11.2. The averages of the standardized U.S. crime data set within the 3 clusters obtained with Euclidean distance. ◘ SMScluscrime

number of all crimes is negative. On the other side, cluster 4 contains the states with high criminality rate. Cluster 2 corresponds to states with a tendency towards burglary, larceny, and auto theft. The souther cluster 3 has large rates of murder and assault.

Comparison

We have seen that each distance leads to another view at the data set. The χ^2 distance compares relative frequencies whereas the Euclidean distance compares the absolute values of the number of each crime. The choice of the method depends in practice mainly on the point of view of the investigator.

EXERCISE 11.11. *Perform the cluster analysis of the U.S. health data set in Table A.19.*

The description of the U.S. health data set is given in Appendix A.19. Basically, it contains the number of deaths in 50 U.S. states classified according to 7 causes of death. We are interested in the numbers of deaths and hence we have decided to perform the analysis using Euclidean analysis on the original data set. The resulting dendrogram is plotted in Figure 11.14.

Cluster 1 contains ME, MA, RI, NY, NJ, PA, IA, MO, SD, NE, MV, FL, and AR. Cluster 2 consists of VT, CT, OH, IN, IL, MI, WI, KS, DE, KY, TN, AL, MS, and OK. Cluster 3 is NH, MN, ND, MD, VA, NC, SC, GA, LA, TX, MT, ID, AZ, NV, WA, OR, and CA and the last cluster 4 consists of WY, CO, NM, UT, AK, and HI. Cluster 4 contains only western states (region code 4). The other three clusters are regionally less homogeneous.

	acc	card	canc	pul	pneu	diab	liv
1	39.56	484.70	210.73	29.35	23.87	16.95	11.78
2	42.48	432.56	189.33	26.41	20.69	16.29	9.99
3	45.55	365.65	168.25	26.16	20.54	13.52	10.48
4	55.37	225.58	111.68	21.37	17.13	10.58	9.38

Table 11.3. The averages of the U.S. health data set within the 4 clusters. Q SMSclushealth

The differences between clusters are summarized in Table 11.3. It seems that most of the differences are due to the number of deaths due to cancer and cardiovascular problems, i.e., to the most common causes of deaths.

In Figure 11.15, we plot the first two principal components. The observations belonging to the four different clusters are plotted using different text size. Obviously, the cluster separated the observations according to their position

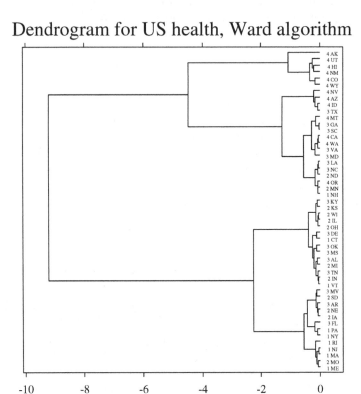

Fig. 11.14. Cluster analysis of U.S. health data set using Ward algorithm and Euclidean distance. ⊙ SMSclushealth

on the horizontal axis of the plot, i.e., according to the value of the first principal component, see also the principal component analysis in Exercise 9.9.

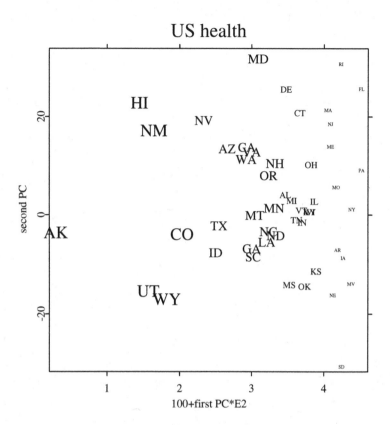

Fig. 11.15. Plot of the first two principal components of the U.S. health data. The size of the symbols is given by the clustering using Ward algorithm and Euclidean distance. ⊡ SMSclushealth

Discriminant Analysis

...if a gentleman walks into my rooms smelling of iodoform, with a black mark of nitrate of silver upon his right fore-finger, and a bulge on the side of his top-hat to show where he has secreted his stethoscope, I must be dull indeed, if I do not pronounce him to be an active member of the medical profession.
Sherlock Holmes in "A Scandal in Bohemia"

Discriminant analysis is used in situations where the clusters are known a priori. The aim of discriminant analysis is to classify an observation, or several observations, into these known groups. For instance, in credit scoring, a bank knows from past experience that there are good customers (who repay their loan without any problems) and bad customers (who have had difficulties repaying their loans). When a new customer asks for a loan, the bank has to decide whether or not to give the loan. The information of the bank is given in two data sets: multivariate observations on the two categories of customers (including age, salary, marital status, the amount of the loan, and the like).

The discrimination rule has to classify the customer into one of the two existing groups, and the discriminant analysis should evaluate the risk of a possible misclassification. Many other examples are described herein. We present ML discrimination and Fisher's linear discrimination function.

In the mathematical formulation of the problem, we try to allocate an observation to one of the populations $\Pi_j, j = 1, 2, ..., J$. A discriminant rule is a separation of the sample space (in general $\mathbb{R}^p$) into disjoint sets R_j such that if a new observation falls into the region R_j, it is identified as a member of population Π_j.

The quality of a discriminant rule can be judged on the basis of the error of misclassification.

If the probability density functions in the populations Π_j are known, we may easily derive a discriminant rule based on the maximum likelihood approach.

Maximum Likelihood Discriminant Rule

Let us assume that each population Π_j, $j = 1, \ldots, J$, can be described by a probability density function (pdf) $f_j(x)$.

The maximum likelihood discriminant rule (ML rule) allocates the new observation x to the population Π_k, maximizing the likelihood $L_k(x) = f_k(x) = \max_{i=1,\ldots,J} f_i(x)$.

Formally, the sets R_j, $j = 1, \ldots, J$, given by the ML discriminant rule are:

$$R_j = \{x : f_j(x) \geq f_i(x) \text{ for } i = 1, \ldots, J\}.$$

In practice, the sets R_j are constructed from estimates of the unknown densities. If the densities are assumed to have a known shape, i.e., normal distribution, it suffices to estimate the unknown parameters; see Exercise 12.1.

Bayes Discriminant Rule

The quality of the ML discriminant rule may be improved if some prior information about the probability of the populations is known. Let π_j denote the prior probability of class j. Note that $\sum_{j=1}^{J} \pi_j = 1$.

The Bayes discriminant rule allocates x to the population Π_k that gives the largest value of $\pi_i f_i(x)$, $\pi_k f_k(x) = \max_{i=1,\ldots,J} \pi_i f_i(x)$. The Bayes discriminant rule can be formally defined by:

$$R_j = \{x : \pi_j f_j(x) \geq \pi_i f_i(x) \text{ for } i = 1, \ldots, J\}.$$

The Bayes rule is identical to the ML discriminant rule if $\pi_j = 1/J$.

Fisher's Linear Discrimination Function

The classical Fisher's linear discriminant rule is based on the maximization of the ratio of the between to the within variance of a projection $a^\top x$.

Suppose we have samples $\mathcal{X}_j$, $j = 1, \ldots, J$, from J populations. Let $\mathcal{Y} = \mathcal{X}a$ and $\mathcal{Y}_j = \mathcal{X}_j a$ denote linear combinations of observations. The within-group sum of squares is given by

$$\sum_{j=1}^{J} \mathcal{Y}_j^\top \mathcal{H}_j \mathcal{Y}_j = \sum_{j=1}^{J} a^\top \mathcal{X}_j^\top \mathcal{H}_j \mathcal{X}_j a = a^\top \mathcal{W}a, \tag{12.1}$$

where $\mathcal{H}_j$ denotes the $(n_j \times n_j)$ centering matrix. The between-group sum of squares is

$$\sum_{j=1}^{J} n_j (\bar{y}_j - \bar{y})^2 = \sum_{j=1}^{J} n_j \{a^\top (\bar{x}_j - \bar{x})\}^2 = a^\top \mathcal{B} a, \qquad (12.2)$$

where $\bar{y}_j$ and $\bar{x}_j$ denote the means of $\mathcal{Y}_j$ and $\mathcal{X}_j$ and $\bar{y}$ and $\bar{x}$ denote the sample means of $\mathcal{Y}$ and $\mathcal{X}$.

Fisher noticed that the vector a that maximizes $a^\top \mathcal{B} a / a^\top \mathcal{W} a$ is the eigenvector of $\mathcal{W}^{-1} \mathcal{B}$ that corresponds to the largest eigenvalue.

Finally, observation x is classified into group j, which is closest to the projected $a^\top x$,

$$R_j = \{x : |a^\top (x - \bar{x}_j)| \leq |a^\top (x - \bar{x}_i)| \text{ for } i = 1, \ldots, J\}.$$

EXERCISE 12.1. *Derive the ML discriminant rule if $\Pi_j = N_p(\mu_j, \Sigma)$, $j = 1, \ldots, J$. Discuss the special case $J = 2$.*

Let us assume that the variance matrix Σ is positive definite. The likelihood of observation x in each of the populations Π_j, $j = 1, \ldots, J$ is

$$L_j(x) = f_j(x) = |2\pi \Sigma|^{-1/2} \exp \left\{ -\frac{1}{2} (x - \mu_j)^\top \Sigma^{-1} (x - \mu_j) \right\}.$$

According to the ML rule, we allocate x to the population Π_j with the largest likelihood. Omitting the constant $|2\pi \Sigma|^{-1/2}$ and taking logarithms, the maximization problem may be equivalently solved by minimizing

$$\delta^2(x, \mu_j) = (x - \mu_j)^\top \Sigma^{-1} (x - \mu_j)$$
$$= \{\Sigma^{-1/2} (x - \mu_j)\}^\top \Sigma^{-1/2} (x - \mu_j).$$

Clearly, $\delta^2(x, \mu_j)$ is the square of the Mahalanobis distance between x and μ_j, see also Exercise 9.7 for the discussion of the Mahalanobis transformation.

Hence, in case of normal distribution with common covariance matrix, the ML rule allocates x to the closest group in the Mahalanobis sense.

For $J = 2$, the observation x is allocated to Π_1 if

$$(x - \mu_1)^\top \Sigma^{-1} (x - \mu_1) \leq (x - \mu_2)^\top \Sigma^{-1} (x - \mu_2).$$

Rearranging terms leads to

$$0 \geq -2\mu_1^\top \Sigma^{-1} x + 2\mu_2^\top \Sigma^{-1} x + \mu_1^\top \Sigma^{-1} \mu_1 - \mu_2^\top \Sigma^{-1} \mu_2$$
$$0 \geq 2(\mu_2 - \mu_1)^\top \Sigma^{-1} x + (\mu_1 - \mu_2)^\top \Sigma^{-1} (\mu_1 + \mu_2)$$
$$0 \leq (\mu_1 - \mu_2)^\top \Sigma^{-1} \{x - \frac{1}{2} (\mu_1 + \mu_2)\}$$
$$0 \leq \alpha^\top (x - \mu),$$

where $\alpha = \Sigma^{-1}(\mu_1 - \mu_2)$ and $\mu = \frac{1}{2}(\mu_1 + \mu_2)$.

It follows that in case of two multinormal populations, the discriminant rule can be written as:

$$R_1 = \{x : \alpha^\top (x - \mu) \geq 0\}.$$

EXERCISE 12.2. *Apply the rule from Exercise 12.1 for $J = 2$ and $p = 1$ and modify it for unequal variances.*

For two univariate normally distributed populations $\Pi_1 = N(\mu_1, \sigma)$ and $\Pi_2 = N(\mu_2, \sigma)$, the ML rule can be written as

$$R_1 = \left\{x : (\mu_1 - \mu_2)\left(x - \frac{\mu_1 + \mu_2}{2}\right) \geq 0\right\}$$

$$R_1 = \left\{x : \text{sign}(\mu_1 - \mu_2)\left(x - \frac{\mu_1 + \mu_2}{2}\right) \geq 0\right\}$$

$$R_1 = \left\{x : \text{sign}(\mu_1 - \mu_2)x \geq \text{sign}(\mu_1 - \mu_2)\frac{\mu_1 + \mu_2}{2}\right\}.$$

Assuming that $\mu_1 < \mu_2$, we obtain

$$R_1 = \left\{x : x \leq \frac{\mu_1 + \mu_2}{2}\right\},$$

i.e., we classify x to R_1 if it is closer to μ_1 than to μ_2.

Assuming that the two normal populations have different variances, $\Pi_1 = N(\mu_1, \sigma_1^2)$ and $\Pi_2 : N(\mu_2, \sigma_2^2)$, we allocate x to R_1 if $L_1(x) > L_2(x)$, where the likelihood is:

$$L_i(x) = (2\pi\sigma_i^2)^{-1/2} \exp\left\{-\frac{1}{2}\left(\frac{x - \mu_i}{\sigma_i}\right)^2\right\}.$$

$L_1(x) \geq L_2(x)$ is equivalent to $L_1(x)/L_2(x) \geq 1$ and we obtain

$$\frac{\sigma_2}{\sigma_1} \exp\left\{-\frac{1}{2}\left[\left(\frac{x - \mu_1}{\sigma_1}\right)^2 - \left(\frac{x - \mu_2}{\sigma_2}\right)^2\right]\right\} \geq 1$$

$$\log\frac{\sigma_2}{\sigma_1} - \frac{1}{2}\left[\left(\frac{x - \mu_1}{\sigma_1}\right)^2 - \left(\frac{x - \mu_2}{\sigma_2}\right)^2\right] \geq 0$$

$$\frac{1}{2}\left[\left(\frac{x - \mu_1}{\sigma_1}\right)^2 - \left(\frac{x - \mu_2}{\sigma_2}\right)^2\right] \leq \log\frac{\sigma_2}{\sigma_1}$$

$$x^2\left(\frac{1}{\sigma_1^2} - \frac{1}{\sigma_2^2}\right) - 2x\left(\frac{\mu_1}{\sigma_1^2} - \frac{\mu_2}{\sigma_2^2}\right) + \left(\frac{\mu_1^2}{\sigma_1^2} - \frac{\mu_2^2}{\sigma_2^2}\right) \leq 2\log\frac{\sigma_2}{\sigma_1}.$$

If $\sigma_1 = \sigma_2$, most of the terms in the above formula disappear and the result simplifies to the discriminant rule obtained in Exercise 12.1.

EXERCISE 12.3. *Calculate the ML discrimination rule based on observations of a one-dimensional variable with an exponential distribution.*

The pdf of the exponential distribution $Exp(\lambda)$ is:

$$f(x) = \lambda \exp\{-\lambda x\} \text{ for } x > 0.$$

Comparing the likelihoods for two populations $\Pi_1 = Exp(\lambda_1)$ and $\Pi_2 = Exp(\lambda_2)$, we allocate the observation x into population Π_1 if

$$L_1(x) \geq L_2(x)$$
$$L_1(x)/L_2(x) \geq 1$$
$$\frac{\lambda_1}{\lambda_2} \exp\{-x(\lambda_1 - \lambda_2)\} \geq 1$$
$$\log \frac{\lambda_1}{\lambda_2} - x(\lambda_1 - \lambda_2) \geq 0$$
$$x(\lambda_1 - \lambda_2) \leq \log \frac{\lambda_1}{\lambda_2}.$$

Assuming that $\lambda_1 < \lambda_2$, we obtain the discriminant rule:

$$R_1 = \left\{ x : x \geq \frac{\log \lambda_1 - \log \lambda_2}{\lambda_1 - \lambda_2} \right\}.$$

The observation x is classified into population Π_1 if it is greater than the constant $(\log \lambda_1 - \log \lambda_2)/(\lambda_1 - \lambda_2)$.

EXERCISE 12.4. *Calculate the ML discrimination rule based on observations of a two-dimensional random vector, where the first component has an exponential distribution and the other has an alternative distribution. What is the difference between the discrimination rule obtained in this exercise and the Bayes discrimination rule?*

Let us assume that the two populations, $\Pi_1 = \{Exp(\lambda_1), Alt(p_1)\}^\top$ and $\Pi_2 = \{Exp(\lambda_2), Alt(p_2)\}^\top$, are characterized by the exponential distribution with parameter λ_j and the alternative distribution with parameter p_j, $j = 1, 2$. The corresponding likelihood can be written as:

$$L_j(x_1, x_2) = \lambda_j \exp(-\lambda_j x_1)\{p_j x_2 + (1 - p_j)(1 - x_2)\}.$$

Since x_2 has the alternative distribution, it can have only two possible outcomes.

Assuming that $x_2 = 1$, we allocate the observation $(x_1, x_2)^\top$ to Π_1 if $L_1(x_1, 1) \geq L_2(x_1, 1)$, i.e.,

$$L_1(x_1, 1)/L_2(x_1, 1) \geq 1$$

$$\frac{\lambda_1 p_1}{\lambda_2 p_2} \exp\left\{-x_1(\lambda_1 - \lambda_2)\right\} \geq 1$$

$$\log \frac{\lambda_1 p_1}{\lambda_2 p_2} - x_1(\lambda_1 - \lambda_2) \geq 0$$

$$x_1(\lambda_1 - \lambda_2) \leq \log \frac{\lambda_1 p_1}{\lambda_2 p_2}$$

Similarly, if $x_2 = 0$, we allocate the observation $(x_1, x_2)^\top$ to Π_1 if

$$x_1(\lambda_1 - \lambda_2) \leq \log \frac{\lambda_1(1 - p_1)}{\lambda_2(1 - p_2)}.$$

Combining both cases and assuming that $\lambda_1 < \lambda_2$, the discriminant rule R_1 can be written as:

$$\left\{ \binom{x_1}{x_2} : x_1 \geq \frac{\lambda_1\{x_2 p_1 + (1 - x_2)(1 - p_1)\} - \lambda_2\{x_2 p_2 + (1 - x_2)(1 - p_2)\}}{\lambda_1 - \lambda_2} \right\}.$$

If the prior probabilities of $\Pi_1 = Exp(\lambda_1)$ and $\Pi_2 = Exp(\lambda_2)$ are π_1 and $\pi_2 = 1 - \pi_1$, respectively, the Bayes rule can be derived by comparing $\pi_i L_i(x)$, $i = 1, 2$, exactly as in Exercise 12.3:

$$R_1 = \left\{ x : x \geq \frac{\log \pi_1 \lambda_1 - \log \pi_2 \lambda_2}{\lambda_1 - \lambda_2} \right\}.$$

Now, it is easy to see that the conditional discriminant rule obtained for the two dimensional random vector under the condition $x_2 = 1$ is equivalent to the Bayes discriminant rule for exponential distribution with $\pi_1 = p_1/(p_1 + p_2)$. Similarly, the conditional discriminant rule if $x_2 = 0$ is a Bayes discriminant rule with $\pi_1 = (1 - p_1)/(2 - p_1 - p_2)$.

EXERCISE 12.5. *Apply the Bayes rule to the car data in Table A.4 in order to discriminate between U.S., Japanese, and European cars. Consider only the variable milage (miles per gallon) and take the relative frequencies as prior probabilities.*

The three regions of origins in the data set are denoted by numbers 1, 2, and 3 standing for U.S., Japanese, and European cars, respectively. Based on the 74 observations given in Table A.4, we will construct a discriminant rule that would allow us to classify a new (75th) car with unknown origin.

Let us start with the maximum likelihood discriminant rule. Usually, the ML rule is based upon assumptions of normality. However, plots of the observed milage suggest that the normality is violated. Hence, instead of mileage measured in miles per gallon, we analyze fuel efficiency measured in liters per

100 kilometers. The averages in U.S., Japan, and Europe are: $\bar{x}_1 = 12.5207$, $\bar{x}_2 = 9.4577$, $\bar{x}_3 = 10.7712$. On average, Japanese cars (group 2) are more fuel efficient than European and U.S. cars.

The ML discriminant rule is calculated according to the description given in Exercise 12.1. In Figure 12.1, we plot the three point clouds corresponding to

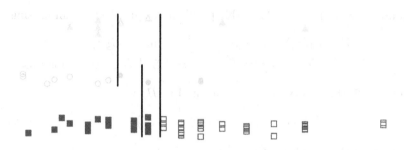

Fig. 12.1. Discrimination of the three regions according to "liters per 100km" with the ML discriminant rule. ◻ SMSdisccar

the three regions and, as vertical lines, we show also the points that separate the discriminant rules R_1, R_2, and R_3. The lowest point cloud (squares) in Figure 12.1 contains U.S. cars, the middle point (circles) cloud the Japanese, and the top point cloud (triangles) the European cars. The correctly classified cars are denoted by empty symbols whereas the filled symbols denote misclassified cars. The counts are given in Table 12.1.

	R_1: U.S.	R_2: JPN	R_3: EUR
Group 1 (U.S.)	33	11	8
Group 2 (Japanese)	2	7	2
Group 3 (European)	3	5	3

Table 12.1. The true region of origins and the region suggested by the ML discriminant rule based on fuel efficiency. The number of correct classifications for each region is given on the diagonal of the table.

The apparent error rate (APER), defined as the percentage of misclassified observations is $(11 + 8 + 2 + 2 + 3 + 5)/79 = 41.89\%$. It seems that the rule is not particularly good since we have less than 60% chance of correct classification. Moreover, this estimate is based on the observations which were used to construct the discriminant rule and it might be way too optimistic.

Let us now consider the Bayes rule which is based on the comparison of the likelihoods weighted by the prior probabilities of the groups. More formally, we allocate the new observation x to the population Π_j maximizing

$$\pi_j L_j(x) = \pi_j f_j(x) = \pi_j |2\pi \Sigma|^{-1/2} \exp\left\{ -\frac{1}{2}(x - \mu_j)^\top \Sigma^{-1}(x - \mu_j) \right\},$$

where $\pi_j, j = 1, \ldots, J$ are the prior probabilities of the respective populations. Similarly as in Exercise 12.1, this problem is equivalent to minimizing

$$\begin{aligned} \delta^2(x, \mu_j, \pi_j) &= (x - \mu_j)^\top \Sigma^{-1}(x - \mu_j) - \log \pi_j \\ &= \{\Sigma^{-1/2}(x - \mu_j)\}^\top \Sigma^{-1/2}(x - \mu_j) - \log \pi_j. \end{aligned}$$

For $J = 2$, the observation x is allocated to Π_1 if

$$(x - \mu_1)^\top \Sigma^{-1}(x - \mu_1) - \log \pi_1 \le (x - \mu_2)^\top \Sigma^{-1}(x - \mu_2) - \log \pi_2.$$

Rearranging terms leads to

$$\begin{aligned} \log \pi_1 - \log \pi_2 &\ge -2\mu_1^\top \Sigma^{-1} x + 2\mu_2^\top \Sigma^{-1} x + \mu_1^\top \Sigma^{-1}\mu_1 - \mu_2^\top \Sigma^{-1}\mu_2 \\ \log \pi_1 - \log \pi_2 &\ge 2(\mu_2 - \mu_1)^\top \Sigma^{-1} x + (\mu_1 - \mu_2)^\top \Sigma^{-1}(\mu_1 + \mu_2) \\ \log \pi_2 - \log \pi_1 &\le (\mu_1 - \mu_2)^\top \Sigma^{-1}\{x - \frac{1}{2}(\mu_1 + \mu_2)\} \\ \log \frac{\pi_2}{\pi_1} &\le \alpha^\top (x - \mu), \end{aligned}$$

where $\alpha = \Sigma^{-1}(\mu_1 - \mu_2)$ and $\mu = \frac{1}{2}(\mu_1 + \mu_2)$. Hence, the Bayes discriminant rule can be written as:

$$R_1 = \left\{ x : \alpha^\top (x - \mu) \ge \log \frac{\pi_2}{\pi_1} \right\}.$$

In our car data example, we use the relative frequencies observed in the data set, $\pi_1 = 0.7027$, $\pi_2 = 0.1486$, $\pi_3 = 0.1486$, as the prior probabilities.

The resulting discriminant rule is graphically displayed in Figure 12.2. Notice that with these weights, it is impossible to classify any new observation as a European car.

The same results are given in Table 12.2. The apparent error rate is equal to 28.38%. Obviously, the Bayes discriminant rule leads to better results since it give large weights to U.S. cars which constitute more than 60% of the entire data set.

EXERCISE 12.6. *Derive simple expressions for matrices W and B and the Fisher discriminant rule in the setup of the Swiss bank notes data set given in Table A.2.*

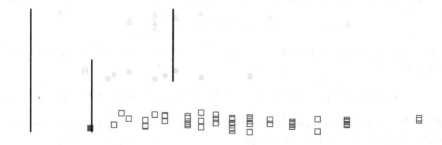

Fig. 12.2. Discrimination of the three regions according to "liters per 100km" using the Bayes rule. ▣ SMSdiscbaycar

	R_1: U.S.	R_2: JPN	R_3: EUR
Group 1 (U.S.)	51	1	0
Group 2 (Japanese)	9	2	0
Group 3 (European)	10	1	0

Table 12.2. The true region of origins and the region suggested by the Bayes discriminant rule based on fuel efficiency. The number of correct classifications for each region is given on the diagonal of the table.

The Swiss bank notes data set, $\mathcal{X}$, contains six measurements taken on 100 genuine and 100 counterfeit bank notes. Let us denote the measurements taken on genuine and counterfeit by $\mathcal{X}_g$ and $\mathcal{X}_f$, respectively. The corresponding linear combinations are $\mathcal{Y} = \mathcal{X}a$, $\mathcal{Y}_g = \mathcal{X}_g a$, and $\mathcal{Y}_f = \mathcal{X}_f a$.

The within-group sum of squares (12.1) satisfies the relation

$$\mathcal{Y}_f^\top \mathcal{H}_f \mathcal{Y}_f + \mathcal{Y}_g^\top \mathcal{H}_g \mathcal{Y}_g = a^\top \mathcal{W}a,$$

where $\mathcal{H}_f$ and $\mathcal{H}_g$ denote the appropriate centering matrices of dimensions $n_f = n_g = 100$. Observe that

$$a^\top \mathcal{W}a = a^\top (\mathcal{X}_f^\top \mathcal{H}_f \mathcal{X}_f + \mathcal{X}_g^\top \mathcal{H}_g \mathcal{X}_f)a$$

and, hence, the matrix $\mathcal{W}$ can be written as:

$$\mathcal{W} = \mathcal{X}_f^\top \mathcal{H}_f \mathcal{X}_f + \mathcal{X}_g^\top \mathcal{H}_g \mathcal{X}_g = \mathcal{H}_f \mathcal{X}_f^\top \mathcal{H}_f \mathcal{X}_f + \mathcal{H}_g \mathcal{X}_g^\top \mathcal{H}_g \mathcal{X}_g$$
$$= n_f \mathcal{S}_f + n_g \mathcal{S}_g = 100(\mathcal{S}_f + \mathcal{S}_g),$$

where $\mathcal{S}_g$ and $\mathcal{S}_f$ denote the empirical covariances w.r.t. the genuine and counterfeit bank notes.

For the between-group sum of squares (12.2) we have

$$a^\top \mathcal{B} a = n_f(\overline{y}_f - \overline{y})^2 + n_g(\overline{y}_g - \overline{y})^2,$$

where $\overline{y}$, $\overline{y}_f$, and $\overline{y}_g$ denote respectively the sample means of $\mathcal{Y}$, $\mathcal{Y}_f$, and $\mathcal{Y}_g$. It follows that

$$a^\top \mathcal{B} a = a^\top \{ n_f(\overline{x}_f - \overline{x})(\overline{x}_f - \overline{x})^\top + n_g(\overline{x}_g - \overline{x})(\overline{x}_g - \overline{x})^\top \} a,$$

where $\overline{x}$, $\overline{x}_f$, and $\overline{x}_g$ denote respectively the column vectors of sample means of $\mathcal{X}$, $\mathcal{X}_f$, and $\mathcal{X}_g$. Hence, we obtain

$$
\begin{aligned}
\mathcal{B} &= n_f(\overline{x}_f - \overline{x})(\overline{x}_f - \overline{x})^\top + n_g(\overline{x}_g - \overline{x})(\overline{x}_g - \overline{x})^\top \\
&= 100\{ (\overline{x}_f - \overline{x})(\overline{x}_f - \overline{x})^\top + (\overline{x}_g - \overline{x})(\overline{x}_g - \overline{x})^\top \} \\
&= 100 \left\{ \left(\overline{x}_f - \frac{\overline{x}_f + \overline{x}_g}{2} \right) \left(\overline{x}_f - \frac{\overline{x}_f + \overline{x}_g}{2} \right)^\top \right. \\
&\quad \left. + \left(\overline{x}_g - \frac{\overline{x}_f + \overline{x}_g}{2} \right) \left(\overline{x}_g - \frac{\overline{x}_f + \overline{x}_g}{2} \right)^\top \right\} \\
&= 25(\overline{x}_f - \overline{x}_g)(\overline{x}_f - \overline{x}_g)^\top.
\end{aligned}
$$

The vector a maximizing the ratio $a^\top \mathcal{B} a / a^\top \mathcal{W} a$ can be calculated as the eigenvector of $\mathcal{W}^{-1}\mathcal{B}$ corresponding to the largest eigenvalue, see Härdle & Simar (2003, theorem 12.4).

For the Swiss bank notes, it is easy to see that the matrix $\mathcal{W}^{-1}\mathcal{B}$ can have at most one nonzero eigenvalue since rank $\mathcal{B} \leq 1$. The nonzero eigenvalue λ_1 can be calculated as:

$$
\begin{aligned}
\lambda_1 = \sum_{j=1}^{p} \lambda_j &= \operatorname{tr} \mathcal{W}^{-1}\mathcal{B} = \operatorname{tr} \mathcal{W}^{-1}25(\overline{x}_f - \overline{x}_g)(\overline{x}_f - \overline{x}_g)^\top \\
&= 25\operatorname{tr}(\overline{x}_f - \overline{x}_g)^\top \mathcal{W}^{-1}(\overline{x}_f - \overline{x}_g) = 25(\overline{x}_f - \overline{x}_g)^\top \mathcal{W}^{-1}(\overline{x}_f - \overline{x}_g).
\end{aligned}
$$

From the equation:

$$\mathcal{W}^{-1}\mathcal{B}\mathcal{W}^{-1}(\overline{x}_f - \overline{x}_g) = 25(\overline{x}_f - \overline{x}_g)^\top \mathcal{W}^{-1}(\overline{x}_f - \overline{x}_g)\mathcal{W}^{-1}(\overline{x}_f - \overline{x}_g)$$

it follows that the eigenvector of $\mathcal{W}^{-1}\mathcal{B}$ corresponding to the largest eigenvalue is $a = \mathcal{W}^{-1}(\overline{x}_f - \overline{x}_g)$. Assuming that $\overline{y}_f > \overline{y}_g$, the corresponding discriminant rule can be formally written as:

$$R_f = \{ x : (\overline{x}_f - \overline{x}_g)^\top \mathcal{W}^{-1}(x - \overline{x}) \geq 0 \}.$$

EXERCISE 12.7. *Compute Fisher's linear discrimination function for the 20 bank notes from Exercise 11.6. Apply it to the entire bank data set. How many observations are misclassified?*

Applying the formulas derived in the previous Exercise 12.6 with $n_f = n_g = 10$, using the randomly chosen observations with indices 7, 8, 16, 39, 71, 73, 89, 94, 94, 100, 110, 121, 129, 131, 149, 150, 154, 161, 163, and 174, we obtain $\overline{x}_g = (214.72, 129.79, 129.64, 8.00, 10.18, 141.48)^\top$, $\overline{x}_f = (214.85, 130.13, 130.13, 10.33, 11.31, 139.53)^\top$, and

$$
\mathcal{W} = \begin{pmatrix}
3.36 & 0.40 & 0.90 & -3.32 & -0.00 & 0.38 \\
0.40 & 1.49 & 0.95 & 0.41 & -0.52 & 0.91 \\
0.90 & 0.95 & 1.91 & 2.43 & -1.38 & 1.31 \\
-3.32 & 0.41 & 2.43 & 18.02 & -10.17 & 2.86 \\
-0.00 & -0.52 & -1.38 & -10.17 & 11.46 & -2.39 \\
0.38 & 0.91 & 1.31 & 2.86 & -2.39 & 3.66
\end{pmatrix}.
$$

The eigenvector of $\mathcal{W}^{-1}\mathcal{B}$ corresponding to the largest eigenvalue can then be calculated as

$$(\overline{x}_f - \overline{x}_g)^\top \mathcal{W}^{-1} = (-1.56, -1.19, 1.38, -1.21, -0.88, 0.87)^\top.$$

The new observation x will be allocated as a counterfeit bank note if $a^\top (x - \overline{x}) \geq 0$. Calculating the Fisher linear discriminant rule for all observations in the Swiss bank notes data set, we obtain altogether six genuine bank notes classified as counterfeit. None of the counterfeit bank notes is classified as genuine. Hence, the estimated error rate is $6/200 = 3\%$. This estimate might be too optimistic since some of the bank notes used for the construction were used also for the evaluation of the rule.

EXERCISE 12.8. *Derive a discriminant rule based on the ML method with* $J = 2$ *minimizing the expected cost misclassification considering the prior probability* $\pi_1 = \frac{1}{3}$ *and the expected cost of misclassification* $C(2|1) = 2C(1|2)$.

The expected cost of misclassification is given by $ECM = C(2|1)p_{21}\pi_1 + C(1|2)p_{12}\pi_2$, where p_{21} is the probability of wrong classification of observation coming from group 1 and p_{12} is the probability of wrong classification of observation coming from group 2.

Assuming that the populations Π_1 and Π_2 are characterized by the probability densities $f_1(.)$ and $f_2(.)$, we can derive the loss $L(R_1)$ as a function of the discriminant rule R_1:

$$
\begin{aligned}
L(R_1) &= C(2|1)\pi_1 p_{21} + C(1|2)\pi_2 p_{12} \\
&= C(2|1)\pi_1 \int_{R_2} f_1(x)dx + C(1|2)\pi_2 \int_{R_1} f_2(x)dx \\
&= C(2|1)\pi_1 \int \{1 - \mathbf{I}(x \in R_1)\} f_1(x)dx + C(1|2)\pi_2 \int \mathbf{I}(x \in R_1) f_2(x)dx \\
&= C(2|1)\pi_1 + \int \mathbf{I}(x \in R_1)\{C(1|2)\pi_2 f_2(x) - C(2|1)\pi_1 f_1(x)\}dx.
\end{aligned}
$$

The loss $L(R_1)$ is obviously minimized if R_1 is chosen so that $x \in R_1$ is equivalent to $C(1|2)\pi_2 f_2(x) - C(2|1)\pi_1 f_1(x) < 0$. Hence, the optimal discriminant rule is:

$$
\begin{aligned}
R_1 &= \{x : C(1|2)\pi_2 f_2(x) - C(2|1)\pi_1 f_1(x) < 0\} \\
&= \{x : C(2|1)\pi_1 f_1(x) > C(1|2)\pi_2 f_2(x)\} \\
&= \left\{ x : \frac{f_1(x)}{f_2(x)} > \frac{C(1|2)\pi_2}{C(2|1)\pi_1} \right\}.
\end{aligned}
$$

Assuming that $\pi_1 = \frac{1}{3}$ and that the expected cost of misclassification $C(2|1) = 2C(1|2)$ leads $\pi_2 = 1 - \pi_1 = 2/3 = 2\pi_1$ and the resulting discriminant rule is:

$$
R_1 = \left\{ x : \frac{f_1(x)}{f_2(x)} > \frac{C(1|2)2\pi_1}{2C(1|2)\pi_1} \right\} = \left\{ x : \frac{f_1(x)}{f_2(x)} > 1 \right\} = \{x : f_1(x) > f_2(x)\},
$$

i.e., we obtain the ML discriminant rule.   ◲ SMSdisfbank

EXERCISE 12.9. *Explain the effect of changing π_1 or $C(1|2)$ on the relative location of the region $R_j, j = 1, 2$ in Exercise 12.8.*

In Exercise 12.8, we have derived the discriminant rule

$$
R_1 = \left\{ x : \frac{f_1(x)}{f_2(x)} > \frac{C(1|2)\pi_2}{C(2|1)\pi_1} \right\}.
$$

Increasing the cost of misclassification $C(1|2)$ would increase the constant in the definition of R_1 and, hence, it would make the region R_1 smaller.

Increasing the prior probability π_1 of the population Π_1 would make the same constant smaller and the region R_1 would grow.

EXERCISE 12.10. *Prove that Fisher's linear discrimination function is identical to the ML rule for multivariate normal distributions with equal covariance matrices $(J = 2)$.*

The ML rule in this situation has been derived in Exercise 12.1,

$$
R_1^{ML} = \{x : \alpha^\top (x - \mu) \geq 0\},
$$

where $\alpha = \Sigma^{-1}(\mu_1 - \mu_2)$ and $\mu = \frac{1}{2}(\mu_1 + \mu_2)$.

Fisher's linear discrimination rule derived for $J = 2$ in Exercise 12.6 is:

$$
R_1^F = \{x : (\overline{x}_1 - \overline{x}_1)^\top \mathcal{W}^{-1}(x - \overline{x}) \geq 0\}.
$$

In the same exercise, we have also shown that $\mathcal{W} = n\mathcal{S}$, where $\mathcal{S}$ denotes the pooled covariance matrix and n the number of observations. Defining the empirical version of α as $\widehat{\alpha} = (\overline{x}_1 - \overline{x}_2)^\top S^{-1}$, we can rewrite the Fisher's discriminant rule as:

$$R_1^F = \{x : \widehat{\alpha}^\top (x - \overline{x}) \geq 0\}.$$

Comparing this expression with the ML discriminant rule, we see that Fisher's rule R_1^F may be interpreted as the empirical version (estimate) of the ML discriminant rule R_1^{ML}.

EXERCISE 12.11. *Suppose that the observations come from three distinct populations, Π_1, Π_2, and Π_3, characterized by binomial distributions:*

$$\Pi_1 : X \sim Bi(10, 0.2) \quad \text{with the prior probability } \pi_1 = 0.5;$$
$$\Pi_2 : X \sim Bi(10, 0.3) \quad \text{with the prior probability } \pi_2 = 0.3;$$
$$\Pi_3 : X \sim Bi(10, 0.5) \quad \text{with the prior probability } \pi_3 = 0.2.$$

Use the Bayes method to determine the discriminant rules R_1, R_2, and R_3.

The corresponding Bayes discriminant rules R_j for $j = 1, 2, 3$ are defined as:

$$R_j = \{x \in \{0, 1, \ldots, 9, 10\} : \pi_j f_j(x) \geq \pi_i f_i(x) \text{ for } i = 1, 2, 3\}.$$

x	$f_1(x)$	$f_2(x)$	$f_3(x)$	$\pi_1 f_1(x)$	$\pi_2 f_2(x)$	$\pi_3 f_3(x)$	$\pi_j f_j(x)$	j
0	0.107374	0.028248	0.000977	0.053687	0.008474	0.000195	0.053687	1
1	0.268435	0.121061	0.009766	0.134218	0.036318	0.001953	0.134218	1
2	0.301990	0.233474	0.043945	0.150995	0.070042	0.008789	0.150995	1
3	0.201327	0.266828	0.117188	0.100663	0.080048	0.023438	0.100663	1
4	0.088080	0.200121	0.205078	0.044040	0.060036	0.041016	0.060036	2
5	0.026424	0.102919	0.246094	0.013212	0.030876	0.049219	0.049219	3
6	0.005505	0.036757	0.205078	0.002753	0.011027	0.041016	0.041016	3
7	0.000786	0.009002	0.117188	0.000393	0.002701	0.023438	0.023438	3
8	0.000074	0.001447	0.043945	0.000037	0.000434	0.008789	0.008789	3
9	0.000004	0.000138	0.009766	0.000002	0.000041	0.001953	0.001953	3
10	0.000000	0.000006	0.000977	0.000000	0.000002	0.000195	0.000195	3

Table 12.3. The values of the likelihood and Bayesian likelihood for three binomial distributions.

The values of $\pi_i f_i(x)$, for $i = 1, \ldots, 3$ and $x = 0, \ldots, 10$ are given in Table 12.3 from which it directly follows that the discriminant rules are:

$$R_1 = \{0, 1, 2, 3\},$$
$$R_2 = \{4\},$$
$$R_3 = \{5, 6, 7, 8, 9, 10\}.$$

EXERCISE 12.12. *Use the Fisher's linear discrimination function on the WAIS data set (Table A.21) and evaluate the results by re-substitution to calculate the probabilities of misclassification.*

The WAIS data set contains results of four subtests of the Wechsler Adult Intelligence Scale for two categories of people. Group 2 contains 12 observations of those presenting a senile factor and group 1 contains 37 people serving as a control.

Applying the formulas derived in Exercise 12.6 and proceeding as in Exercise 12.7, we obtain the eigenvector

$$(\bar{x}_2 - \bar{x}_1)^\top \mathcal{W}^{-1} = (-0.0006, -0.0044, -0.0002, -0.0095)^\top.$$

Calculating the Fisher's discriminant rule from all observations leads to 4 misclassified observations in group 2 and 8 misclassified observations in group 1.

Hence, the apparent error rate (APER) is equal to $(4 + 8)/49 = 24.49\%$. The disadvantage of this measure of the quality of the discriminant rule is that it is based on the same observations that were used to construct the rule.

In order to obtain a more appropriate estimate of the misclassification probability, we may proceed in the following way:

1. Calculate the discrimination rule from all but one observation.

2. Allocate the omitted observation according to the rule from step 1.

3. Repeat steps 1 and 2 for all observations and count the number of correct and wrong classifications.

The estimate of the misclassification rate based on this procedure is called the actual error rate (AER).

Running the algorithm for the WAIS data set, we misclassify 4 observations in group 2 and 11 observations in group 1. The AER is $(4 + 11)/49 = 30.61\%$.

Hence, if a new patient arrives, he will be correctly classified with probability approximately 70%.   ◖ SMSdisfwais

13

Correspondence Analysis

The method was no doubt suggested to Clay's ingenious mind by the colour of his accomplice's hair.
Sherlock Holmes in "The Red-Headed League"

Contingency tables contain information about the joint distribution of statistical variables. For a large number of classes (for each variable) the resulting $n \times p$ frequency matrix can be hard to interpret. Correspondence analysis is a tool for developing simple indices that show us relations between row and column categories.

These indices tell us, for example, which column categories have more weight in a row category and vice versa. A typical example is the statistical analysis of consumer preferences.

Suppose that one has recorded the frequencies of newspaper distribution across regions. If the number of newspapers and regions is big, then one sits in front of a huge $n \times p$ matrix with numbers from which we have to tell which region prefers which newspaper. Correspondence analysis provides a way out of this: reducing the dimensionality of the table via factors helps to concentrate on the most important variables.

The basic idea is to extract the indices in a decreasing order of importance so that the main information of the table can be summarized in spaces with smaller dimensions. If only two factors (indices) are used, the results can be shown in two-dimensional graphs. The dimension reduction techniques are similar to the principal component method but, due to the different character of the categorical data, we decompose a measure of dependency (χ^2-statistic) between the variables rather than the variance.

A contingency table $\mathcal{X}(n \times p)$ consists of frequencies of joint occurrence of row and column events—the entry x_{ij} in the table $\mathcal{X}$ is the number of observations in a sample that simultaneously fall in the ith row category and the jth column category. The symbol $x_{i\bullet} = \sum_{j=1}^{n} x_{ij}$ denotes the number of observations falling into the ith row category. Similarly, $x_{\bullet j} = \sum_{i=1}^{n} x_{ij}$. The total number of observations is $x_{\bullet\bullet} = \sum_{i=1}^{n} x_{i\bullet} = \sum_{j=1}^{n} x_{\bullet j}$. For simplification, define the matrices $\mathcal{A}\,(n \times n)$ and $\mathcal{B}\,(p \times p)$ as

$$\mathcal{A} = \mathrm{diag}(x_{i\bullet}) \quad \text{and} \quad \mathcal{B} = \mathrm{diag}(x_{\bullet j}). \tag{13.1}$$

These matrices provide the marginal row frequencies $a(n \times 1) = (x_{1\bullet}, \ldots, x_{n\bullet})^{\top}$ and the marginal column frequencies $b(p \times 1)(x_{\bullet 1}, \ldots, x_{\bullet p})^{\top}$:

$$a = \mathcal{A}1_n \quad \text{and} \quad b = \mathcal{B}1_p. \tag{13.2}$$

E_{ij} is the estimated expected value in the (i, j)th category under the assumption of independence, i.e.,

$$E_{ij} = \frac{x_{i\bullet}\,x_{\bullet j}}{x_{\bullet\bullet}}. \tag{13.3}$$

Technically speaking, the basic idea is to decompose the χ^2-statistic of dependence:

$$t = \sum_{i=1}^{n}\sum_{j=1}^{p}(x_{ij} - E_{ij})^2/E_{ij}. \tag{13.4}$$

Under the hypothesis of independence of the row and column categories, the statistic t has a $\chi^2_{(n-1)(p-1)}$ distribution.

The correspondence analysis is targeted toward the analysis of the contributions to the χ^2-statistic (13.4):

$$c_{ij} = (x_{ij} - E_{ij})/E_{ij}^{1/2}, \tag{13.5}$$

which may be viewed as a measure of the departure of the observed x_{ij} from independence. The desired lower-dimensional decomposition is then produced by the singular value decomposition (SVD) of the matrix $\mathcal{C} = (c_{ij})_{i=1,\ldots,n;j=1,\ldots,p}$. The exact expressions for the row and column factors (r_k and s_k, respectively) are given in Exercise 13.2. Their mean and variance and the relationship to the χ^2-statistic (13.4) are investigated in Exercises 13.3 and 13.4.

Both the basic properties of the factors and some applications of correspondence analysis are demonstrated in the following exercises.

EXERCISE 13.1. *Show that the matrices $\mathcal{A}^{-1}\mathcal{X}\mathcal{B}^{-1}\mathcal{X}^{\top}$ and $\mathcal{B}^{-1}\mathcal{X}^{\top}\mathcal{A}^{-1}\mathcal{X}$ have an eigenvalue equal to 1 and that the corresponding eigenvectors are proportional to $(1, \ldots, 1)^{\top}$.*

It suffices to show that for $\mathcal{A}^{-1}\mathcal{X}\mathcal{B}^{-1}\mathcal{X}^{\top}$. The second equation follows by exchanging rows and columns of the contingency table $\mathcal{X}$. Eigenvalue λ and eigenvector γ are solutions of the equation

$$\mathcal{A}^{-1}\mathcal{X}\mathcal{B}^{-1}\mathcal{X}^\top\gamma = \lambda\gamma \qquad (13.6)$$

and it remains to show that (13.6) is satisfied for $\lambda = 1$ and $\gamma = (1,\dots,1)^\top = 1_n$:

$$\mathcal{A}^{-1}\mathcal{X}\mathcal{B}^{-1}\mathcal{X}^\top \begin{pmatrix} 1 \\ 1 \\ \vdots \\ 1 \end{pmatrix} = \begin{pmatrix} \frac{1}{x_{1\bullet}} & 0 & \cdots & 0 \\ 0 & \frac{1}{x_{2\bullet}} & \cdots & 0 \\ \vdots & \vdots & \ddots & \vdots \\ 0 & \cdots & 0 & \frac{1}{x_{n\bullet}} \end{pmatrix} \begin{pmatrix} x_{11} & x_{12} & \cdots & x_{1p} \\ x_{21} & x_{22} & \cdots & x_{2p} \\ \vdots & \vdots & \ddots & \vdots \\ x_{n1} & x_{n2} & \cdots & x_{np} \end{pmatrix}$$

$$\times \begin{pmatrix} \frac{1}{x_{\bullet 1}} & 0 & \cdots & 0 \\ 0 & \frac{1}{x_{\bullet 2}} & \cdots & 0 \\ \vdots & \vdots & \ddots & \vdots \\ 0 & \cdots & 0 & \frac{1}{x_{\bullet p}} \end{pmatrix} \begin{pmatrix} x_{11} & x_{21} & \cdots & x_{n1} \\ x_{12} & x_{22} & \cdots & x_{n2} \\ \vdots & \vdots & \ddots & \vdots \\ x_{1p} & x_{2p} & \cdots & x_{np} \end{pmatrix} \begin{pmatrix} 1 \\ 1 \\ \vdots \\ 1 \end{pmatrix}$$

$$= \begin{pmatrix} E_{11} & E_{12} & \cdots & E_{1p} \\ E_{21} & E_{22} & \cdots & E_{2p} \\ \vdots & \vdots & \ddots & \vdots \\ E_{n1} & E_{n2} & \cdots & E_{np} \end{pmatrix} \begin{pmatrix} x_{11} & x_{21} & \cdots & x_{n1} \\ x_{12} & x_{22} & \cdots & x_{n2} \\ \vdots & \vdots & \ddots & \vdots \\ x_{1p} & x_{2p} & \cdots & x_{np} \end{pmatrix} \begin{pmatrix} 1 \\ 1 \\ \vdots \\ 1 \end{pmatrix}$$

$$= \left(\sum_{k=1}^{p} E_{ik}x_{jk} \right)_{i=1,\dots,n;j=1,\dots,n} \begin{pmatrix} 1 \\ 1 \\ \vdots \\ 1 \end{pmatrix}$$

$$= \left(\sum_{j=1}^{n}\sum_{k=1}^{p} E_{ik}x_{jk} \right)_{i=1,\dots,n}$$

$$= \left(\sum_{k=1}^{p} E_{ik}x_{\bullet k} \right)_{i=1,\dots,n}$$

$$= \left(\sum_{k=1}^{p} \frac{x_{ik}x_{\bullet k}}{x_{i\bullet}x_{\bullet k}} \right)_{i=1,\dots,n}$$

$$= \begin{pmatrix} 1 \\ 1 \\ \vdots \\ 1 \end{pmatrix}.$$

Hence, (13.6) is satisfied for $\lambda = 1$ and $\gamma = 1_n$ and we have proven the statement proposed in Exercise 13.1.

EXERCISE 13.2. *Let δ_k and γ_k denote the kth eigenvectors of $\mathcal{C}^\top\mathcal{C}$ and $\mathcal{C}\mathcal{C}^\top$, respectively. Verify the relations:*

$$\mathcal{C}\sqrt{b} = 0 \quad and \quad \mathcal{C}^\top\sqrt{a} = 0, \tag{13.7}$$

$$\delta_k^\top\sqrt{b} = 0 \quad and \quad \gamma_k^\top\sqrt{a} = 0, \tag{13.8}$$

$$r_k^\top a = 0 \quad and \quad s_k^\top b = 0. \tag{13.9}$$

Notice that the second part of all equations follows by applying the first part to the contingency table $\mathcal{X}^\top$.

The ith element of the vector $\mathcal{C}_{(n\times p)}\sqrt{b}_{(p\times 1)}$ is $\sum_{j=1}^p \frac{x_{ij}-E_{ij}}{\sqrt{E_{ij}}}\sqrt{x}_{\bullet j}$, for $i = 1,...,n$. Using simple algebra we write

$$
\begin{aligned}
\mathcal{C}\sqrt{b} &= \left(\sum_{j=1}^p \frac{x_{ij}-E_{ij}}{\sqrt{E_{ij}}}\sqrt{x}_{\bullet j}\right)_{i=1,...,n} \\
&= \left(\sum_{j=1}^p \frac{x_{ij} - \frac{x_{i\bullet}x_{\bullet j}}{x_{\bullet\bullet}}}{\sqrt{\frac{x_{i\bullet}x_{\bullet j}}{x_{\bullet\bullet}}}}\sqrt{x}_{\bullet j}\right)_{i=1,...,n} \\
&= \left(\sum_{j=1}^p \frac{x_{ij}x_{\bullet\bullet}-x_{i\bullet}x_{\bullet j}}{x_{\bullet\bullet}}\frac{\sqrt{x_{\bullet\bullet}}}{\sqrt{x_{i\bullet}x_{\bullet j}}}\sqrt{x}_{\bullet j}\right)_{i=1,...,n} \\
&= \left(\sum_{j=1}^p \frac{x_{\bullet\bullet}x_{ij}}{\sqrt{x_{\bullet\bullet}}\sqrt{x_{i\bullet}}} - \sum_{j=1}^p \frac{x_{i\bullet}x_{\bullet j}}{\sqrt{x_{\bullet\bullet}}\sqrt{x_{i\bullet}}}\right)_{i=1,...,n} \\
&= \left(\frac{\sqrt{x_{\bullet\bullet}}}{\sqrt{x_{i\bullet}}}\sum_{j=1}^p x_{ij} - \frac{\sqrt{x_{i\bullet}}}{\sqrt{x_{\bullet\bullet}}}\sum_{j=1}^p x_{\bullet j}\right)_{i=1,...,n} \\
&= \left(\frac{\sqrt{x_{\bullet\bullet}}}{\sqrt{x_{i\bullet}}}x_{i\bullet} - \frac{\sqrt{x_{i\bullet}}}{\sqrt{x_{\bullet\bullet}}}x_{\bullet\bullet}\right)_{i=1,...,n} \\
&= \left(\sqrt{x_{\bullet\bullet}x_{i\bullet}} - \sqrt{x_{\bullet\bullet}x_{i\bullet}}\right)_{i=1,...,n} \\
&= 0_n.
\end{aligned}
$$

This proves the first part of (13.7). The second part follows from the symmetry of the situation.

The symbol $\delta_k^\top$ in relation (13.8) denotes the kth eigenvector of $\mathcal{C}^\top\mathcal{C}$ and $\gamma_k^\top$ is the kth eigenvector of $\mathcal{C}\mathcal{C}^\top$. From the properties of SVD (Härdle & Simar 2003, chapter 8) we know the relationship between δ_k and γ_k:

$$\delta_k = \frac{1}{\sqrt{\lambda_k}}\mathcal{C}^\top\gamma_k \quad and \quad \gamma_k = \frac{1}{\sqrt{\lambda_k}}\mathcal{C}\delta_k. \tag{13.10}$$

Applying the above proved formula (13.7) leads directly

$$\delta_k^\top\sqrt{b} = \frac{1}{\sqrt{\lambda_k}}\gamma_k^\top\mathcal{C}\sqrt{b} = \frac{1}{\sqrt{\lambda_k}}\gamma_k^\top 0 = 0$$

and

$$\gamma_k^\top \sqrt{a} = \frac{1}{\sqrt{\lambda_k}} \delta_k^\top C^\top \sqrt{a} = \frac{1}{\sqrt{\lambda_k}} \delta_k^\top 0 = 0.$$

The row coordinates r_k and the column coordinates s_k are defined as

$$r_k = \mathcal{A}^{-\frac{1}{2}} \mathcal{C} \delta_k$$
$$s_k = \mathcal{B}^{-\frac{1}{2}} \mathcal{C}^\top \gamma_k. \tag{13.11}$$

Using this definition and (13.7) it follows that

$$r_k^\top a = \delta_k^\top \mathcal{C}^\top \mathcal{A}^{-\frac{1}{2}} a = \delta_k^\top \mathcal{C}^\top \sqrt{a} = \delta_k^\top 0 = 0$$

and

$$s_k^\top b = \gamma_k^\top \mathcal{C} \mathcal{B}^{-\frac{1}{2}} b = \gamma_k^\top \mathcal{C} \sqrt{b} = \gamma_k^\top 0 = 0.$$

The vectors of row and column coordinates, r_k and s_k, are the row and column factors.

EXERCISE 13.3. *Rewrite the χ^2-statistic (13.4) in terms of the matrix $\mathcal{C}$. Describe the relationship of the χ^2-statistic to the SVD of $\mathcal{C}$.*

The SVD of $\mathcal{C}$ yields $\mathcal{C} = \Gamma \Lambda \Delta^\top$ with $\Lambda = \mathrm{diag}(\lambda_1^{1/2}, \ldots, \lambda_R^{1/2})$, where $\lambda_1, \ldots, \lambda_R$ are the nonzero eigenvalues of both $\mathcal{C}^\top \mathcal{C}$ and $\mathcal{C} \mathcal{C}^\top$ (Härdle & Simar 2003, chapter 8).

Now, it is easy to see that

$$t = \sum_{i=1}^{n} \sum_{j=1}^{p} (x_{ij} - E_{ij})^2 / E_{ij} = \sum_{i=1}^{n} \sum_{j=1}^{p} c_{ij}^2 = \mathrm{tr}(\mathcal{C} \mathcal{C}^\top) = \sum_{k=1}^{R} \lambda_k.$$

Hence, the SVD of the matrix $\mathcal{C}$ decomposes the χ^2-statistic t. In Exercise 13.4, we will show that also the variances of the row and column factors provide a decomposition of the χ^2-statistic.

EXERCISE 13.4. *Calculate the means and variances of the row and column factors r_k and s_k.*

Using the relation (13.9), it is easy to see that the means (weighted by the row and column marginal frequencies) are:

$$\bar{r}_k = \frac{1}{x_{\bullet\bullet}} r_k^\top a = 0,$$

$$\bar{s}_k = \frac{1}{x_{\bullet\bullet}} s_k^\top b = 0.$$

Hence, both row and column factors are centered.

For the variances of r_k and s_k we have the following:

$$Var(r_k) = \frac{1}{x_{\bullet\bullet}} \sum_{i=1}^{n} x_{i\bullet} r_{ki}^2 = r_k^\top \mathcal{A} r_k / x_{\bullet\bullet} = \delta_k^\top \mathcal{C}^\top \mathcal{C} \delta_k / x_{\bullet\bullet} = \frac{\lambda_k}{x_{\bullet\bullet}},$$

$$Var(s_k) = \frac{1}{x_{\bullet\bullet}} \sum_{j=1}^{p} x_{\bullet j} s_{kj}^2 = s_k^\top \mathcal{B} s_k / x_{\bullet\bullet} = \gamma^\top \mathcal{C} \mathcal{C}^\top \gamma_k / x_{\bullet\bullet} = \frac{\lambda_k}{x_{\bullet\bullet}}.$$

Hence, the proportion of the variance explained by the kth factor is

$$Var(r_k) / \sum_{i=1}^{R} Var(r_k) = \lambda_k / \sum_{i=1}^{R} \lambda_i.$$

The variance of the kth row factor, $Var(r_k)$, can be further decomposed into the absolute single row contributions defined as

$$C_a(i, r_k) = \frac{x_{i\bullet} r_{ki}^2}{\lambda_k}, \text{ for } i = 1, \ldots, n, \ k = 1, \ldots, R.$$

Similarly, the proportions

$$C_a(j, s_k) = \frac{x_{\bullet j} s_{kj}^2}{\lambda_k}, \text{ for } j = 1, \ldots, p, \ k = 1, \ldots, R$$

are the absolute contributions of column j to the variance of the column factor s_k. These absolute contributions may help to interpret the row and column factors obtained by the correspondence analysis.

EXERCISE 13.5. *Do a correspondence analysis for the car marks data in Table A.5. Explain how this data set can be considered as a contingency table.*

The car marks data set consists of averaged marks. The numbers could be seen as "number of points" corresponding to the quality of cars (the worse the more points). In this way, the entries in the data set can be interpreted as counts and the data set as a contingency table.

Correspondence analysis is based on SVD of matrix $\mathcal{C}$. The eigenvalues tell us the proportion of explained variance. From Table 13.1 we can see that the first two eigenvalues account for 93% of the variance. Here, representation in two dimensions is satisfactory.

Figure 13.1 shows the projections of the rows (the 23 types of cars) and columns (the 8 features). The projections on the first 3 axis along with their absolute contributions to the variance of the axis are given in Table 13.2 for the cars and in Table 13.3 for features.

nr. of factors	eigenvalues	cumulated percentage
1	31.0730	0.8133
2	4.5016	0.9311
3	1.1900	0.9623
4	0.0000	0.9623
5	0.5806	0.9775
6	0.1849	0.9823
7	0.3454	0.9914
8	0.3298	1.0000

Table 13.1. Eigenvalues and cumulated percentage of explained variance for the car marks data.

Cars	r_1	r_2	r_3	$C_a(i, r_1)$	$C_a(i, r_2)$	$C_a(i, r_3)$
Audi	−0.1862	0.0536	0.0114	0.0272	0.0156	0.0027
BMW	−0.4385	0.0650	−0.0702	0.1374	0.0208	0.0919
Cit	0.1498	0.0267	0.0042	0.0205	0.0045	0.0004
Ferr	−0.4400	−0.2143	0.0128	0.1663	0.2725	0.0037
Fiat	0.2356	0.0385	0.0781	0.0502	0.0092	0.1442
Ford	0.1161	−0.0470	0.0432	0.0105	0.0119	0.0380
Hyun	0.1421	−0.0182	0.0212	0.0153	0.0017	0.0089
Jagu	−0.4657	−0.1493	−0.0029	0.1633	0.1159	0.0002
Lada	0.2162	−0.0192	−0.0319	0.0448	0.0024	0.0255
Mazd	0.0971	−0.0659	0.0671	0.0079	0.0250	0.0979
Merc	−0.3406	0.1659	−0.0425	0.0806	0.1320	0.0327
Mit	−0.0349	0.0072	0.0249	0.0010	0.0003	0.0127
Nis	0.1937	−0.0060	−0.0143	0.0308	0.0002	0.0044
OpCo	0.1045	0.0882	0.0108	0.0078	0.0392	0.0022
OpVe	−0.1142	0.0463	0.0338	0.0093	0.0105	0.0212
Peug	0.0889	0.0072	−0.0012	0.0065	0.0003	0.0000
Rena	0.0532	−0.0062	0.0323	0.0022	0.0002	0.0215
Rov	−0.1454	−0.0341	−0.0199	0.0171	0.0065	0.0083
Toy	0.0537	−0.0272	0.0545	0.0022	0.0040	0.0601
Tra	0.2918	−0.0501	−0.1061	0.0937	0.0191	0.3234
VwGo	−0.2156	0.1833	−0.0043	0.0343	0.1708	0.0004
VwPa	−0.0303	0.1441	0.0094	0.0007	0.1024	0.0016
War	0.2493	−0.0669	−0.0577	0.0702	0.0349	0.0981

Table 13.2. Coefficients and absolute contributions for the cars in car marks data.

Feature	s_1	s_2	s_3	$C_a(j,s_1)$	$C_a(j,s_2)$	$C_a(j,s_3)$
Econ.	−0.2810	0.0023	−0.0821	0.1923	0.0000	0.4292
Service	0.1239	−0.0553	0.0271	0.0348	0.0478	0.0433
Value	0.2149	−0.0407	−0.0070	0.1077	0.0267	0.0030
Price	−0.4254	0.0376	0.0582	0.4384	0.0236	0.2146
Design	0.1553	0.1024	0.0545	0.0571	0.1714	0.1836
Sport	0.1587	0.1436	−0.0431	0.0646	0.3653	0.1244
Safety	0.1722	−0.1121	−0.0046	0.0721	0.2110	0.0013
Easy	−0.1263	−0.1040	0.0033	0.0329	0.1540	0.0006

Table 13.3. Coefficients and absolute contributions for features in car marks data.

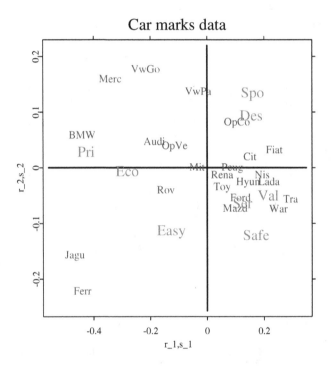

Fig. 13.1. Projections of rows and columns for car marks data. ⊙ SMScorrcarm

Figure 13.1 shows that price on the left side and value on the right side are most strongly responsible for the variation on the first axis. The second axis can be described as a contrast between *sport* and *easy and safe*. This interpretation is confirmed in Table 13.3, where in the first column factor s_1, the difference between the coefficient of price (-0.42535) and value (0.21488) is the largest, and in the second column factor s_2, the difference between the coefficient of sport (0.14364) and safety (-0.1121) is the largest. These two axes are quite sensible since expensive cars (with high marks in price) tend to depreciate faster (with low marks in value), and sport cars tend to be less safe and less easily handled.

In Figure 13.1, Mitsubishi, Toyota, Renault and Peugeot are quite close to the center, which means they are kind of average cars (with average marks in the 8 features). On the left we see the more expensive cars and on the right the cheaper ones. Cars in the lower sector are more safe and easily handled than those in the upper sector. Among all cars, Ferrari plays an important role on each axis. On the first axis it is opposed to Trabant and Wartburg, which are the cheapest (lowest marks in price). On the second axis it is opposed to Volkswagen Golf and Mercedes. About half of the cars are concentrated in the right part of the picture and not far from the origin, these the most common types of cars. They typically have strong marks in service and value (located very close to service and value), and a little far from economy which means they are not very economical, and far from price which means that they are cheap.

EXERCISE 13.6. *Compute the χ^2-statistic and test independence for the French baccalauréat data.*

The χ^2-statistic of independence compares observed counts x_{ij} to their estimated (under the hypothesis of independence) expected values E_{ij} (13.3):

$$t = \sum_{i=1}^{n} \sum_{j=1}^{p} (x_{ij} - E_{ij})^2 / E_{ij}. \qquad (13.12)$$

Under the hypothesis of independence, t has the $\chi^2_{(n-1)(p-1)}$ distribution.

For the French baccalauréat data, the test statistic is $t = 6670.6$ and the 0.95 quantile of the χ^2-distribution $\chi^2_{(n-1)(p-1)} = 199.24$ ▢ SMSchi2bac. The test statistic is larger then the critical value and we reject independence between the row and column categories.

EXERCISE 13.7. *Prove that $\mathcal{C} = \mathcal{A}^{-1/2}(\mathcal{X} - E)\mathcal{B}^{-1/2}\sqrt{x_{\bullet\bullet}}$ and $E = ab^\top x_{\bullet\bullet}^{-1}$ and verify:*

$$r_k = \sqrt{\frac{x_{\bullet\bullet}}{\lambda_k}} \mathcal{A}^{-1} \mathcal{X} s_k, \tag{13.13}$$

$$s_k = \sqrt{\frac{x_{\bullet\bullet}}{\lambda_k}} \mathcal{B}^{-1} \mathcal{X}^\top r_k. \tag{13.14}$$

Some properties of the row and column coordinates r_k and s_k and of the matrix $\mathcal{C}$ were discussed already in Exercise 13.2. Using the definitions of $\mathcal{A}$, $\mathcal{B}$, $\mathcal{C}$, and E_{ij}, we have

$$\mathcal{A}^{-1/2}(\mathcal{X} - E)\mathcal{B}^{-1/2}\sqrt{x_{\bullet\bullet}}$$

$$= \sqrt{x_{\bullet\bullet}} \begin{pmatrix} x_{1\bullet}^{-\frac{1}{2}} & 0 & \cdots & 0 \\ 0 & x_{2\bullet}^{-\frac{1}{2}} & \cdots & 0 \\ \vdots & \vdots & \ddots & \vdots \\ 0 & \cdots & 0 & x_{n\bullet}^{-\frac{1}{2}} \end{pmatrix} \begin{pmatrix} x_{11} - E_{11} & x_{12} - E_{12} & \cdots & x_{1p} - E_{1p} \\ x_{21} - E_{21} & x_{22} - E_{22} & \cdots & x_{2p} - E_{2p} \\ \vdots & \vdots & \ddots & \vdots \\ x_{n1} - E_{n1} & x_{n2} - E_{n2} & \cdots & x_{np} - E_{np} \end{pmatrix}$$

$$\times \begin{pmatrix} x_{\bullet 1}^{-\frac{1}{2}} & 0 & \cdots & 0 \\ 0 & x_{\bullet 2}^{-\frac{1}{2}} & \cdots & 0 \\ \vdots & \vdots & \ddots & \vdots \\ 0 & \cdots & 0 & x_{\bullet p}^{-\frac{1}{2}} \end{pmatrix}$$

$$= \sqrt{x_{\bullet\bullet}} \begin{pmatrix} \frac{x_{11} - E_{11}}{\sqrt{x_{1\bullet} x_{\bullet 1}}} & \frac{x_{12} - E_{12}}{\sqrt{x_{1\bullet} x_{\bullet 2}}} & \cdots & \frac{x_{1p} - E_{1p}}{\sqrt{x_{1\bullet} x_{\bullet p}}} \\ \frac{x_{21} - E_{21}}{\sqrt{x_{2\bullet} x_{\bullet 1}}} & \frac{x_{22} - E_{22}}{\sqrt{x_{2\bullet} x_{\bullet 2}}} & \cdots & \frac{x_{2p} - E_{2p}}{\sqrt{x_{2\bullet} x_{\bullet p}}} \\ \vdots & \vdots & \ddots & \vdots \\ \frac{x_{n1} - E_{n1}}{\sqrt{x_{n\bullet} x_{\bullet 1}}} & \frac{x_{n2} - E_{n2}}{\sqrt{x_{n\bullet} x_{\bullet 2}}} & \cdots & \frac{x_{np} - E_{np}}{\sqrt{x_{n\bullet} x_{\bullet p}}} \end{pmatrix}$$

$$= \left(\sqrt{x_{\bullet\bullet}} \frac{x_{ij} - E_{ij}}{\sqrt{x_{i\bullet} x_{\bullet j}}} \right)_{i=1,\ldots,n; j=1,\ldots,p} = \left(\frac{x_{ij} - E_{ij}}{\sqrt{\frac{x_{i\bullet} x_{\bullet j}}{x_{\bullet\bullet}}}} \right)_{i=1,\ldots,n; j=1,\ldots,p}$$

$$= \left(\frac{x_{ij} - E_{ij}}{\sqrt{E_{ij}}} \right)_{i=1,\ldots,n; j=1,\ldots,p} = \mathcal{C}.$$

The relation $E = ab^\top x_{\bullet\bullet}^{-1}$ is very easy to show since

$$\frac{ab^\top}{x_{\bullet\bullet}} = \frac{1}{x_{\bullet\bullet}} \begin{pmatrix} x_{1\bullet} \\ x_{2\bullet} \\ \vdots \\ x_{n\bullet} \end{pmatrix} (x_{\bullet 1}, x_{\bullet 2}, \ldots, x_{n\bullet})$$

$$= \frac{1}{x_{\bullet\bullet}} \begin{pmatrix} x_{1\bullet}x_{\bullet 1} & x_{1\bullet}x_{\bullet 2} & \cdots & x_{1\bullet}x_{\bullet p} \\ x_{2\bullet}x_{\bullet 1} & x_{2\bullet}x_{\bullet 2} & \cdots & x_{2\bullet}x_{\bullet p} \\ \vdots & \vdots & \ddots & \vdots \\ x_{n\bullet}x_{\bullet 1} & x_{n\bullet}x_{\bullet 2} & \cdots & x_{n\bullet}x_{\bullet p} \end{pmatrix} = E.$$

It follows from definition (13.11) of s_k and from the relation (13.10) between γ_k and δ_k that

$$s_k = \mathcal{B}^{-1/2}\mathcal{C}^\top \gamma_k = \sqrt{\lambda_k}\mathcal{B}^{-1/2}\delta_k.$$

Next, using the definition (13.11) of r_k and applying the above proved properties and (13.9), we have

$$r_k = \mathcal{A}^{-1/2}\mathcal{C}\delta_k = \sqrt{x_{\bullet\bullet}}\mathcal{A}^{-1/2}\mathcal{A}^{-1/2}(\mathcal{X}-E)\mathcal{B}^{-1/2}\delta_k$$
$$= \sqrt{\frac{x_{\bullet\bullet}}{\lambda_k}}\mathcal{A}^{-1}(\mathcal{X}-E)s_k = \sqrt{\frac{x_{\bullet\bullet}}{\lambda_k}}\mathcal{A}^{-1}\left(\mathcal{X}s_k - \frac{ab^\top s_k}{x_{\bullet\bullet}}\right)$$
$$= \sqrt{\frac{x_{\bullet\bullet}}{\lambda_k}}\mathcal{A}^{-1}\mathcal{X}s_k.$$

The expression for s_k follows exactly in the same way.

EXERCISE 13.8. *Do the full correspondence analysis of the U.S. crime data in Table A.18, and determine the absolute contributions for the first three axes. How can you interpret the third axis? Try to identify the states with one of the four regions to which it belongs. Do you think the four regions have a different behavior with respect to crime?*

The results of the correspondence analysis for the U.S. crime data are presented in Table 13.4 containing the projections and absolute contributions of the rows (states) and in Table 13.5 which contains the corresponding projections and absolute contributions of the columns (crimes).

The third axis could be interpreted as contrast between *robbery* versus *burglary*. The states with largest contributions to the 3rd axis are MA in contrast to IL, MI, and MD.

The differences between different regions can be best assessed in a graphics. Figure 13.2 shows the projections of all states and crimes where each region is colored differently. The biggest differences are between states from Northeast (squares) and South (triangles). The distributions of the crimes in Midwest (triangles) and West (crosses) is very similar, the points are relatively close to the origin.

State	r_1	r_2	r_3	$C_a(i,r_1)$	$C_a(i,r_2)$	$C_a(i,r_3)$
ME	−0.1188	−0.0382	0.1062	0.0059	0.0012	0.0151
NH	−0.0639	0.0242	0.1573	0.0016	0.0005	0.0308
VT	−0.0778	−0.1068	0.2051	0.0026	0.0098	0.0578
MA	0.3142	0.2536	0.2139	0.0840	0.1088	0.1240
RI	0.1334	0.2381	0.1228	0.0173	0.1093	0.0466
CT	0.0683	0.0849	0.1301	0.0037	0.0114	0.0427
NY	0.3812	−0.0012	−0.1769	0.1585	0.0000	0.1085
NJ	0.2003	0.1111	0.0149	0.0325	0.0199	0.0006
PA	0.2300	0.0569	−0.0004	0.0258	0.0031	0.0000
OH	0.0834	0.0941	−0.0465	0.0056	0.0143	0.0056
IN	0.0489	0.0816	0.0039	0.0018	0.0099	0.0000
IL	0.1756	0.0415	−0.1926	0.0265	0.0029	0.1014
MI	0.0991	−0.0506	−0.1442	0.0123	0.0064	0.0828
WI	−0.2485	0.1085	−0.0626	0.0380	0.0144	0.0077
MN	−0.0621	0.1099	−0.0253	0.0028	0.0175	0.0015
IA	−0.2700	0.0779	−0.0680	0.0416	0.0069	0.0084
MO	0.1541	0.0076	−0.0255	0.0227	0.0001	0.0020
ND	−0.3916	0.1048	−0.1064	0.0595	0.0085	0.0140
SD	−0.2841	−0.0295	−0.0421	0.0377	0.0008	0.0026
NE	−0.0718	0.0516	−0.0487	0.0030	0.0031	0.0044
KS	−0.1629	0.0007	−0.0459	0.0220	0.0000	0.0056
DE	0.0392	0.0333	0.0305	0.0015	0.0021	0.0029
MD	0.1912	−0.0271	−0.2101	0.0386	0.0015	0.1483
VA	−0.0642	−0.0259	−0.0442	0.0031	0.0010	0.0047
WV	−0.0634	−0.1672	0.0255	0.0013	0.0174	0.0006
NC	0.0344	−0.3622	0.0569	0.0007	0.1567	0.0062
SC	0.0396	−0.1880	0.1168	0.0011	0.0491	0.0303
GA	−0.0052	−0.0828	−0.0041	0.0000	0.0105	0.0000
FL	0.0080	−0.1259	−0.0194	0.0000	0.0381	0.0015
KY	0.1314	−0.0094	0.0744	0.0097	0.0000	0.0100
TN	0.2057	−0.1591	0.1108	0.0231	0.0274	0.0213
AL	0.1021	−0.2626	0.1161	0.0057	0.0750	0.0235
MS	−0.0162	−0.3623	0.0515	0.0000	0.0772	0.0025
AR	−0.0220	−0.2719	0.1117	0.0003	0.0811	0.0219
LA	0.1515	−0.1232	−0.0191	0.0173	0.0227	0.0009
OK	−0.0427	−0.0422	0.0531	0.0012	0.0024	0.0061
TX	0.0313	−0.0667	−0.0004	0.0009	0.0082	0.0000
MT	−0.2471	0.0595	−0.0339	0.0400	0.0046	0.0024
ID	−0.3161	−0.0051	−0.0575	0.0717	0.0000	0.0075
WY	−0.2884	0.0157	−0.0447	0.0562	0.0003	0.0043
CO	−0.0183	0.0296	0.0164	0.0004	0.0021	0.0010
NM	−0.0631	−0.0487	0.0493	0.0038	0.0045	0.0075
AZ	−0.1042	−0.0097	−0.0091	0.0146	0.0003	0.0004
UT	−0.2381	0.0833	−0.0466	0.0542	0.0132	0.0066
NV	0.0480	0.0278	0.0219	0.0030	0.0020	0.0020
WA	−0.1148	−0.0005	0.0305	0.0146	0.0000	0.0033
OR	−0.1266	−0.0141	−0.0127	0.0171	0.0004	0.0005
CA	0.0295	0.0095	0.0014	0.0013	0.0003	0.0000
AK	0.0057	0.0849	0.0210	0.0000	0.0124	0.0012
HI	−0.1047	0.1307	0.0737	0.0131	0.0406	0.0207

Table 13.4. Coefficients and absolute contributions for regions according to U.S. crimes. ◻ SMScorrcrime

Crime	s_1	s_2	s_3	$C_a(j,s_1)$	$C_a(j,s_2)$	$C_a(j,s_3)$
murder	0.1727	−0.4860	0.0643	0.0023	0.0366	0.0010
rape	0.0661	−0.1874	−0.0079	0.0008	0.0124	0.0000
robbery	0.5066	−0.0261	−0.4045	0.2961	0.0016	0.6009
assault	0.1807	−0.3933	0.0116	0.0503	0.4731	0.0007
burglary	0.0620	−0.0631	0.0830	0.0406	0.0837	0.2320
larceny	−0.1199	0.0222	−0.0345	0.3176	0.0217	0.0835
auto theft	0.2644	0.2113	0.0785	0.2923	0.3710	0.0820

Table 13.5. Coefficients and absolute contributions for U.S. crimes.
Q SMScorrcrime

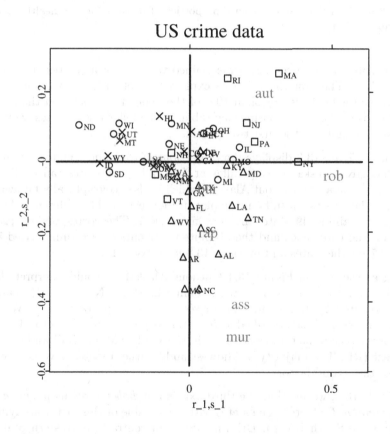

Fig. 13.2. Projection of rows (states) and columns (crimes) in U.S. crime data. Northeast (square), Midwest (circle), South (triangle) and West (cross).
Q SMScorrcrime

EXERCISE 13.9. *Repeat Exercise 13.8 with the U.S. health data in Table A.19. Only analyze the columns indicating the number of deaths per state.*

λ_j	percentage of variance	cumulated percentage
255.390	0.6046	0.6046
75.097	0.1778	0.7824
41.518	0.0983	0.8807
19.749	0.0468	0.9275
19.126	0.0453	0.9728
11.512	0.0273	1.0000
0.000	0.0000	1.0000

Table 13.6. Eigenvalues and explained proportion of variance for U.S. health Data.
Q SMScorrhealth

The eigenvalues and percentages of explained variance for all states are given in Table 13.6. The first three factors explain 88% of the total variance. As the third factor explains less than 10% of the dependency between the rows and columns of the given contingency table, in the following analyses we will concentrate mainly on the first two factors.

The plot in Figure 13.3 displays the projections of rows and columns. It suggests that AK (Alaska) is very different from all other states (an outlier). Repeating the analysis without Alaska—which is also geographically far away from the other states—results in the plot in Figure 13.4. The differences between the remaining 49 states are now more clear. The corresponding projection on the three axes and the absolute contributions are summarized in Table 13.7 for the states and in Table 13.8 for causes of death.

Looking at the plot in Figure 13.4 (without Alaska) we could interpret the first axis as an *accident*(+) factor with dominating states NV(+), NM(+) and WY(+) versus RI(−). This first factor seems to be important in the West. The second axis may be described as *liver* versus *pneumonia flu* factor. Large values of the second factor are observed in the West (NM and NV) and in the Northeast (RI). The majority of Midwest and Southern states have negative values of the second factor.

From Table 13.8, we see that the third axis is the *diabetes* versus *pulmonary and pneumonia flu* factor. The states with large value of this factor are lying mainly in the South LA(+), DE(+), MS(+) in contrast to Western States CO(−), OR(−), and AZ(−).

The regions have clearly different behavior with respect to causes of death. We could even say that the axes of the graph divide the states into four groups which correspond to the four U.S. regions. The biggest differences are observed between Western and Midwestern states.

State	r_1	r_2	r_3	$C_a(i, r_1)$	$C_a(i, r_2)$	$C_a(i, r_3)$
ME	−0.0508	0.0365	−0.0321	0.0081	0.0143	0.0200
NH	−0.0287	0.0302	−0.0115	0.0022	0.0084	0.0022
VT	−0.0096	0.0091	−0.0409	0.0003	0.0008	0.0302
MA	−0.0891	0.0201	−0.0342	0.0247	0.0043	0.0223
RI	−0.1154	0.0803	0.0354	0.0427	0.0703	0.0247
CT	−0.0634	0.0297	0.0009	0.0116	0.0087	0.0000
NY	−0.1018	0.0113	−0.0233	0.0335	0.0014	0.0108
NJ	−0.0984	0.0390	0.0098	0.0299	0.0159	0.0018
PA	−0.1007	0.0131	0.0198	0.0336	0.0019	0.0080
OH	−0.0791	0.0217	0.0136	0.0184	0.0047	0.0034
IN	−0.0526	−0.0142	0.0146	0.0079	0.0019	0.0038
IL	−0.0853	−0.0002	−0.0028	0.0213	0.0000	0.0001
MI	−0.0602	0.0181	−0.0057	0.0100	0.0031	0.0006
WI	−0.0840	−0.0237	0.0114	0.0203	0.0055	0.0023
MN	−0.0396	−0.0317	−0.0211	0.0042	0.0091	0.0073
IA	−0.0597	−0.0503	−0.0283	0.0113	0.0274	0.0156
MO	−0.0439	−0.0179	−0.0147	0.0061	0.0035	0.0042
ND	0.0097	−0.0553	0.0358	0.0003	0.0281	0.0213
SD	0.0070	−0.1107	−0.0317	0.0002	0.1326	0.0196
NE	−0.0414	−0.0701	−0.0423	0.0053	0.0516	0.0339
KS	−0.0211	−0.0450	−0.0183	0.0013	0.0206	0.0061
DE	−0.0405	0.0739	0.0668	0.0046	0.0525	0.0777
MD	−0.0408	0.0710	0.0303	0.0043	0.0444	0.0147
VA	−0.0181	0.0074	−0.0066	0.0008	0.0005	0.0007
WV	−0.0293	−0.0298	0.0013	0.0028	0.0098	0.0000
NC	0.0096	−0.0212	0.0171	0.0002	0.0040	0.0048
SC	0.0300	−0.0355	0.0474	0.0023	0.0108	0.0348
GA	0.0450	−0.0255	0.0164	0.0051	0.0056	0.0042
FL	−0.0388	0.0605	−0.0042	0.0052	0.0428	0.0004
KY	0.0040	−0.0191	0.0048	0.0000	0.0037	0.0004
TN	−0.0109	−0.0322	−0.0009	0.0003	0.0100	0.0000
AL	0.0101	−0.0012	0.0441	0.0003	0.0000	0.0334
MS	0.0502	−0.0671	0.0641	0.0071	0.0430	0.0710
AR	−0.0123	−0.0431	0.0132	0.0005	0.0201	0.0034
LA	0.0293	−0.0241	0.0938	0.0023	0.0052	0.1423
OK	0.0688	−0.0537	0.0268	0.0142	0.0293	0.0132
TX	0.0789	−0.0181	0.0374	0.0142	0.0025	0.0196
MT	0.1231	−0.0023	−0.0216	0.0407	0.0000	0.0077
ID	0.1303	−0.0223	−0.0297	0.0393	0.0039	0.0126
WY	0.3139	−0.0452	0.0095	0.1962	0.0138	0.0011
CO	0.1482	−0.0078	−0.0822	0.0449	0.0004	0.0848
NM	0.2959	0.1168	0.0364	0.1756	0.0930	0.0163
AZ	0.1107	0.0604	−0.0645	0.0301	0.0305	0.0629
UT	0.1280	−0.0434	0.0267	0.0273	0.0107	0.0073
NV	0.1778	0.1030	−0.0097	0.0733	0.0836	0.0013
WA	0.0346	0.0305	−0.0416	0.0030	0.0080	0.0269
OR	0.0198	0.0082	−0.0612	0.0011	0.0006	0.0620
CA	0.0278	0.0576	−0.0561	0.0020	0.0286	0.0491
HI	0.0744	0.0707	0.0298	0.0093	0.0284	0.0091

Table 13.7. Coefficients and absolute contributions for regions in U.S. health data set. 🔲 SMScorrhealth

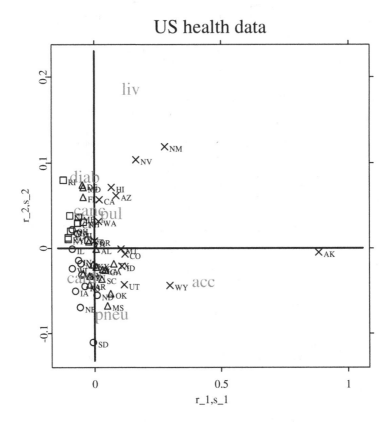

Fig. 13.3. Projection of rows (states) and columns (causes of death) for U.S. health data with Alaska. Northeast (square), Midwest (circle), South (triangle) and West (cross). ⊙ SMScorrhealth

Cause of death	s_1	s_2	s_3	$C_a(j, s_1)$	$C_a(j, s_2)$	$C_a(j, s_3)$
accident	0.2990	−0.0333	0.0500	0.7453	0.0314	0.1283
cardiovascular	−0.0372	−0.0274	0.0013	0.1072	0.1980	0.0008
cancer	−0.0218	0.0520	0.0068	0.0165	0.3180	0.0099
pulmonary	0.1370	0.0456	−0.1070	0.0967	0.0364	0.3627
pneumonia flu	0.0708	−0.0711	−0.0953	0.0204	0.0700	0.2273
diabetes	−0.0050	0.0899	0.1100	0.0000	0.0795	0.2153
liver	0.0826	0.1969	−0.0669	0.0138	0.2666	0.0557

Table 13.8. Coefficients and absolute contributions for causes of death in the U.S. health data set. ⊙ SMScorrhealth

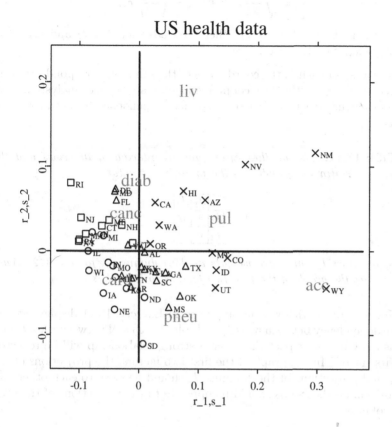

Fig. 13.4. Projection of rows (states) and columns (causes of death) for U.S. health data without Alaska. Northeast (square), Midwest (circle), South (triangle) and West (cross). ◘ SMScorrhealth

EXERCISE 13.10. *Consider a $(n \times n)$ contingency table being a diagonal matrix $\mathcal{X}$. What do you expect the factors r_k, s_k to be like?*

If $\mathcal{X}$ is a diagonal matrix then both the column totals $x_{i\bullet}$ and row totals $x_{\bullet i}$ for $i = 1, \dots, n$ are equal to the diagonal elements x_{ii}. It follows that $\mathcal{X} = \mathcal{A} = \mathcal{B}$. Now, we can apply the relations (13.13) and (13.14) between r_k and s_k from Exercise 13.7 and we obtain:

$$r_k = \sqrt{\frac{x_{\bullet\bullet}}{\lambda_k}} \mathcal{A}^{-1} \mathcal{X} s_k = \sqrt{\frac{x_{\bullet\bullet}}{\lambda_k}} s_k$$

and

$$s_k = \sqrt{\frac{x_{\bullet\bullet}}{\lambda_k}} \mathcal{B}^{-1} \mathcal{X}^\top r_k = \sqrt{\frac{x_{\bullet\bullet}}{\lambda_k}} r_k.$$

Plugging the first formula into the other one leads that $r_k = s_k$ and $x_{\bullet\bullet}/\lambda_k = 1$, i.e., $\lambda_k = x_{\bullet\bullet}$ for all $k = 1, \ldots, n$.

In other words, for each k, the coordinates of the kth row correspond perfectly to the coordinates of the kth column and correspondence analysis always discovers the true structure if there is a perfect dependency between rows and columns.

EXERCISE 13.11. *Assume that after some reordering of the rows and the columns, the contingency table has the following structure:*

$$\mathcal{X} = \begin{array}{c|c|c} & J_1 & J_2 \\ \hline I_1 & * & 0 \\ \hline I_2 & 0 & * \end{array}.$$

That is, the rows I_i only have weights in the columns J_i, for $i = 1, 2$. What do you expect the graph of the first two factors to look like?

A contingency table with a structure given in Exercise 13.11 displays strong negative dependency between rows I_1 and columns J_2 and between rows I_2 and columns J_1. One can expect that such a strong relationship will be reflected in the first factor. In the graph of the first two factors, the projections of the rows I_1 and projections of the columns J_1 should lie close to each other and their position on the x-axis should be opposite to the projections of the rows I_2 and columns J_2.

As an illustration, we calculate the factors for a $(2n \times 2n)$ contingency table $\mathcal{X}$ containing only ones in the blocks on the diagonal

$$\mathcal{X} = \begin{pmatrix} 1_n 1_n^\top & 0_n 0_n^\top \\ 0_n 0_n^\top & 1_n 1_n^\top \end{pmatrix}.$$

Clearly, $E_{ij} = n^2/2n^2 = 1/2$ and

$$C = \frac{1}{2} \begin{pmatrix} 1_n 1_n^\top & -1_n 1_n^\top \\ -1_n 1_n^\top & 1_n 1_n^\top \end{pmatrix} = \begin{pmatrix} 1_n \\ -1_n \end{pmatrix} \frac{1}{2} \begin{pmatrix} 1_n^\top & -1_n^\top \end{pmatrix}.$$

Matrix C has only one nonzero eigenvalue and the representation in one dimension describes all dependencies in the contingency table. The projections of the first n rows coincide with the projections of the first n columns and have opposite sign than the projections of the remaining rows and columns.

In practice, the output of correspondence analysis will depend on the data contained in the given contingency table and it might differ a lot from our expectations.

EXERCISE 13.12. *Redo Exercise 13.11 using the following contingency table:*

$$
\mathcal{X} =
\begin{array}{c|c|c|c}
 & J_1 & J_2 & J_3 \\
\hline
I_1 & * & 0 & 0 \\
\hline
I_2 & 0 & * & 0 \\
\hline
I_3 & 0 & 0 & *
\end{array}.
$$

In a contingency table with the above structure, one could expect that the first two factors will be driven by the block diagonal structure. Two factors should suffice to display clearly the strong negative dependency between the different blocks of the variables. In the graph of the first two factors, we should see three groups of points, one corresponding to rows I_1 and columns J_1, second group to rows I_2 and columns J_2 and third group to rows I_3 and columns J_3.

As in Exercise 13.11, we calculate the factors for an idealized $(3n \times 3n)$ contingency table $\mathcal{X}$ containing ones in the $(n \times n)$ blocks on the diagonal

$$
\mathcal{X} =
\begin{pmatrix}
1_n 1_n^\top & 0_n 0_n^\top & 0_n 0_n^\top \\
0_n 0_n^\top & 1_n 1_n^\top & 0_n 0_n^\top \\
0_n 0_n^\top & 0_n 0_n^\top & 1_n 1_n^\top
\end{pmatrix}.
$$

Here, $E_{ij} = n^2/3n^2 = 1/3$ and

$$
\mathcal{C} = \frac{1}{3}
\begin{pmatrix}
2(1_n 1_n^\top) & -1_n 1_n^\top & -1_n 1_n^\top \\
-1_n 1_n^\top & 2(1_n 1_n^\top) & -1_n 1_n^\top \\
-1_n 1_n^\top & -1_n 1_n^\top & 2(1_n 1_n^\top)
\end{pmatrix}
$$

$$
= \frac{1}{3}
\begin{pmatrix}
1_n & 0_n \\
-(1/2)1_n & (3/4)^{1/2}1_n \\
-(1/2)1_n & -(3/4)^{1/2}1_n
\end{pmatrix}
\begin{pmatrix}
2 & 0 \\
0 & 2
\end{pmatrix}
\begin{pmatrix}
1_n^\top & -(1/2)1_n^\top & -(1/2)1_n^\top \\
0_n^\top & (3/4)^{1/2}1_n^\top & -(3/4)^{1/2}1_n^\top
\end{pmatrix}.
$$

Matrix $\mathcal{C}$ has two nonzero eigenvalues and the representation in two dimensions describes all dependencies in the contingency table. The projections of the first n rows coincide with the projections of the first n columns, second n rows have the same coordinates as the second n columns and the last n rows overlap with the last n columns. Notice that the first factor explains the same amount of dependency as the second factor. Also the distances between the projections for all three groups are identical.

Again, the exact shape of the 2-dimensional graph will strongly depend on the data and, depending on the structure inside the blocks lying on the diagonal, it might lead to other results.

EXERCISE 13.13. *Consider the French food data in Table A.9. Given that all of the variables are measured in the same units (French Francs), explain how this table can be considered as a contingency table. Perform a correspondence analysis and compare the results to those obtained in the NPCA analysis in Härdle & Simar (2003, chapter 9).*

The amount of money spent by a certain family on a certain kind of food can be rephrased as, e.g., number of one-franc notes falling into that category. Hence, we can say that the entries in the French food data set are counts and the data set can be interpreted as a contingency table.

λ_j	percentage of variance	cumulated percentage
852.44	0.6606	0.6606
319.78	0.2478	0.9084
61.04	0.0473	0.9557
31.89	0.0247	0.9804
18.23	0.0141	0.9945
7.01	0.0055	1.0000
0.00	0.0000	1.0000

Table 13.9. Eigenvalues and cumulated percentage of explained variance for the French food data. ⬛ SMScorrfood

From Table 13.9, we can see that the first two eigenvalues account for 91% of the variance. Representation in two dimensions will be satisfactory. Figure 13.5 plots the projections of the rows (12 types of families) and columns (7 kinds of food). The projections on the first three axes along with their

Type of family	r_1	r_2	r_3	$C_a(i, r_1)$	$C_a(i, r_2)$	$C_a(i, r_3)$
MA2	−0.0977	−0.1443	0.0418	0.0420	0.2443	0.1072
EM2	0.0414	−0.0158	0.0319	0.0078	0.0030	0.0638
CA2	0.0756	−0.0909	−0.0093	0.0351	0.1355	0.0074
MA3	−0.1298	−0.0461	0.0151	0.0808	0.0272	0.0153
EM3	−0.0798	−0.0115	0.0312	0.0308	0.0017	0.0657
CA3	0.1580	−0.0464	−0.0336	0.1772	0.0408	0.1121
MA4	−0.1529	0.0240	−0.0265	0.1272	0.0084	0.0534
EM4	−0.0509	−0.0079	0.0143	0.0153	0.0010	0.0170
CA4	0.1680	−0.0175	−0.0300	0.1953	0.0056	0.0871
MA5	−0.1695	0.0298	−0.0404	0.1833	0.0151	0.1454
EM5	−0.0277	0.1215	−0.0206	0.0053	0.2731	0.0412
CA5	0.1091	0.1046	0.0493	0.0996	0.2442	0.2844

Table 13.10. Coefficients and absolute contributions for row factors of the French food data. ⬛ SMScorrfood

absolute contribution to the variance of the axes are given in Table 13.10 for the families and in Table 13.11 for the food. The row labels describe the type of family using the following code: MA denotes manual worker, EM denotes employees, and CA denotes manager families. The number denotes the number of children.

Food category	s_1	s_2	s_3	$C_a(j, s_1)$	$C_a(j, s_2)$	$C_a(j, s_3)$
bread	−0.1862	0.0437	−0.0536	0.2179	0.0320	0.2525
vegetables	0.0077	0.0030	0.0638	0.0001	0.1251	0.0032
fruits	0.0352	0.1355	0.0074	0.1140	0.0030	0.2782
meat	0.0808	0.0272	0.0153	0.0355	0.0455	0.0297
poultry	0.1224	−0.0166	−0.0448	0.1694	0.0083	0.3173
milk	−0.1875	0.1517	0.0369	0.1773	0.3095	0.0957
wine	−0.2345	−0.1856	0.0179	0.2852	0.4766	0.0233

Table 13.11. Coefficients and absolute contributions for column factors of the French food data. Q SMScorrfood

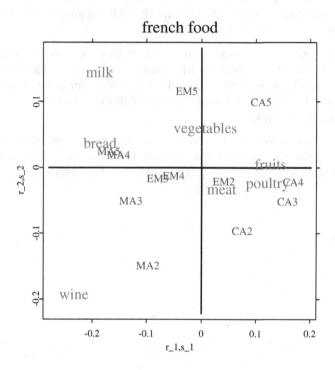

Fig. 13.5. Factorial decomposition of the French food data. Q SMScorrfood

Figure 13.5 shows that wine on the left side and fruits and poultry on the right side are most strongly responsible for the variation on the first axis. The second axis describes an opposition between milk and wine. These interpretations are confirmed in Table 13.11, where in the first column factor s_1, the difference between the coefficient of wine (-0.2345) and fruits (0.1267) is the largest, and in the second column factor s_2, the difference between the coefficient of wine (-0.1856) and milk (0.1517) is the largest.

The relationship between the row and the column categories can be assessed by looking at the position of the row and column projections in Figure 13.5. On the x-axis, the employee families are lying close to the origin and seem to have a general food structure consisting mainly of meat and vegetables. On the left, we observe the poorer group consisting of manual workers close to wine, bread, and milk. On the right we find the richer manager families which are projected close to fruits and poultry.

The position of the projections on the y-axis corresponds to the number of children in the family. The families with many children lie in the upper part of the graph together with the projections of the column categories milk, vegetables, and bread. The families with less children seem to be related to the column category wine.

The results of the correspondence analysis are in a good agreement with the results of the principal component analysis of the same data set in Härdle & Simar (2003, example 9.6) although the method is based on a different look at the data.

Canonical Correlation Analysis

A glance at our friend here reveals the rounded head of the Celt, which carries inside it the Celtic enthusiasm and power of attachment.
Dr. Mortimer in "The Hound of the Baskervilles"

The association between two sets of variables may be quantified by canonical correlation analysis (CCA). Given a set of variables $X \in \mathbb{R}^q$ and another set $Y \in \mathbb{R}^p$, one asks for the linear combination $a^\top X$ that "best matches" a linear combination $b^\top Y$. The best match in CCA is defined through maximal correlation. The task of CCA is therefore to find $a \in \mathbb{R}^q$ and $b \in \mathbb{R}^p$ so that the correlation $\rho(a, b) = \rho_{a^\top X, b^\top Y}$ is maximized. These best-matching linear combinations $a^\top X$ and $b^\top Y$ are then called canonical correlation variables; their correlation is the canonical correlation coefficient. The coefficients a and b of the canonical correlation variables are the canonical vectors.

Let us assume that the two random vectors under investigation, X and Y, have the following covariance structure

$$Var \begin{pmatrix} X \\ Y \end{pmatrix} = \begin{pmatrix} \Sigma_{XX} & \Sigma_{XY} \\ \Sigma_{YX} & \Sigma_{YY} \end{pmatrix}.$$

The algorithm of CCA consists of calculating the matrix

$$\mathcal{K} = \Sigma_{XX}^{-1/2} \Sigma_{XY} \Sigma_{YY}^{-1/2}$$

and its SVD

$$\mathcal{K} = \Gamma \Lambda \Delta^\top.$$

The diagonal elements of the matrix Λ are the canonical correlation coefficients. The canonical correlation vectors can be obtained as

$$a_i = \Sigma_{XX}^{-1/2} \gamma_i,$$
$$b_i = \Sigma_{YY}^{-1/2} \delta_i,$$

and the canonical correlation variables are

$$\eta_i = a_i^\top X,$$
$$\varphi_i = b_i^\top Y.$$

It can be easily verified that

$$\text{Var} \begin{pmatrix} \eta \\ \varphi \end{pmatrix} = \begin{pmatrix} \mathcal{I}_k & \Lambda \\ \Lambda & \mathcal{I}_k \end{pmatrix}.$$

EXERCISE 14.1. *Calculate the canonical variables for the complete car marks data set. Interpret the coefficients.*

As in Härdle & Simar (2003, example 14.1), we split the observed variables into two logical subsets: $X = $ (price, value)$^\top$ and $Y = $ (economy, service, design, sportiness, safety, easy handling)$^\top$.

The empirical covariance matrix is

$$S = \begin{pmatrix} 1.41 & -1.11 & 0.78 & -0.71 & -0.90 & -1.04 & -0.95 & 0.18 \\ -1.11 & 1.19 & -0.42 & 0.82 & 0.77 & 0.90 & 1.12 & 0.11 \\ 0.78 & -0.42 & 0.75 & -0.23 & -0.45 & -0.42 & -0.28 & 0.28 \\ -0.71 & 0.82 & -0.23 & 0.66 & 0.52 & 0.57 & 0.85 & 0.14 \\ -0.90 & 0.77 & -0.45 & 0.52 & 0.72 & 0.77 & 0.68 & -0.10 \\ -1.04 & 0.90 & -0.42 & 0.57 & 0.77 & 1.05 & 0.76 & -0.15 \\ -0.95 & 1.12 & -0.28 & 0.85 & 0.68 & 0.76 & 1.26 & 0.22 \\ 0.18 & 0.11 & 0.28 & 0.14 & -0.10 & -0.15 & 0.22 & 0.32 \end{pmatrix}.$$

In this case, the first random vector has only two components. Hence, we can obtain only two pairs of canonical variables. The corresponding canonical correlations are $r_1 = 0.98$ and $r_2 = 0.89$. The relationship between both pairs of canonical variables seems to be quite strong.

The first pair of canonical vectors, corresponding to r_1, is

$$a_1 = (-0.33, 0.59)^\top,$$
$$b_1 = (-0.43, 0.19, 0.00, 0.46, 0.22, 0.38)^\top,$$

and the second pair of canonical vectors

$$a_2 = (1.602, 1.686)^\top,$$
$$b_2 = (0.568, 0.544, -0.012, -0.096, -0.014, 0.915)^\top.$$

These coefficients lead to the canonical variables

$$\eta_1 = -0.33x_1 + 0.59x_2,$$
$$\varphi_1 = -0.43y_1 + 0.19y_2 + 0.46y_4 + 0.22y_5 + 0.38y_6,$$

and

$$\eta_2 = 1.602x_1 + 1.686x_2,$$
$$\varphi_2 = 0.568y_1 + 0.544y_2 - 0.012y_3 - 0.096y_4 - 0.014y_5 + 0.915y_6.$$

From the first canonical variables, we see that x_1 (price) is positively related to y_1 (economy), and negatively related to the remaining characteristics of a car (service, sportiness, safety and easy handling). The variable x_2 (value) is negatively related to y_1 (economy), and positively related to the other characteristics.

The canonical variable η_1 can be interpreted as a value index of the car. On the one side, we observe cars with good (low) price and bad (high) appreciation of value such as Trabant and Wartburg and on the other side, we see cars with high price and good (low) appreciation of value such as BMW, Jaguar, Ferrari and Mercedes. Similarly, φ_1 can be interpreted as a quality index consisting of variables such as service and safety. The value and quality indeces are highly correlated with the canonical correlation coefficient 0.98. We can see this correlation in Figure 14.1.

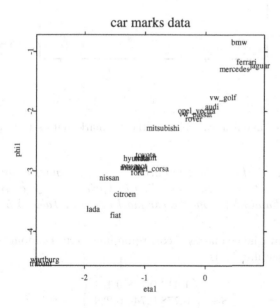

Fig. 14.1. Scatterplot of first canonical variables for the car marks data set.
Q SMScancarm1

The second pair of canonical variables provides more insight into the relationship between the two sets of variables. η_2 has low values for cars with good marks both in price and value, e.g., VW and Opel. On the right hand side, we should see cars with bad marks in these two variables such as Ferrari and Wartburg. The canonical variable φ_2 consists mainly of variables economy and service. The position of cars is displayed on Figure 14.2.

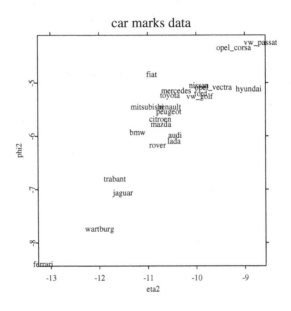

Fig. 14.2. Second canonical variables for the car marks data set. Q SMScancarm2

EXERCISE 14.2. *Perform the canonical correlation analysis for the following subsets of variables:* $\mathcal{X}$ *corresponding to {price} and* $\mathcal{Y}$ *corresponding to {economy, easy handling} from the car marks data in Table A.5.*

The estimated covariance matrix S corresponding to the random vector (price, economy, easy handling)$^\top$ is

$$S = \begin{pmatrix} 1.412 & 0.778 & 0.181 \\ 0.778 & 0.746 & 0.284 \\ 0.181 & 0.284 & 0.318 \end{pmatrix}.$$

The canonical vectors maximizing the correlation between linear combinations of {price} and {economy, easy handling} are $a = -0.84155$ and $b = (-1.3378, 0.58526)^\top$. The canonical variables are thus $\eta = -0.84155 \times$ price and $\varphi = -1.3378 \times$ economy$+0.58526 \times$ easy handling. In this example, we obtain only one pair of canonical variables. We observe that the price has negative influence on the canonical variable η which means that price is positively related to economy and negatively related to easy handling. The canonical correlation coefficient is $r = 0.78718$.

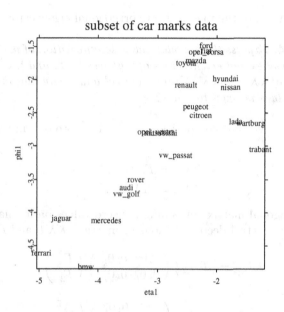

Fig. 14.3. Scatterplot of the first pair of canonical variables for a subset of the car marks data set. Ⓠ SMScancarm

From Figure 14.3, we can that see the relationship between the two canonical variables is not so strong as in Exercise 14.1 where more variables from the same data set are analyzed.

EXERCISE 14.3. *Use the SVD of matrix $\mathcal{K}$ to show that the canonical variables η_1 and η_2 are not correlated.*

Recall that the canonical vectors are defined as $a_i^\top = \Sigma_{XX}^{-\frac{1}{2}} \gamma_i$, where γ_i are eigenvectors of matrix $\mathcal{K}\mathcal{K}^\top$ with $\mathcal{K} = \Sigma_{XX}^{-\frac{1}{2}} \Sigma_{XY} \Sigma_{YY}^{-\frac{1}{2}}$.

To show that the correlation between the first two canonical variables $\eta_i = a_i^\top X$, $i = 1, 2$ is equal to zero, it is sufficient to show that the covariance between these random variables is zero:

$$\begin{aligned}
\mathrm{Cov}(\eta_1, \eta_2) &= \mathrm{Cov}(\gamma_1^\top \Sigma_{XX}^{-\frac{1}{2}} X, \gamma_2^\top \Sigma_{XX}^{-\frac{1}{2}} X) \\
&= \gamma_1^\top \Sigma_{XX}^{-\frac{1}{2}} \mathrm{Cov}(X, X) \Sigma_{XX}^{-\frac{1}{2}} \gamma_2 \\
&= \gamma_1^\top \Sigma_{XX}^{-\frac{1}{2}} \Sigma_{XX} \Sigma_{XX}^{-\frac{1}{2}} \gamma_2 \\
&= \gamma_1^\top \gamma_2 \\
&= 0
\end{aligned}$$

because the columns of the matrix Γ are orthogonal eigenvectors.

EXERCISE 14.4. *Express the singular value decomposition of matrices $\mathcal{K}$ and $\mathcal{K}^\top$ using eigenvalues and eigenvectors of matrices $\mathcal{K}^\top \mathcal{K}$ and $\mathcal{K}\mathcal{K}^\top$, show that the eigenvalues of $\mathcal{K}\mathcal{K}^\top$ and $\mathcal{K}^\top \mathcal{K}$ are identical and verify that the number of nonzero eigenvalues is equal to* $\mathrm{rank}(\Sigma_{XY})$.

Using the singular value decomposition $\mathcal{K} = \Gamma \Lambda \Delta^\top$, we obtain the decompositions

$$\mathcal{K}\mathcal{K}^\top = \Gamma \Lambda^2 \Gamma^\top,$$
$$\mathcal{K}^\top \mathcal{K} = \Delta \Lambda^2 \Delta^\top,$$

where Λ is a diagonal matrix containing nonzero values on its diagonal. This implies that the spectral decompositions of matrices $\mathcal{K}\mathcal{K}^\top$ and $\mathcal{K}^\top \mathcal{K}$ can be written as

$$\mathcal{K}\mathcal{K}^\top = \begin{pmatrix} \Gamma & \Gamma_2 \end{pmatrix} \begin{pmatrix} \Lambda^2 & 0_k 0_k^\top \\ 0_k 0_k^\top & 0_k 0_k^\top \end{pmatrix} \begin{pmatrix} \Gamma^\top \\ \Gamma_2^\top \end{pmatrix}$$

and

$$\mathcal{K}^\top \mathcal{K} = \begin{pmatrix} \Delta & \Delta_2 \end{pmatrix} \begin{pmatrix} \Lambda^2 & 0_k 0_k^\top \\ 0_k 0_k^\top & 0_k 0_k^\top \end{pmatrix} \begin{pmatrix} \Delta^\top \\ \Delta_2^\top \end{pmatrix},$$

i.e., we see that the nonzero eigenvalues of the two matrices are identical.

We remark that the number of zero eigenvalues depends on the dimension of matrix $\mathcal{K}$. The number of nonzero eigenvalues of both $\mathcal{K}\mathcal{K}^\top$ and $\mathcal{K}^\top \mathcal{K}$ is identical and it is equal to the dimension of the matrix Λ and hence also to

$$\mathrm{rank}(\mathcal{K}) = \mathrm{rank}\left(\Sigma_{XX}^{-\frac{1}{2}} \Sigma_{XY} \Sigma_{YY}^{-\frac{1}{2}}\right) = \mathrm{rank}(\Sigma_{XY})$$

because the matrices Σ_{XX} and Σ_{YY} have full rank.

EXERCISE 14.5. *What will be the result of CCA for $Y = X$?*

We know that the variance matrix of the canonical variables is equal to

$$Var \begin{pmatrix} \eta \\ \varphi \end{pmatrix} = \begin{pmatrix} \mathcal{I}_k & \Lambda \\ \Lambda & \mathcal{I}_k \end{pmatrix}.$$

Defining $\eta = \Sigma_{XX}^{-1/2} X = \Sigma_{YY}^{-1/2} Y = \varphi$ leads to the desired correlation structure with all canonical correlation coefficients equal to one.

EXERCISE 14.6. *What will be the results of CCA for $Y = 2X$ and for $Y = -X$?*

Similarly as in the previous Exercise 14.5, we define $\eta = \Sigma_{XX}^{-1/2} X$ and $\varphi = \Sigma_{YY}^{-1/2} Y$ in order to obtain the perfect correlation structure

$$Var \begin{pmatrix} \eta \\ \varphi \end{pmatrix} = \begin{pmatrix} \mathcal{I}_k & \mathcal{I}_k \\ \mathcal{I}_k & \mathcal{I}_k \end{pmatrix}.$$

EXERCISE 14.7. *What results do you expect if you perform CCA for X and Y such that $\Sigma_{XY} = 0_p 0_q^\top$? What if $\Sigma_{XY} = \mathcal{I}_p$?*

CCA for two uncorrelated sets of variables, with $\Sigma_{XY} = 0_p 0_q^\top$, would lead to zero canonical correlation coefficients. The canonical variables would be the Mahalanobis transformations of the original variables and, due to the zero correlation, the assignment of the variables to the pairs could be arbitrary.

The assumption $\Sigma_{XY} = \mathcal{I}_p$ means that both vectors have the same dimension. The canonical correlation coefficients and canonical variables cannot be deduced only from this information, we would need to know also the variance matrices of X and Y. We can say only that each component of X is positively related to the same component of Y. Thus, we can expect that both canonical variables η and φ will be calculated as weighted averages of the variables X and Y, respectively, with all weights positive.

Multidimensional Scaling

> It was a nice question, for the Cape de Verds were about 500 miles to
> the north of us, and the African coast about 700 miles to the east. On
> the whole, as the wind was coming round to north, we thought that
> Sierra Leone might be best, ...
> James Armitage in "The Adventure of the "Gloria Scott""

Multidimensional scaling (MDS) is a mathematical tool that uses proximities
between observations to produce their spatial representation. In contrast to
the techniques considered so far, MDS does not start from the raw multivariate
data matrix $\mathcal{X}$, but from an $(n \times n)$ dissimilarity or distance matrix, $\mathcal{D}$, with
the elements δ_{ij} and d_{ij}, respectively. Hence, the underlying dimensionality
of the data under investigation is in general not known.

MDS is a data reduction technique because it is concerned with the problem
of finding a set of points in low dimension that represents the configuration
of data in high dimension.

The metric MDS solution may result in projections of data objects that con-
flict with the ranking of the original observations. The nonmetric MDS solves
this problem by iterating between a monotonizing algorithmic step and a least
squares projection step. The examples presented in this chapter are based on
reconstructing a map from a distance matrix and on marketing concerns such
as ranking the outfit of cars.

The Euclidean distance between the ith and jth points, d_{ij}, is defined as

$$d_{ij}^2 = \sum_{k=1}^{p}(x_{ik} - x_{jk})^2,$$

where p is the dimension of the observations. Multidimensional scaling aims
to find the original Euclidean coordinates from a given distance matrix $\mathcal{D} =
(d_{ij})_{i,j=1,\ldots,n}$.

With a_{ij} defined as $-d_{ij}^2/2$ and

$$a_{i\bullet} = \frac{1}{n} \sum_{j=1}^{n} a_{ij}, \quad a_{\bullet j} = \frac{1}{n} \sum_{i=1}^{n} a_{ij}, \quad \text{and} \quad a_{\bullet\bullet} = \frac{1}{n^2} \sum_{i=1}^{n} \sum_{j=1}^{n} a_{ij}, \quad (15.1)$$

we get

$$b_{ij} = a_{ij} - a_{i\bullet} - a_{\bullet j} + a_{\bullet\bullet}, \quad (15.2)$$

where $b_{ij} = x_i^\top x_j$. The inner product matrix $\mathcal{B} = (b_{ij})$ can be expressed as

$$\mathcal{B} = \mathcal{X}\mathcal{X}^\top, \quad (15.3)$$

where $\mathcal{X} = (x_1, \ldots, x_n)^\top$ is the $(n \times p)$ matrix of coordinates. The matrix $\mathcal{B}$ is symmetric, positive semidefinite, and of rank p; hence it has p non-negative eigenvalues and $n - p$ zero eigenvalues and thus a spectral decomposition

$$\mathcal{B} = \Gamma \Lambda \Gamma^\top, \quad (15.4)$$

which allows us to obtain the matrix of coordinates $\mathcal{X}$ containing the point configuration in $\mathbb{R}^p$ as

$$\mathcal{X} = \Gamma \Lambda^{\frac{1}{2}}. \quad (15.5)$$

Nonmetric Solution

The idea of a nonmetric MDS is to demand a less-rigid relationship between the final configuration of the points and the distances. In nonmetric MDS, it is assumed only that this relationship can be described by some monotone function.

More formally, let us assume that we want to find a configuration of points corresponding to a given dissimilarities δ_{ij}. In nonmetric MDS, we attempt to find a configuration of points in a lower-dimensional space such that their Euclidean distances are $f(\delta_{ij})$, where $f(.)$ is some increasing function.

The most common approach is the iterative Shepard-Kruskal algorithm. In the first step, we calculate Euclidean distance d_{ij} from an initial configuration of the points. In the second step, we calculate the so-called disparities $\widehat{d}_{ij}$ such that they are a monotone function of the given dissimilarities δ_{ij} and the quality of the configuration of the points is measured by the STRESS measure:

$$\text{STRESS} = \left(\frac{\sum_{i<j}(d_{ij} - \widehat{d}_{ij})^2}{\sum_{i<j} d_{ij}^2} \right)^{1/2}. \quad (15.6)$$

In the third step, based on the differences between d_{ij} and $\widehat{d}_{ij} = \widehat{f}(\delta_{ij})$, we define a new position of the points:

$$x_{ik}^{\text{NEW}} = x_{ik} + \frac{\alpha}{n-1} \sum_{\substack{j=1 \\ j \neq i}}^{n} \left(1 - \widehat{d}_{ij}/d_{ij}\right)(x_{jk} - x_{ik}),$$

where α determines the step width of the iteration. In the fourth step, the STRESS measure is used to decide whether the change as a result of the last iteration is sufficiently small or if the iterative procedure has to be continued.

EXERCISE 15.1. *Apply the MDS method to the Swiss bank note data. What do you expect to see?*

We apply MDS on the 200×200 matrix $\mathcal{D}$ of Euclidean distance between all Swiss bank notes. We know that these distances are Euclidean. Therefore, we try to reconstruct the original configuration of points using metric MDS.

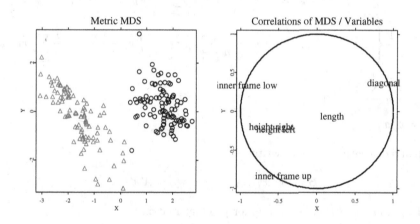

Fig. 15.1. MDS for Swiss bank notes. ⓠ SMSmdsbank

The results of metric MDS are displayed in Figure 15.1. One would expect that our results would be very similar to the principal component analysis (Härdle & Simar 2003, chapter 9).

The correlations of the projections with the original variables look indeed quite similar. Contrary to our expectations, the scatterplot of the two-dimensional projections look rather different. One can see that the separation of the two point clouds is much better in the MDS method. The reason could be that principal components are based only on an estimated of a covariance matrix which is wrong if the data set consists of more subgroups. MDS is based only on the distance matrix, it is based only on the distances between observations and it does not assume any covariance structure.

EXERCISE 15.2. *Using (15.1), show that b_{ij} in (15.2) can be written in the form (15.3).*

In the following calculation, we shall use the relations $a_{ij} = -d_{ij}^2/2$ and $d_{ij}^2 = x_i^\top x_i + x_j^\top x_j - 2x_i^\top x_j$ and we assume that the observations are centered, i.e., $\sum x_i = 0_p$.

$$b_{ij} = a_{ij} - a_{i\bullet} - a_{\bullet j} + a_{\bullet\bullet}$$

$$= -\frac{1}{2}\left\{ d_{ij}^2 - \frac{1}{n}\sum_{k=1}^{n} d_{ik}^2 - \frac{1}{n}\sum_{k=1}^{n} d_{kj}^2 + \frac{1}{n^2}\sum_{k=1}^{n}\sum_{l=1}^{n} d_{kl}^2 \right\}$$

$$= -\frac{1}{2}\left\{ x_i^\top x_i + x_j^\top x_j - 2x_i^\top x_j - \frac{1}{n}\sum_{k=1}^{n}(x_k^\top x_k + x_j^\top x_j - 2x_k^\top x_j) \right.$$

$$\left. -\frac{1}{n}\sum_{k=1}^{n}(x_i^\top x_i + x_k^\top x_k - 2x_i^\top x_k) + \frac{1}{n^2}\sum_{k=1}^{n}\sum_{l=1}^{n}(x_k^\top x_k + x_l^\top x_l - 2x_k^\top x_l) \right\}$$

$$= -\frac{1}{2}\left\{ x_i^\top x_i + x_j^\top x_j - 2x_i^\top x_j - \frac{1}{n}\sum_{k=1}^{n} x_k^\top x_k - x_j^\top x_j - x_i^\top x_i - \frac{1}{n}\sum_{k=1}^{n} x_k^\top x_k \right.$$

$$\left. +\frac{2}{n}\sum_{k=1}^{n} x_k^\top x_k \right\}$$

$$= x_i^\top x_j.$$

In matrix notation, we can write

$$\mathcal{B} = (x_i^\top x_j)_{i=1,\dots,n;j=1,\dots,n} = \mathcal{X}\mathcal{X}^\top$$

and the matrix $\mathcal{B}$ is called the inner product matrix.

EXERCISE 15.3. *Show that*

1. $b_{ii} = a_{\bullet\bullet} - 2a_{i\bullet}$; $b_{ij} = a_{ij} - a_{i\bullet} - a_{\bullet j} + a_{\bullet\bullet}$; $i \neq j$,
2. $\mathcal{B} = \sum_{i=1}^{p} x_i x_i^\top$,
3. $\sum_{i=1}^{n} \lambda_i = \sum_{i=1}^{n} b_{ii} = \frac{1}{2n}\sum_{i,j=1}^{n} d_{ij}^2$.

The first part of this question was verified in the previous exercise. The formula for b_{ii} follows immediately by setting $i = j$ in (15.2).

Also from the previous exercise, we know that $\mathcal{B} = \mathcal{X}\mathcal{X}^\top$. Let us now investigate the matrix $\mathcal{B}$ elementwise.

$$B = (x_i^\top x_j)_{i=1,\dots,n;j=1,\dots,n}$$
$$= \left(\sum_{k=1}^{p} x_{ik}x_{jk}\right)_{i=1,\dots,n;j=1,\dots,n}$$
$$= \sum_{k=1}^{p} (x_{ik}x_{jk})_{i=1,\dots,n;j=1,\dots,n}$$
$$= \sum_{k=1}^{p} x_{[k]}x_{[k]}^\top,$$

where $x_{[k]}$ denotes the kth column of the matrix $\mathcal{X}$.

The sum of eigenvalues is equal to the trace of the matrix $\mathcal{B}$,

$$\sum_{i=1}^{p} \lambda_i = \sum_{i=1}^{n} b_{ii}$$
$$= \sum_{i=1}^{p} (a_{\bullet\bullet} - 2a_{i\bullet})$$
$$= -\frac{1}{2}\sum_{i=1}^{p}\left\{\frac{1}{n^2}\sum_{k}\sum_{l} d_{kl}^2 - \frac{2}{n}\sum_{k} d_{ik}^2\right\}$$
$$= -\frac{1}{2n}\sum_{k}\sum_{l} d_{kl}^2 + \frac{1}{n}\sum_{i}\sum_{k} d_{ik}^2$$
$$= \frac{1}{2n}\sum_{i}\sum_{j} d_{ij}^2$$

EXERCISE 15.4. *Redo a careful analysis of the car marks data based on the following dissimilarity matrix:*

j	1	2	3	4
i	Nissan	Kia	BMW	Audi
1 Nissan	-			
2 Kia	2	-		
3 BMW	4	6	-	
4 Audi	3	5	1	-

The dissimilarity matrix contains obviously only ranks of dissimilarity. Applying metric MDS would not be appropriate in this situation. Nonmetric MDS, on the other hand, does not assume that the distance matrix is Euclidean. It only assumes that the dissimilarities are monotone functions of the Euclidean distances and it uses the iterative Shepard-Kruskal algorithm to find a configuration of points in two dimensions that satisfy this monotonicity.

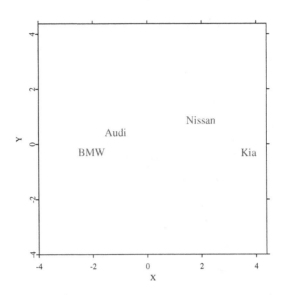

Fig. 15.2. Nonmetric MDS for four cars. ▢ SMSnmdscarm

The outcome of the Shepard-Kruskal algorithm is given in Figure 15.2. It is important that both axes have the same scale, different scales could lead to wrong interpretations.

Audi and BMW are lying very close to each other in opposition to Kia. In between, we find Nissan who seems to lie a bit closer to Kia than to Audi and BMW.

EXERCISE 15.5. *Apply the MDS method to the U.S. health data. Is the result in accordance with the geographic location of the U.S. states?*

The results of both the metric and nonmetric MDS are displayed in Figure 15.3. The metric MDS on the left hand side is used as the first iteration for the Shepard-Kruskal algorithm. The last iteration of the algorithm is displayed on the right hand side of the graphics.

We can see that standardization leads to a more informative and a much better scaled plot than the original data set.

Metric MDS applied on the original data set shows large difference between Texas and all other states. Nonmetric MDS shifts the positions of the states slightly and one can see California and New York emerging. These states,

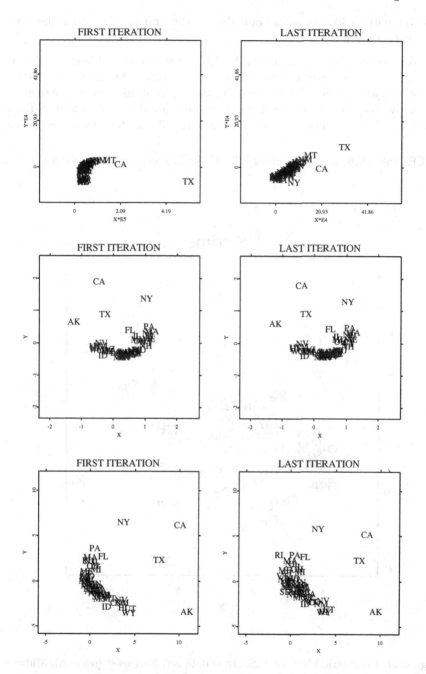

Fig. 15.3. Nonmetric MDS for the original, the 0–1 scaled, and the standardized U.S. health data set. ◻ SMSnmdsushealth

together with Arkansas, stand out also on the graphics based on the standardized data set.

We can also see some geographical East/West structure inside the big cloud containing the majority of the states. Closer to New York, we see Eastern states such as Florida or New Jersey. On the opposite side, closer to Arkansas, we see Western states such as Utah, Idaho or Nevada. California and Texas stand out of this structure and seem to be very different from the other states.

EXERCISE 15.6. *Redo Exercise 15.5 with the U.S. crime data set (Table A.18).*

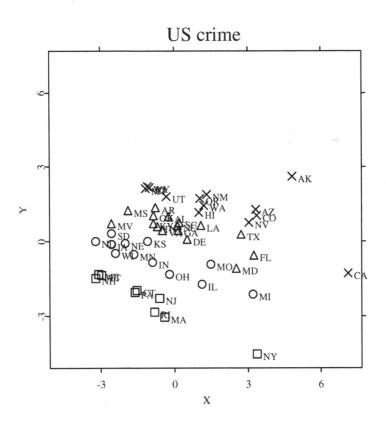

Fig. 15.4. Nonmetric MDS for U.S. crime data set. Northeast (squares), Midwest (circles), South (triangles) and West (crosses). **Q** SMSnmdsuscrime

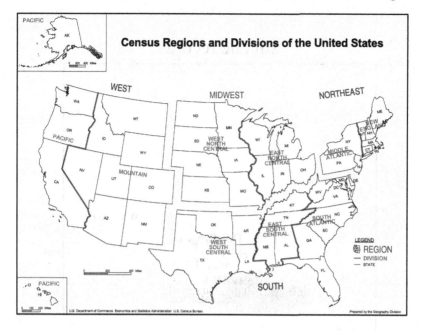

Fig. 15.5. U.S. states. Source: U.S. Census Bureau.

The results of nonmetric MDS are displayed in Figure 15.4. We standardize the data set by subtracting the sample mean and dividing by its standard deviation.

Similarly as in the previous exercise, we see that New York, California, and Arkansas stand somewhat aside. The other states seem to form a group of similar and neighboring states. The four census regions are clearly separated in the direction of the vertical axis. The West (denoted by crosses) lies in the upper part of Figure 15.4. Northeastern states (squares) are located in the lower part of the graphics. The South (triangles) seem to be more similar to West whereas Midwest lies closer to Northeast.

EXERCISE 15.7. *Perform the MDS analysis on the athletic records data in Table A.1. Can you see which countries are "close to each other"?*

Applying the nonmetric MDS in Figure 15.6, we see a cloud containing most of the countries. At some distance we observe four outliers: Netherlands, Mauritius, West Samoa and Cook Islands. Closer look at the original data set reveals that West Samoa and Cook Islands are very bad in all disciplines and that Mauritius and Netherlands are very bad in 200 meters.

It seems that the horizontal direction of the scatterplot corresponds to the overall performance of each country with Cook Islands as the worst and the

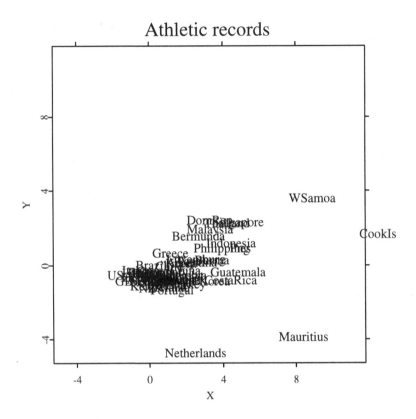

Fig. 15.6. Nonmetric MDS for the the standardized athletic records data set.
Q SMSnmdsathletic

USA as the best country. Neighboring countries seem to be usually quite close
to each other.

EXERCISE 15.8. *Repeat Exercise 15.7 without the outlying countries: Netherlands, West Samoa, Mauritius, and Cook Islands.*

In Figure 15.7, we can see the structure of the athletic records more clearly.
The countries with the best athletic records, such as USA, Italy, USSR, and
GB are located on the left. The countries with worse national athletic records
can be found on the right hand side. These countries are also more spread out
since, for example, Dominican Republic is quite good in short distance while
Costa Rica performs well in marathon.

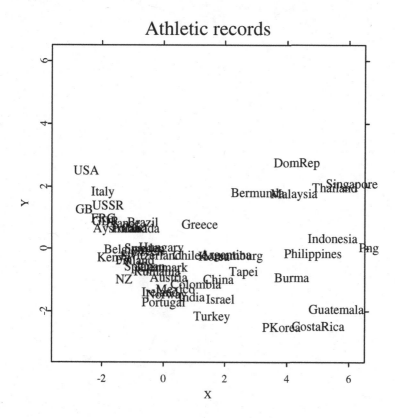

Fig. 15.7. Nonmetric MDS for the the standardized athletic records data set without the four most outlying countries. ⊡ SMSnmdsathlesub

In this exercise, the removal of the outliers leads to a better graphical display for the remaining countries.

Conjoint Measurement Analysis

It only remains, therefore, to discover what is wanted by this German who writes upon Bohemian paper, and prefers wearing mask to showing his face.
Sherlock Holmes in "A Scandal in Bohemia"

Conjoint measurement analysis is a technique to investigate the utilities attributes to certain factor levels. It is heavily used in marketing and in the analysis of consumer preferences. The statistical core of conjoint measurement analysis is ANOVA in a linear model with a specially constrained design matrix $\mathcal{X}$.

We observe the factors (elements of $\mathcal{X}$) and the preferences $\mathcal{Y}$. The aim is to estimate the part-worth that is the contribution of each factor level to its preference.

In the metric solution, the distance between any two adjacent preference orderings corresponds to the same difference in utility, i.e., the utility gain between products ranked 1st and 2nd is the same as the gain between say the 4th- and 5th-ranked product.

In the nonmetric solution, one adjusts the estimated utilities by the PAV (pool-adjacent-violators) algorithm and iterates in order to minimize a stress measure.

Design of Data Generation

A stimulus is defined as a combination of the factor levels of different components.

The profile method investigates the utility of each stimulus, i.e., we need to ask the tester for the utility of each combination of the components. For example, three components with four levels each would lead to $4 \cdot 4 \cdot 4 = 64$ different stimuli.

If the number of components and their levels is increasing, the number of stimuli might soon become too large for a questionnaire. In such a situation, we can investigate only selected subset of stimuli. One possibility is the two-factor method, which considers only pairwise combinations of the components. For three components with four levels, we would observe only $4 \cdot 4 + 4 \cdot 4 + 4 \cdot 4 = 48$ stimuli.

The utilities of stimuli are then decomposed into the part-worths of the factor levels by the standard ANOVA procedure.

Estimation of Preferences

The estimation procedure is formulated here only for data collected by the profile method. The necessary modifications for other data setups are straightforward.

The conjoint measurement problem for one individual may be rewritten as a linear regression model:

$$Y = \mathcal{X}\beta + \varepsilon$$

with $\mathcal{X}$ being a design matrix with dummy variables. If the profile method is used, the row dimension of $\mathcal{X}$ is $K = \prod_{j=1}^{J} L_j$ (the number of stimuli) and the column dimension $D = \sum_{j=1}^{J} L_j - J + 1$.

In practice we have more than one person to answer the utility rank question for the different factor levels. The design matrix is then obtained by stacking the design matrix n times. Hence, for n persons we have a design matrix:

$$\mathcal{X}^* = 1_n \otimes \mathcal{X} = \left. \begin{pmatrix} \mathcal{X} \\ \vdots \\ \vdots \\ \mathcal{X} \end{pmatrix} \right\} n - \text{times}$$

with dimensions $(nK)(L - J)$ (where $L = \sum_{j=1}^{J} L_j$) and $Y^* = (Y_1^{\top}, ..., Y_n^{\top})^{\top}$.

The linear model can now be written as:

$$Y^* = \mathcal{X}^*\beta + \varepsilon^*. \tag{16.1}$$

The solution to the least squares problem, leading to estimates of the vector of the part-worths β, is in fact provided by the standard analysis of variance (ANOVA) technique.

Nonmetric Solution

Often, the utilities are not measured on a metric scale. In this situation, we may use the monotone ANOVA (Kruskal 1965) based on a monotone transformation $\hat{Z} = f(\hat{Y})$ to the observed stimulus utilities Y.

The transformation $\hat{Z}_k = f(\hat{Y}_k)$ of the fitted values $\hat{Y}_k$ is introduced to guarantee monotonicity of preference orderings. The relationship is now monotone, but model (16.1) may now be violated. Hence, as in (15.6) in Chapter 15, the procedure is iterated until the STRESS measure

$$\text{STRESS} = \frac{\sum_{k=1}^{K}(\hat{Z}_k - \hat{Y}_k)^2}{\sum_{k=1}^{K}(\hat{Y}_k - \bar{\hat{Y}})^2}$$

is minimized over β and the monotone transformation $f(.)$.

EXERCISE 16.1. *Compute the part-worths for the following table of rankings*

		X_2	
		1	2
	1	1	2
X_1	2	4	3
	3	6	5

The given table contains the respondents rankings of the utilities of the stimuli given by all six combinations of $L_1 = 3$ levels of X_1 and $L_2 = 2$ levels of X_2.

The design matrix $\mathcal{X}$ has $K = L_1 L_2 = 6$ rows and $D = L_1 + L_2 - 2 + 1 = 4$ linearly independent columns. The design matrix is not unique and its choice depends largely on the desired interpretation of the coefficients. It is possible to increase the number of columns of $\mathcal{X}$ if we add linear constraints on the parameters β of the linear model.

For example, we can parametrize the model by calculating the overall mean utility, $\hat{\mu} = (1 + 2 + 4 + 3 + 6 + 5)/6 = 3.5$. The part-worths of X_1 can be described by the parameter vector $\beta_1 = (\beta_{11}, \beta_{12}, \beta_{13})^\top$ satisfying the constraint $1_3^\top \beta_1 = \beta_{11} + \beta_{12} + \beta_{13} = 0$. The part worths are given by $\beta_2 = (\beta_{22}, \beta_{22})^\top$ such that $\beta_{22} + \beta_{22} = 0$.

Formally, this linear model can be written in the matrix form:

$$
Y = \begin{pmatrix} Y_1 \\ Y_2 \\ Y_3 \\ Y_4 \\ Y_5 \\ Y_6 \end{pmatrix} = \begin{pmatrix} 1 \\ 2 \\ 4 \\ 3 \\ 6 \\ 5 \end{pmatrix} = \begin{pmatrix} 1\,1\,0\,0\,1\,0 \\ 1\,0\,1\,0\,0\,1 \\ 1\,0\,0\,1\,1\,0 \\ 1\,1\,0\,0\,0\,1 \\ 1\,0\,1\,0\,1\,0 \\ 1\,0\,0\,1\,0\,1 \end{pmatrix} \begin{pmatrix} \mu \\ \beta_1 \\ \beta_2 \end{pmatrix} + \varepsilon = \mathcal{X}\beta + \varepsilon,
$$

under two linear constraints $1_3^\top \beta_1 = 0$ and $1_2^\top \beta_2$. The estimation procedure is demonstrated in Table 16.1, where we provide also the mean utilities $\bar{p}_{x_1\bullet}$ and $\bar{p}_{x_2\bullet}$ for all levels of the factor X_1 and X_2, respectively.

		X_2			
		1st level	2nd level	$\bar{p}_{x_1\bullet}$	$\widehat{\beta}_{1l}$
	1st level	1	2	1.5	−2
X_1	2nd level	4	3	3.5	0
	3rd level	6	5	5.5	2
	$\bar{p}_{x_2\bullet}$	3.67	3.33	3.5	
	$\widehat{\beta}_{2l}$	0.17	−0.17		

Table 16.1. Metric solution for the example.

The coefficients (part-worths) were calculated as the difference of the marginal mean utility and the overall mean utility, $\widehat{\beta}_{ji} = \bar{p}_{x_{ji}} - \mu$. The resulting part worths,
$$\widehat{\beta}_{11} = -2 \quad \widehat{\beta}_{21} = 0.17$$
$$\widehat{\beta}_{12} = 0 \quad \widehat{\beta}_{22} = -0.17,$$
$$\widehat{\beta}_{13} = 2$$
model the utility for each stimulus. For example, the estimated utility for the stimulus given by 1st level of X_1 and 2nd level of X_2 is $\widehat{Y}_2 = \widehat{\mu} + \widehat{\beta}_{11} + \widehat{\beta}_{22} = 3.5 - 2 - 0.17 = 1.33$.

EXERCISE 16.2. *Rewrite the design matrix $\mathcal{X}(K \times (D+2))$ given in Exercise 16.1 and the parameter vector β without the parameter for the overall mean effect μ and without the additional constraints on the parameters, i.e., find a design matrix $\mathcal{X}'(K \times D)$ such that $\mathcal{X}\beta = \mathcal{X}'\beta'$.*

The design matrix $\mathcal{X}(6 \times 6)$ proposed in Exercise 16.1 allowed to interpret the model parameters as the overall mean utility μ and the part-worths β_{ji} as the deviation from μ. As the design matrix is not uniquely given, we can choose it to suit different interpretations. Often, some of the factor levels is considered as a reference level and the model parameters are the constructed to describe the differences with respect to the reference. This leads to the model:

$$Y = \begin{pmatrix} Y_1 \\ Y_2 \\ Y_3 \\ Y_4 \\ Y_5 \\ Y_6 \end{pmatrix} = \begin{pmatrix} 1 \\ 2 \\ 4 \\ 3 \\ 6 \\ 5 \end{pmatrix} = \begin{pmatrix} 1 & 0 & 0 & 0 \\ 1 & 0 & 0 & 1 \\ 1 & 1 & 0 & 0 \\ 1 & 1 & 0 & 1 \\ 1 & 0 & 1 & 0 \\ 1 & 0 & 1 & 1 \end{pmatrix} \begin{pmatrix} \beta'_{11} \\ \beta'_{12} \\ \beta'_{13} \\ \beta'_{22} \end{pmatrix} + \varepsilon = \mathcal{X}'\beta' + \varepsilon,$$

where the parameter β'_{11} is the reference stimulus corresponding to the combination of the 1st level of X_1 and 1st level X_2. The remaining parameters then measure the part-worths with respect to the reference level.

The parameter estimates may be obtained similarly as in Exercise 16.1 from the marginal mean utilities,

$$\widehat{\beta}_{11} = 1.67, \quad \widehat{\beta}_{12} = 2, \quad \widehat{\beta}_{13} = 4, \quad \widehat{\beta}_{22} = -0.34.$$

Hence, the utility of the second stimulus, given by 1st level of X_1 and 2nd level of X_2 is $\widehat{Y}_2 = \beta'_{11} + \beta'_{22} = 1.67 - 0.34 = 1.33$, the same value as in Exercise 16.1. For the utility of the last stimulus, given by 3rd level of X_1 and 2nd level of X_2 we would obtain $\widehat{Y}_6 = \beta'_{11} + \beta'_{13} + \beta'_{22} = 1.67 + 4 - 0.34 = 5.33$.

EXERCISE 16.3. *Is it possible that different rankings lead to identical part-worths?*

Yes, this can happen. It suffices to realize that the parameter estimates are based only on the marginal mean utilities. Modifying the data set in a way that does not change this means leads to the same results. An example is given in Table 16.2.

		X_2			
		1st level	2nd level	$\bar{p}_{x_1 \bullet}$	$\widehat{\beta}_{1l}$
	1st level	2	1	1.5	-2
X_1	2nd level	4	3	3.5	0
	3rd level	5	6	5.5	2
	$\bar{p}_{x_2 \bullet}$	3.67	3.33	3.5	
	$\widehat{\beta}_{2l}$	0.17	-0.17		

Table 16.2. Metric solution for the counterexample demonstrating the nonuniqueness of the problem.

The resulting coefficients (part-worths), calculated as the difference of the marginal mean utility and the overall mean utility,

$$\begin{aligned} \widehat{\beta}_{11} &= -2 & \widehat{\beta}_{21} &= 0.17 \\ \widehat{\beta}_{12} &= 0 & \widehat{\beta}_{22} &= -0.17, \\ \widehat{\beta}_{13} &= 2 \end{aligned}$$

are identical to the coefficients obtained from different rankings in Exercise 16.1.

EXERCISE 16.4. *Compute the design matrix in the setup of Exercise 16.1 for $n = 3$ persons ranking the same products with X_1 and X_2.*

As described in the introduction to this Chapter, the design matrix $\mathcal{X}^*$ is obtained by stacking three identical individual design matrices $\mathcal{X}$. Denoting by Y the vector of the all $nK = 18$ rankings of the 6 stimuli by the 3 people, we can write the model as:

$$Y = \begin{pmatrix} 1\,1\,0\,0\,1\,0 \\ 1\,0\,1\,0\,0\,1 \\ 1\,0\,0\,1\,1\,0 \\ 1\,1\,0\,0\,0\,1 \\ 1\,0\,1\,0\,1\,0 \\ 1\,0\,0\,1\,0\,1 \\ 1\,1\,0\,0\,1\,0 \\ 1\,0\,1\,0\,0\,1 \\ 1\,0\,0\,1\,1\,0 \\ 1\,1\,0\,0\,0\,1 \\ 1\,0\,1\,0\,1\,0 \\ 1\,0\,0\,1\,0\,1 \\ 1\,1\,0\,0\,1\,0 \\ 1\,0\,1\,0\,0\,1 \\ 1\,0\,0\,1\,1\,0 \\ 1\,1\,0\,0\,0\,1 \\ 1\,0\,1\,0\,1\,0 \\ 1\,0\,0\,1\,0\,1 \end{pmatrix} \begin{pmatrix} \mu \\ \beta_1 \\ \beta_2 \end{pmatrix} + \varepsilon = \begin{pmatrix} \mathcal{X} \\ \mathcal{X} \\ \mathcal{X} \end{pmatrix} \beta + \varepsilon = \mathcal{X}^*\beta + \varepsilon,$$

where $1_3^\top \beta_1 = 0$ and $1_2^\top \beta_2$. The parameters μ, β, β_1, and β_2 have the same interpretation and dimension as in Exercise 16.1.

EXERCISE 16.5. *Compare the predicted and observed utilities for the example analyzed in Exercise 16.1.*

The observed and predicted rankings, denoted respectively by Y_k and $\widehat{Y}_k$, $k = 1,\ldots,6$ are given in Table 16.3.

Stimulus	X_1	X_2	Y_k	$\widehat{Y}_k$	$Y_k - \widehat{Y}_k$	$(Y_k - \widehat{Y}_k)^2$
1	1	1	1	1.67	−0.67	0.44
2	1	2	2	1.33	0.67	0.44
3	2	1	4	3.67	0.33	0.11
4	2	2	3	3.33	−0.33	0.11
5	3	1	6	5.67	0.33	0.11
6	3	2	5	5.33	−0.33	0.11
$\sum$	-	-	21	21	0	1.33

Table 16.3. Deviations between model and data.

We observe that largest deviations in Table 16.3 occur for the first level of X_1. However, we would need larger sample of respondents for meaningful analysis of the part-worths.

EXERCISE 16.6. *Compute the part-worths on the basis of the following tables of rankings observed on $n = 3$ persons:*

$$
\begin{array}{c|cc}
 & \multicolumn{2}{c}{X_2} \\
 & 1 & 2 \\
\hline
 & 1 & 2 \\
X_1 & 4 & 3 \\
 & 6 & 5
\end{array}
\;,\quad
\begin{array}{c|cc}
 & \multicolumn{2}{c}{X_2} \\
 & 1 & 3 \\
\hline
X_1 & 4 & 2 \\
 & 5 & 6
\end{array}
\;,\quad
\begin{array}{c|cc}
 & \multicolumn{2}{c}{X_2} \\
 & 3 & 1 \\
\hline
X_1 & 5 & 2 \\
 & 6 & 4
\end{array}
\;.
$$

The analysis can be carried out similarly as in Exercise 16.1. We obtain Table 16.4 summarizing the results.

| | X_2 | | | |
	1st level	2nd level	$\bar{p}_{x_1 \bullet}$	$\widehat{\beta}_{1l}$
1st level	1,1,3	2,3,1	1.83	−1.67
X_1 2nd level	4,4,5	3,2,2	3.33	−0.17
3rd level	6,5,6	5,6,4	5.33	1.83
$\bar{p}_{x_2 \bullet}$	3.89	3.11	3.5	
$\widehat{\beta}_{2l}$	0.39	−0.39		

Table 16.4. Metric solution for the example with $n = 3$.

For computer implementation of this procedure, it is better to use the parametrization in terms of parameters β' described in Exercise 16.2. The corresponding parameter estimates calculated by the appropriate statistical software are:

$$\widehat{\beta}'_1 = 2.22, \quad \widehat{\beta}'_2 = 1.50, \quad \widehat{\beta}'_3 = 3.50, \quad \widehat{\beta}'_4 = -0.78,$$

and it is easy to see that these values correspond exactly to the values calculated by hand in Table 16.4.

The main advantage of performing the analysis on a computer is that a reasonable software implementations of the two-way ANOVA give us also statistical tests of significance of the X_1 and X_2 factors.

The hypothesis H_0^1 : "no effect of X_1", tested by the usual F-test, leads to p-value 0.0001. The hypothesis H_0^2: "no effect of X_2" leads, in the same way, to the p-value 0.1062. Hence, the effect of X_1 on the product utilities is statistically significant whereas the effect of X_2 is not.   ◨ SMSconjexmp

EXERCISE 16.7. *Suppose that in the car example a person has ranked cars by the profile method on the following characteristics: X_1=motor, X_2=safety, and X_3=doors.*

The preferences are given in the following tables:

X_1	X_2	X_3	preference
1	1	1	1
1	1	2	3
1	1	3	2
1	2	1	5
1	2	2	4
1	2	3	6

X_1	X_2	X_3	preference
2	1	1	7
2	1	2	8
2	1	3	9
2	2	1	10
2	2	2	12
2	2	3	11

X_1	X_2	X_3	preference
3	1	1	13
3	1	2	15
3	1	3	14
3	2	1	16
3	2	2	17
3	2	3	18

Estimate and analyze the part-worths.

There are $k = 18$ observations corresponding to 3 levels of X_1, 2 levels of X_2, and 3 levels of X_3. Due to the profile method, we have observations for all $3 \cdot 2 \cdot 3 = 18$ possible combinations (stimuli) of the factor levels.

		X_2			
		1st level	2nd level	$\bar{p}_{x_1\bullet}$	$\widehat{\beta}_{1l}$
	1st level	1,3,2	5,4,6	3.5	−6
X_1	2nd level	7,8,9	10,12,11	9.5	0
	3rd level	13,15,14	16,17,18	15.5	6
				$\bar{p}_{x_3\bullet}$	$\widehat{\beta}_{3l}$
	1st level	1,7,13	5,10,16	8.67	−0.83
X_3	2nd level	3,8,15	4,12,17	9.83	0.33
	3rd level	2,9,14	6,11,18	10.00	0.50
$\bar{p}_{x_2\bullet}$		8	11	9.5	
$\widehat{\beta}_{2l}$		−1.5	1.5		

Table 16.5. Metric solution for the ranking of the cars.

The tests of significance of the factors can be carried out by the usual F-test. For the significance of X_1, X_2, and X_3, we respectively obtain p-values 0.0000, 0.2445, and 0.9060.

We conclude that factor X_1, motor, has significant influence on the consumer preferences. The part-worth of the remaining two factors under consideration, safety (X_2) and doors (X_3), are not statistically significant. ⊙ SMSconjcars

Applications in Finance

"It is interesting, chemically, no doubt," I answered, "but practically
—"

Dr. Watson in "Study in Scarlet"

Multivariate statistical analysis is frequently used in quantitative finance, risk management, and portfolio optimization. A basic rule says that one should diversify in order to spread financial risk. The question is how to assign weights to the different portfolio positions. Here we analyze a so-called mean-variance optimization that leads to weights that minimize risk given a budget constraint. Equivalently, we may optimize the weights of a portfolio for maximal return given a bound on the risk structure. The discussion naturally leads to links to the capital asset pricing model (CAPM).

Financial data sets are of multivariate nature because they contain information about the joint development of assets, derivatives, important market indicators, and the likes.

A typical investor question is how much he should invest in what type of asset. Suppose that p_{ij} denotes the price of the jth asset in the ith time period. The return from this asset is then $x_{ij} = (p_{ij} - p_{i-1,j})/p_{ij}$.

Let us assume that the random vector X of returns of selected p assets has p-dimensional probability distribution $X\,(\mu, \Sigma)$. The return of a given portfolio is the weighted sum of the individual returns:

$$Q = c^\top X,$$

where c denotes the proportions of the assets in the portfolio, $c^\top 1_p = 1$. Each asset contributes with a weight c_j, $j = 1, \ldots, p$, to the portfolio. The performance of the portfolio $c^\top X$ is a function of both the stochastic random vector X and the weights $c = (c_1, \ldots, c_p)^\top$. The mean return of the portfolio is defined as the expected value of $Q = c^\top X$, whereas the variance $Var(Q) = c^\top \Sigma c$ measures the risk of the portfolio.

Given a certain level of risk, an investor wants to know how much he should . invest in what asset. Put into mathematical terms this is equivalent to asking how to choose the vector c of asset weights in order to optimize a certain portfolio risk measure.

The first part of the exercises will analyze the minimization of $Var(Q) = c^\top \Sigma c$ with respect to c. We then consider the relation to the CAPM model.

Efficient Portfolios

The variance efficient portfolio, defined as the portfolio with minimum risk (measured by the variance), is derived in Theorem 17.1.

THEOREM 17.1. *Assume that the returns X have multivariate distribution (μ, Σ) and that $\Sigma > 0$.*

The variance efficient portfolio weights are $c = \{1_p^\top \Sigma^{-1} 1_p\}^{-1} \Sigma^{-1} 1_p$.

Including in the portfolio a riskless asset with a fixed return and zero variance allows one to derive a portfolio with a given mean return, $EQ = \bar{\mu}$, and minimum variance. Such a portfolio is called the mean-variance efficient.

THEOREM 17.2. *Assume that a riskless asset has constant return r and that the remaining returns $X = (X_1, \dots, X_p)^\top$ have multivariate distribution (μ, Σ), $\Sigma > 0$.*

The weights of mean-variance efficient portfolio are

$$c = \{\mu^\top \Sigma^{-1} (\mu - r 1_p)\}^{-1} \bar{\mu} \Sigma^{-1} (\mu - r 1_p)$$

for the risky assets X and $c_r = 1 - 1_p^\top c$ for the riskless asset.

In practice, the variance matrix Σ is estimated from the past returns. However, this approach assumes that the covariance structure is stable over time. In practice, one can expect that this assumption might be broken; see Franke, Härdle & Hafner (2004) for an overview of the commonly used modern methods.

Capital Asset Pricing Model

The capital asset pricing model (CAPM) investigates the relation between a mean-variance efficient portfolio and an asset uncorrelated with this portfolio. This is typically a market index or the riskless interest rate. Starting from the mean-variance efficient portfolio weights given in Theorem 17.2, we can arrive at $\mu = r 1_p + \Sigma c \{c^\top \Sigma c\}^{-1} (\bar{\mu} - r)$; see Härdle & Simar (2003, section 17.4). Setting $\beta = \Sigma c \{c^\top \Sigma c\}^{-1}$, we arrive at the well-known CAPM model:

$$\mu = r1_p + \beta(\overline{\mu} - r),$$

where r is the return of the riskless asset or the index and $\overline{\mu}$ is the expected return of the market. The difference $\overline{\mu} - r$ is the risk premium. The beta factors $\beta = (\beta_1, \ldots, \beta_p)^\top$ are a measure of the relative performance (or sensitivity) of the p assets with respect to the market risk. The econometric interpretation of the CAPM model says that the expected return of any asset is a sum of the return of the riskless asset plus the risk premium determined by the asset beta factor (Franke et al. 2004).

We start with two exercises on matrix inversion. The inversion techniques are used later in the construction of efficient portfolios.

EXERCISE 17.1. *Derive the inverse of $(1 - \rho)\mathcal{I}_p + \rho 1_p 1_p^\top$.*

In Exercise 2.8, we have already shown that

$$(\mathcal{A} + aa^\top)^{-1} = \mathcal{A}^{-1} - \frac{\mathcal{A}^{-1}aa^\top\mathcal{A}^{-1}}{1 + a^\top\mathcal{A}^{-1}a}.$$

Setting $\mathcal{A} = (1 - \rho)\mathcal{I}_p$ and $a = (\text{sign } \rho)\sqrt{\rho}1_p$, we easily obtain:

$$\{(1 - \rho)\mathcal{I}_p + \rho 1_p 1_p^\top\}^{-1}$$

$$= \{(1 - \rho)\mathcal{I}_p\}^{-1} - \frac{\{(1 - \rho)\mathcal{I}_p\}^{-1}\rho 1_p 1_p^\top\{(1 - \rho)\mathcal{I}_p\}^{-1}}{1 + \rho 1_p^\top\{(1 - \rho)\mathcal{I}_p\}^{-1}1_p}$$

$$= \frac{\mathcal{I}_p}{1 - \rho} - \frac{\rho 1_p 1_p^\top}{(1 - \rho)^2\{1 + (1 - \rho)^{-1}\rho 1_p^\top 1_p\}}$$

$$= \frac{\mathcal{I}_p}{1 - \rho} - \frac{\rho 1_p 1_p^\top}{(1 - \rho)(1 - \rho + \rho p)} = \frac{\mathcal{I}_p}{1 - \rho} - \frac{\rho 1_p 1_p^\top}{(1 - \rho)\{1 + \rho(p - 1)\}}.$$

Notice that the above derivation applies only if $\rho \neq 1$.

EXERCISE 17.2. *For which values of ρ is the matrix $\mathcal{Q} = (1 - \rho)\mathcal{I}_p + \rho 1_p 1_p^\top$ positive definite?*

The eigenvalues are found by solving $|\mathcal{Q} - \lambda\mathcal{I}_p| = 0$. According to the expression $|\mathcal{A} + aa^\top| = |\mathcal{A}||1 + a^\top\mathcal{A}^{-1}a|$ derived in Exercise 2.8, we can write:

$$|\Sigma - \lambda\mathcal{I}_p| = |(1 - \rho - \lambda)\mathcal{I}_p + \rho 1_p 1_p^\top|$$

$$= |(1 - \rho - \lambda)\mathcal{I}_p||1 + \rho 1_p^\top\{(1 - \rho - \lambda)\mathcal{I}_p\}^{-1}1_p|$$

$$= (1 - \rho - \lambda)|1 + \rho p(1 - \rho - \lambda)^{-1}|.$$

Hence, the eigenvalues are $\lambda = 1 - \rho$ and $\lambda = 1 - \rho + \rho p = 1 + \rho(p - 1)$.

The matrix $\mathcal{Q}$ is positive definite if and only if all its eigenvalues are positive. This implies that $\mathcal{Q} > 0$ if $1 - \rho > 0$ and $1 + \rho(p - 1) > 0$, i.e., $\rho < 1$ and $\rho > -(p - 1)^{-1}$.

EXERCISE 17.3. *Calculate the variance efficient portfolio of equally corre-lated assets with equal variances.*

According to Theorem 17.1, the assets have to be weighted by

$$c = \{1_p^\top \Sigma^{-1} 1_p\}^{-1} \Sigma^{-1} 1_p.$$

In our case, the variance matrix of the returns can be written as:

$$\Sigma = \sigma^2 \mathcal{Q} = \sigma^2 \begin{pmatrix} 1 & \rho & \cdots & \rho \\ \rho & 1 & \cdots & \rho \\ \vdots & \vdots & \ddots & \vdots \\ \rho & \rho & \cdots & 1 \end{pmatrix},$$

where $-(p-1)^{-1} < \rho < 1$ guarantees that the matrix Σ is positive definite, see Exercise 17.2.

According to Exercise 17.1, the inverse is

$$\Sigma^{-1} = \sigma^{-2} \mathcal{Q}^{-1} = \frac{\mathcal{I}_p}{\sigma^2(1-\rho)} - \frac{\rho 1_p 1_p^\top}{\sigma^2(1-\rho)\{1+(p-1)\rho\}}$$

and it follows that

$$\begin{aligned} \Sigma^{-1} 1_p &= \frac{1_p}{\sigma^2(1-\rho)} - \frac{\rho 1_p 1_p^\top 1_p}{\sigma^2(1-\rho)\{1+(p-1)\rho\}} \\ &= \frac{[\{1+(p-1)\rho\} - \rho p] 1_p}{\sigma^2(1-\rho)\{1+(p-1)\rho\}} = \frac{(1-\rho) 1_p}{\sigma^2(1-\rho)\{1+(p-1)\rho\}} \\ &= \frac{1_p}{\sigma^2\{1+(p-1)\rho\}} \end{aligned}$$

which yields

$$1_p^\top \Sigma^{-1} 1_p^\top = \frac{p}{\sigma^2\{1+(p-1)\rho\}}.$$

The weights of the variance efficient portfolio are thus

$$c = \{1_p^\top \Sigma^{-1} 1_p\}^{-1} \Sigma^{-1} 1_p = \frac{1}{p} 1_p,$$

i.e., all assets are equally weighted.

EXERCISE 17.4. *Calculate the variance efficient portfolio of equally corre-lated assets.*

Let $\sigma_i^2 > 0$ be the variance of ith asset X_i, $i = 1, \ldots, p$, and define $\mathcal{D} = \text{diag}(\sigma_1^2, \ldots, \sigma_p^2)$. The variance matrix of X can be written as $\Sigma = \mathcal{D}^{1/2} \mathcal{Q} \mathcal{D}^{1/2}$, where $\mathcal{Q}$ is the correlation matrix defined in Exercise 17.3.

Obviously, the inverse of the variance matrix is $\Sigma^{-1} = \mathcal{D}^{-1/2}\mathcal{Q}^{-1}\mathcal{D}^{-1/2}$. Expressing the inverse $\mathcal{Q}^{-1}$ as in Exercise 17.1, we have

$$\Sigma^{-1}1_p = \mathcal{D}^{-1/2}\mathcal{Q}^{-1}\mathcal{D}^{-1/2}1_p$$

$$= \frac{\mathcal{D}^{-1}1_p}{1-\rho} - \frac{\rho \mathcal{D}^{-1/2}1_p 1_p^{\top}\mathcal{D}^{-1/2}1_p}{(1-\rho)\{1+(p-1)\rho\}}$$

$$= \left\{ \frac{\sigma_i^{-2}}{1-\rho} - \frac{\rho\sigma_i^{-1}\sum_{j=1}^{p}\sigma_j^{-1}}{(1-\rho)\{1+(p-1)\rho\}} \right\}_{i=1,\ldots,p}.$$

Hence, the weight of the ith asset in the variance efficient portfolio can be expressed as:

$$c_i = \left\{ \sum_{j=1}^{p}\sigma_j^{-2} - \frac{\rho\left(\sum_{j=1}^{p}\sigma_j^{-1}\right)^2}{\{1+(p-1)\rho\}} \right\}^{-1} \left\{ \sigma_i^{-2} - \frac{\rho\sigma_i^{-1}\sum_{j=1}^{p}\sigma_j^{-1}}{\{1+(p-1)\rho\}} \right\}.$$

EXERCISE 17.5. *How does the result of Exercise 17.4 look like if $\rho = 0$.*

Setting $\rho = 0$ in the variance efficient portfolio weights derived in Exercise 17.4 leads to

$$c_i = \frac{\sigma_i^{-2}}{\sum_{j=1}^{p}\sigma_j^{-2}}.$$

Hence, the weight of the ith asset in the variance efficient portfolio is decreasing function of its variance. This results corresponds to Härdle & Simar (2003, corollary 17.3).

EXERCISE 17.6. *Derive the variance efficient portfolio for IBM, PanAm, and the Digital Equipment company using the returns given in Table A.13.*

The empirical covariance matrix is:

$$S = \begin{pmatrix} 0.0035 & 0.0017 & 0.0026 \\ 0.0017 & 0.0174 & 0.0035 \\ 0.0026 & 0.0035 & 0.0098 \end{pmatrix}.$$

Using S as estimate of the unknown variance matrix Σ and applying Theorem 17.1 leads the estimated variance efficient portfolio weights

$$c = (0.829, 0.092, 0.078)^{\top}.$$

The large majority should be invested in IBM which, as we see from the empirical covariance matrix, has also the smallest variance.

In Figure 17.1, we compare the returns of the variance efficient and equally weighted portfolio. Both plots are using the same scale and, as expected, the variability of the variance efficient portfolio is obviously much smaller.

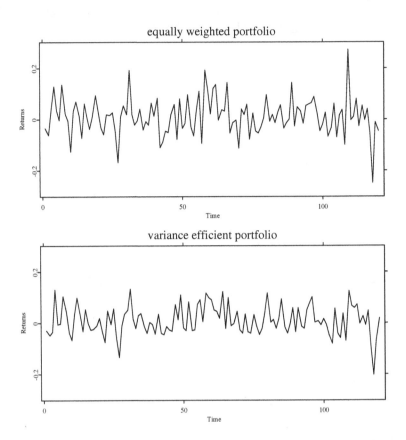

Fig. 17.1. Returns of the equally weighted and variance efficient portfolio for IBM, PanAm and DEC. ⊙ SMSportfol

EXERCISE 17.7. *The empirical covariance between the 120 returns of IBM and PanAm in Exercise 17.6 is 0.0017. Test if the true covariance is zero.*

The value $s_{IBM,PanAm} = 0.0017$ seems to be quite small but we have to keep in mind that it depends on the scale of measurement. In this case, it is better to work with correlations which are scale independent and zero covariance is equivalent to zero correlation.

The empirical correlation matrix of the variables analyzed in Exercise 17.6 is

$$\mathcal{R} = \begin{pmatrix} 1.0000 & 0.2126 & 0.4441 \\ 0.2126 & 1.0000 & 0.2654 \\ 0.4441 & 0.2654 & 1.0000 \end{pmatrix}.$$

The empirical correlation of IBM and PanAm returns is $r_{IBM,PanAm} = 0.2126$.

The significance of the correlation coefficient can be tested using Fisher's Z-transformation, see also Exercise 3.5. Under the null hypothesis, $H_0 : \rho_{IBM,PanAm} = 0$, the random variable

$$W = \frac{1}{2} \log \left(\frac{1 + r_{IBM,PanAm}}{1 - r_{IBM,PanAm}} \right)$$

has asymptotically normal distribution with expected value $EW = 0$ and variance $Var\, W = (n - 3)^{-1}$.

Comparing the value

$$\frac{\sqrt{n - 3}}{2} \log \left(\frac{1 + r_{IBM,PanAm}}{1 - r_{IBM,PanAm}} \right) = \frac{\sqrt{117}}{2} \log \frac{1.2126}{0.7874} = 2.3352$$

to the appropriate quantile of the standard normal distribution, $u_{0.975} = 1.96$, we reject the null hypothesis.

Hence, on probability level $1 - \alpha = 95\%$, we conclude that the covariance between IBM and PanAm returns is significantly positive.

EXERCISE 17.8. *Explain why in both the equally and optimally weighted portfolio plotted on Figure 17.1 in Exercise 17.6 have negative returns just before the end of the series, regardless of whether they are optimally weighted or not!*

In the NYSE returns data set given in Table A.13, we can clearly see that at the end of the data set, all considered stocks have negative returns. In such situation, it is clear that any positively weighted portfolio ends up in a loss.

The worst results can be seen in the third row from the end of the data set. Since the data set contains monthly returns and it stops in December 1987, the worst results are achieved in October 1987. Actually, the stock market crash of October 19th 1987 was one of the largest market crashes in history. On this so-called Black Monday, the Dow-Jones index lost 22.6% of its value (Sobel 1988).

EXERCISE 17.9. *Could some of the weights in Exercise 17.6 be negative?*

The efficient portfolio weights, $c = \{1_p^{\top} \Sigma^{-1} 1_p\}^{-1} \Sigma^{-1} 1_p$, are given in Theorem 17.1. Clearly, the denominator $1_p^{\top} \Sigma^{-1} 1_p$ is always positive since the variance matrix Σ is positive definite. Thus, the weight of the ith asset $c_i < 0$ if and only if the ith element of the vector $\Sigma^{-1} 1_p < 0$.

Noticing that the vector $\Sigma^{-1} 1_p$ contains the sums of row elements of the matrix Σ^{-1}, we just need to design a suitable positive definite matrix. For example,

$$\Sigma^{-1} = \begin{pmatrix} 1.0 & -0.8 & -0.4 \\ -0.8 & 1.0 & 0.2 \\ -0.4 & 0.2 & 1.0 \end{pmatrix}$$

is a positive definite matrix with negative row sums. The corresponding variance matrix of the asset returns would be:

$$\Sigma = (\Sigma^{-1})^{-1} = \begin{pmatrix} 3.3333 & 2.5000 & 0.8333 \\ 2.5000 & 2.9167 & 0.4167 \\ 0.8333 & 0.4167 & 1.2500 \end{pmatrix}.$$

It is now easy to see that the variance efficient portfolio weights are indeed $c = (-0.2, 0.4, 0.8)^{\top}$ with $c_1 = -0.2 < 0$.

Hence, we conclude that the variance efficient portfolio weights (see Theorem 17.1 and Exercise 17.6) could be negative for certain covariance structure of the asset returns.

EXERCISE 17.10. *In the CAPM the β value tells us about the performance of the portfolio relative to the riskless asset. Calculate the β value for each single stock price series relative to the "riskless" asset IBM.*

We have already seen in Exercise 17.6 that IBM returns have smallest variance. Hence, it makes sense to use IBM as a replacement of the market index in CAPM.

Let us denote the returns of the index (IBM) by r_i, $i = 1, \ldots, n$. The coefficient β_j corresponding to the jth asset returns x_{ij}, $j = 1, \ldots, p$, can be estimated using the following linear model:

$$x_{ij} = \alpha_i + \beta_j r_i + \varepsilon_i,$$

where ε_i are iid random variables with zero mean and variance σ^2. As shown in Exercise 3.7, the estimates of β_i by the least squares method can be calculated as:

$$\widehat{\beta}_i = \frac{s_{X_i,R}}{s_{RR}},$$

where $s_{X_i,R}$ denotes the empirical covariance of the ith asset returns and the market index and $s_{R,R}$ is the empirical variance of the market index.

The betas can now be calculated from the covariance matrix given in Exercise 17.6:

$$\widehat{\beta}_2 = 0.0017/0.0035 = 0.49,$$
$$\widehat{\beta}_3 = 0.0026/0.0035 = 0.74.$$

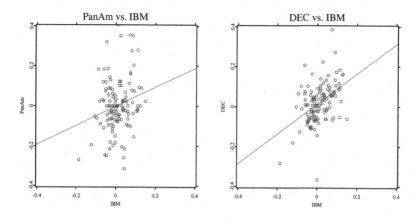

Fig. 17.2. Returns of PanAm and DEC plotted against IBM returns with the corresponding regression lines. **Q** SMScapmnyse

The estimated regression lines are plotted in Figure 17.2. The larger sensitivity of DEC is clearly visible.

The estimated betas suggest that both companies, PanAm ($\widehat{\beta}_2$) and DEC ($\widehat{\beta}_3$) are both less sensitive to market changes than the "index" IBM.

Highly Interactive, Computationally Intensive Techniques

Then we have stopped all the holes. And now we must be silent and wait.
Sherlock Holmes in "The Red-Headed League"

Modern statistics is impossible without computers. The introduction of modern computers in the last quarter of the 20th century created the subdiscipline "computational statistics." This new science has subsequently initiated a variety of new computer-aided techniques. Some of these techniques, such as brushing of scatter plots, are highly interactive and computationally intensive.

Computer-aided techniques can help us to discover dependencies among the variables without formal tools and are essential, especially when we consider extremely high-dimensional data. For example, visual inspection and interactive conditioning via the brush helps us to discover lower-dimensional relations between variables. Computer-aided techniques are therefore at the heart of multivariate statistical analysis.

In this chapter we first present simplicial depth, a generalization of the data depth allowing straightforward definition of the multivariate median. Next, projection pursuit is a semiparametric technique based on "interesting" one-dimensional projection. A multivariate nonparametric regression model is underlying sliced inverse regression, a technique that leads to a dimensionality reduction of the space of the explanatory variables. The technique of support vector machines (SVM) is motivated by nonlinear classification (discrimination problems). The last technique presented in this chapter, classification and regression trees (CART), is a decision tree procedure developed by Breiman, Friedman, Olshen & Stone (1984).

Simplicial Depth

In p-dimensional space, the depth of a data point x is defined as the number of convex hulls formed from all possible selections of $p+1$ points covering x. The multivariate median is defined as the point with the largest depth

$$x_{\text{med}} = \arg\max_i \#\{k_0, \ldots, k_p \in \{1, \ldots, n\} : x_i \in \text{hull}(x_{k_0}, \ldots, x_{k_p})\}.$$

Unfortunately, with increasing dimension p and number of observations n, the calculation of the multivariate median becomes very time-consuming.

Exploratory Projection Pursuit

The projection pursuit searches for interesting directions in a p-dimensional data set by maximizing a chosen index. In Chapter 9, the method of principal components is based on the maximization of variance. In Chapter 14, the method of canonical correlations maximizes the correlation between linear combinations of two subgroups of the observed variables.

Assume that the p-dimensional random vector X has zero mean, $EX = 0_p$, and unit variance $\text{Var}(X) = \mathcal{I}_p$. Such a covariance structure can be achieved by the Mahalanobis transformation. Let $\hat{f}_{h,\alpha}$ denote the kernel density estimator of the pdf of the projection $\alpha^\top X$, where h denotes the kernel estimator bandwidth. Friedman & Tukey (1974) proposed the index

$$I_{FT,h}(\alpha) = n^{-1} \sum_{i=1}^{n} \hat{f}_{h,\alpha}(\alpha^\top X_i),$$

leading to the maximization of $\int f^2(z)dz$. An alternative index can be based on the entropy measure $\int f(z) \log f(z)dz$ or on the Fisher information $\int \{f'(z)\}^2/f(z)dz$. Jones & Sibson (1987) suggested approximating the deviations from the normal density by

$$I_{JS}(\alpha) = \{\kappa_3^2(\alpha^\top X) + \kappa_4^2(\alpha^\top X)/4\}/12,$$

where $\kappa_3(\alpha^\top X) = E\{(\alpha^\top X)^3\}$ and $\kappa_4(\alpha^\top X) = E\{(\alpha^\top X)^4\}-3$ are cumulants of $\alpha^\top X$. The maximization of these indices usually leads to the least-normal-looking view of the data set.

Sliced Inverse Regression

Given a response variable Y and a (random) vector $X \in \mathbb{R}^p$ of explanatory variables, the idea of sliced inverse regression (Duan & Li 1991) is to find a smooth regression function that operates on a variable set of projections:

$$Y = m(\beta_1^\top X, \dots, \beta_k^\top X, \varepsilon),$$

where $\beta_1, \dots, \beta_k$ are unknown projection vectors, $k \leq p$ is unknown, $m : \mathbb{R}^{k+1} \to \mathbb{R}$ is an unknown function, and ε is the random error with $E(\varepsilon|X) = 0$. The unknown β_is are called effective dimension-reduction directions (EDR directions). The span of EDR directions is denoted as an effective dimension reduction space (EDR space).

The EDR space can be identified by considering the inverse regression (IR) curve $m_1(y) = E(Z|Y = y)$ of the standardized variable $Z = \Sigma^{-1/2}(X - EX)$. The SIR algorithm exploits the fact that the conditional expectation $m_1(y)$ is moving in $\text{span}(\eta_1, \dots, \eta_k)$. The EDR directions $\widehat{\beta}_i$, $i = 1, \dots, k$ are calculated from the eigenvectors $\widehat{\eta}_i$ of $\text{Var}\{m_1(y)\}$. The eigenvalues of $\text{Var}\{m_1(y)\}$ show which of the EDR directions are important (Cook & Weisberg 1991, Li 1991, Hall & Li 1993).

SIR Algorithm

1. Standardize x by calculating $z_i = \widehat{\Sigma}^{-1/2}(x_i - \overline{x})$.

2. Divide the range of the response y_i into S disjoint intervals (slices) H_s, $s = 1, \dots, S$. The number of observations within slice H_s is $n_s = \sum_{i=1}^n I_{H_s}(y_i)$.

3. Compute the mean of z_i over all slices, $\overline{z}_s = n_s^{-1} \sum_{i=1}^n z_i I_{H_s}(y_i)$ as a crude estimate of the IR curve $m_1(y)$.

4. Calculate the estimate for the conditional variance of the IR curve: $\widehat{V} = n^{-1} \sum_{s=1}^S n_s \overline{z}_s \overline{z}_s^\top$.

5. Identify the eigenvalues $\widehat{\lambda}_i$ and eigenvectors $\widehat{\eta}_i$ of $\widehat{V}$.

6. Put the standardized EDR directions $\widehat{\eta}_i$ back to the original scale: $\widehat{\beta}_i = \widehat{\Sigma}^{-1/2}\widehat{\eta}_i$.

SIR II Algorithm

In some cases, the EDR directions are hard to find using the SIR algorithm. The SIR II algorithm overcomes this difficulty by considering the conditional variance $\text{Var}(X|y)$ instead of the IR curve $m_1(y)$.

In practice, it is recommended to use SIR and SIR II jointly (Cook & Weisberg 1991, Li 1991, Schott 1994) or to investigate higher-order conditional moments. For further reading, see Kötter (1996).

CART

CART is based on sequential splitting of the data space into a binary tree. At each node, the split is determined by minimization of an impurity measure. For regression trees this impurity measure is, e.g., the variance; for classification trees, it is, e.g., the misclassification error. An example is the classification of patients into low- and high-risk patients.

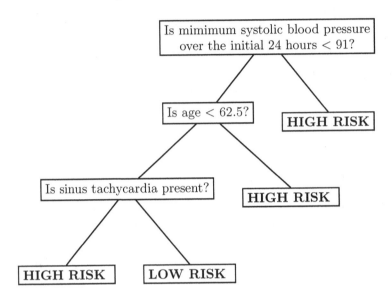

Left branches of the tree correspond to positive answers, right branches to negative answers to questions like "$X_j \leq a$." Here X_j denotes one of the many variables recorded for each patient and a is the threshold that has to be computed by minimizing the (chosen) impurity measure at each node.

An important characteristic is that CART always splits one of the coordinate axes, i.e., in only one variable. A simple classification into two groups that lie above and below a diagonal of a square will be hard for CART. We study this effect in the following exercises.

The splitting procedure is defined via the Gini, the twoing, or the least squares criterion. The Gini method typically performs best. Industries using CART include telecommunications, transportation, banking, financial services, health care, and education.

Support Vector Machines

The theoretical basis of the SVM methodology is provided by the statistical learning theory (Vapnik 2000). The basic idea of the SVM classification is to find a separating hyperplane $x^\top w + b = 0$ corresponding to the largest possible margin between the points of different classes. The classification error $\xi_i \geq 0$ of the ith observation is defined as the distance from the misclassified point x_i to the canonical hyperplane $x^\top w + b = \pm 1$ bounding its class:

$$x_i^\top w + b \geq 1 - \xi_i \text{ in group 1,} \tag{18.1}$$
$$x_i^\top w + b \leq -1 + \xi_i \text{ in group 2.} \tag{18.2}$$

Using this notation, the margin between the points of different classes, i.e., the distance between the canonical hyperplanes, is equal to $2/\|w\|$. The problem of penalized margin maximization can now be formulated as a constrained minimization of the expression:

$$\frac{1}{2}\|w\|^2 + c \left(\sum_{i=1}^n \xi_i \right)^\nu \tag{18.3}$$

under constraints (18.1) and (18.2), where c and $\nu \geq 1$ are parameters controlling the behavior of the algorithm.

Nonlinear classification is achieved by mapping the data into a high-dimensional feature space and finding a linear separating hyperplane in this feature space. This can be easily achieved by using a kernel function in the dual formulation of the minimization problem (18.3). Throughout the rest of this chapter, we will use the stationary Gaussian kernel with an anisotropic radial basis,

$$K(x_i, x_j) = \exp\{-(x_i - x_j)^\top r^2 \Sigma^{-1}(x_i - x_j)/2\},$$

where Σ is taken as the empirical variance matrix and r is a constant.

For more insight into the SVM methodology, we refer to Vapnik (2000).

EXERCISE 18.1. *Construct a configuration of points in $\mathbb{R}^2$ such that the point with coordinates given by the univariate medians, $(x_{med,1}, x_{med,2})^\top$, is not in the center of the scatterplot.*

In Figure 18.1, we plot an example with 11 points. Ten points are lying roughly on the unit circle while the 11th point lies somewhere between them. The depth of the 11 points is given at the location of each point in Figure 18.1. Given point, say x, lying on the diameter of the circle should be covered only by the triangles (convex hulls of $p + 1 = 3$ points) containing x. The number of such triangles is clearly

$$\binom{10}{2} = \frac{10!}{8!2!} = \frac{90}{2} = 45$$

and, in Figure 18.1, we observe that this is indeed the depth of the points lying on the diameter of the circle.

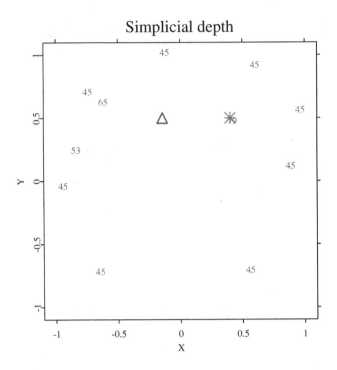

Fig. 18.1. The deepest point (star) and the coordinatewise median (triangle) of the simulated data set. The numbers are giving the simplicial depth of the points.
Q SMSsimpdsimu

The deepest point, the multivariate median, is denoted by the star. The triangle shows the location of the coordinatewise median. Clearly, the coordinatewise median does not lie close to any observation.

EXERCISE 18.2. *Calculate the Simplicial Depth for the Swiss bank notes data set (Table A.2) and compare the results to the univariate medians. Calculate the Simplicial Depth again for the genuine and counterfeit bank notes separately.*

The Swiss bank notes data set has altogether two hundred 6-dimensional observations. In order to calculate the depth of each point, we should check if each of these points lies inside a convex hull formed by every possible $p+1 = 7$ points. From 200 points, we can select 7 distinct points in altogether $\binom{200}{7} =$ 2283896214600 ways. Clearly, the evaluation of this relatively small data set might take a while even if some smart numerical algorithm is used.

In order to demonstrate the concept of the Simplicial Depth, we calculate the depth only on the 20 Swiss bank notes selected in Exercise 11.6, see also Figure 11.5 for the numbers of the selected observations. To increase the speed of calculation even further, we calculate the simplicial depth only in the 2-dimensional space given by the first two principal components.

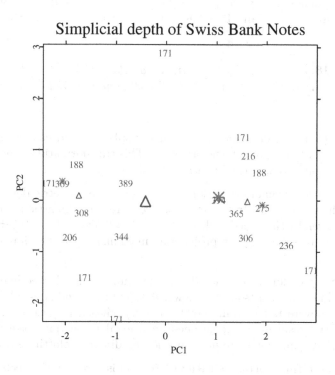

Fig. 18.2. The deepest points (big star) and the coordinatewise medians (big triangle) of the first two PCs of the Swiss bank notes. Smaller symbols show the deepest points and coordinatewise medians for genuine and counterfeit banknotes. The numbers are giving the simplicial depth of the points. ▣ SMSsimpdbank

The simplicial depth of the selected 20 points is plotted in Figure 18.2. The smallest possible data depth if given by $\binom{19}{2} = 171$. The largest possible depth would be $\binom{20}{3} = 1140$ if some of the points would like in the convex hull of all possible combinations of 3 points.

The deepest point, $(1.05, 0.07)^{\top}$, in Figure 18.2 is denoted by the big red star. Smaller stars denote the deepest points calculated separately for the genuine and counterfeit bank notes. The large triangle denotes the coordinatewise median, $(-0.41, -0.01)^\top$, of all observations—notice that it is lying quite far from the deepest point and it even has opposite sign.

The coordinatewise medians calculated only for the 10 genuine and 10 counterfeit bank notes are plotted as small triangles in Figure 18.2. The differences between the deepest point and the coordinatewise median are clearly visible: the deepest point is always one of the points given in the data set whereas the coordinatewise median often lies quite far away even from the closest observation.

EXERCISE 18.3. *Apply the EPP technique on the Swiss bank notes data set (Table A.2) and compare the results to the PC analysis and Fisher's linear discriminant rule.*

The first step in projection pursuit is usually sphering and centering of the data set by the Mahalanobis transformation. This transformation removes the effect of location, scale, and correlation structure.

The search of the optimal projection is based on nonparametric density estimators of the projections. In this exercise, we were using Quartic kernel with bandwidth given by the Scott's rule-of-thumb, $h = 2.62n^{-1/5}$, see Härdle et al. (2004). We were searching for projections maximizing the Friedman-Tukey index.

In Figure 18.3, we plot the estimated densities minimizing (dashed line, upper dotplot) and maximizing (solid line, lower dotplot) the Friedman-Tukey index (the extremes were taken from 10000 randomly chosen projections). In the dotplots of the resulting extreme one-dimensional projections, the genuine and counterfeit bank notes are distinguished by different plotting symbols.

The most interesting Friedman-Tukey projection is given by the vector:

$$(0.9083, -0.2494, -0.0368, 0.2568, -0.2126, 0.0181)^\top,$$

i.e., the largest weight is assigned to the the first variable, the length of the bank note. In the least interesting projection (dashed line in Figure 18.3), it would be impossible to separate the genuine and counterfeit bank notes although we see some outlying group of counterfeit bank notes on the right hand side. In the lower dotplot, the separation between the counterfeit and genuine bank notes seems to be much better. However, the best separation by far is achieved by the Fisher's linear discriminant rule plotted in Figure 18.4.

The Fisher's LDA projection is given by the coefficients:

$$a = (-0.1229, -0.0307, 0.0009, 0.0057, 0.0020, -0.0078)^\top$$

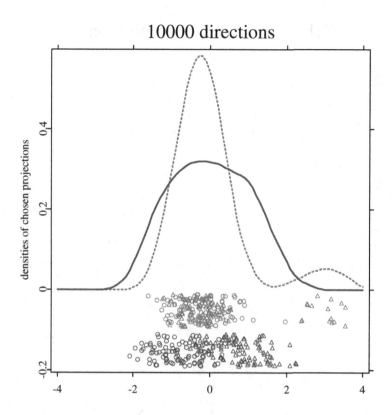

Fig. 18.3. The least (dashed line, upper dotplot) and the most informative (solid line, lower dotplot) from 10000 randomly chosen directions. The Friedman-Tukey index. ▢ SMSeppbank

and the projected points are displayed in the upper dotplot in Figure 18.4. The corresponding kernel density estimate is given by the dashed line. The clear separation of the two groups confirms the optimality of the Fisher's projection (note that the prior knowledge of the two groups was used in the construction of this projection).

The principal components projection of the same (sphered and centered) data set is given by the linear combination:

$$v = (0.6465, 0.3197, 0.0847, -0.5688, -0.1859, 0.3383)^\top$$

and the resulting one-dimensional projection is plotted as the solid line and the lower dotplot in Figure 18.4. However, the Mahalanobis transformation

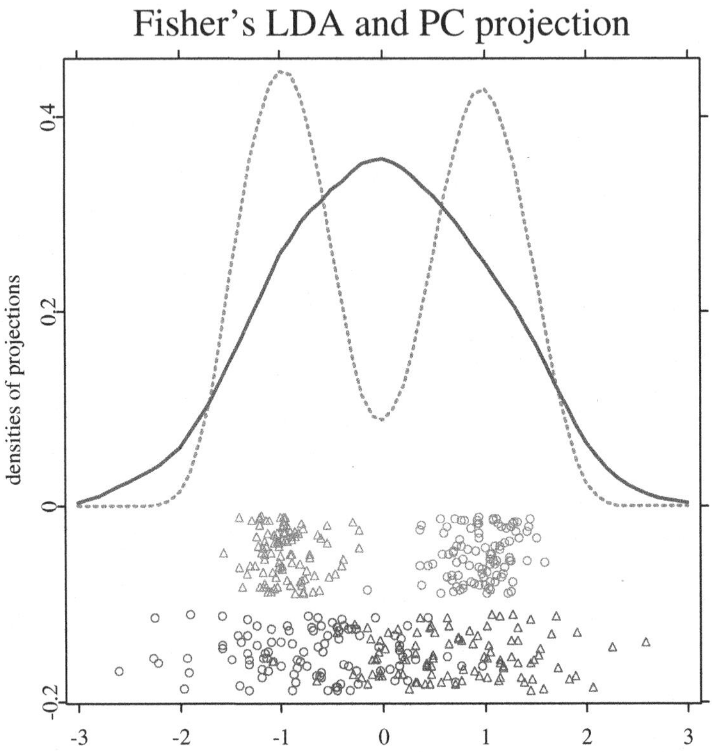

Fig. 18.4. The Fisher's LDA projection (dashed line, upper dotplot) first PC (solid line, lower dotplot) for Swiss bank notes. `SMSdisfbank2`

used for sphering and centering of the data set "guarantees" that the PC transformation has no chance of producing an interesting result, see Exercise 9.7.

Comparing the projections plotted in Figures 18.3 and 18.4, we can say the PC projection and EPP lead to similar results and both assign the largest weight to the first variable. The Fisher's discriminant rule, using prior knowledge of the group membership of the observations, shows the best possible separation of genuine and counterfeit bank notes.

From the computational point of view, the PC and Fisher's discriminant projections are very simple to implement. The implementation of the exploratory projection pursuit is much more involved and it requires choices of additional parameters such as the kernel function or bandwidth. For large and high-dimensional data sets, the computation might take very long, the numerical algorithm does not have to find the global maximum and it even is not guaranteed that such a unique maximum exists.

EXERCISE 18.4. *Apply the SIR technique to the U.S. companies data (Table A.17) with $Y =$ "market value" and $X =$ "all other variables". Which EDR directions do you find?*

The U.S. companies data set contains 6 variables measured on 79 U.S. companies. Apart of the response variable, market value, the data set contains information on assets, sales, profits, cash flow and number of employees. As described in the introduction to this chapter, SIR attempts to find lower dimensional projections of the five explanatory variables with a strong (possibly nonlinear) relationship to the market value.

This data set has been already investigated in Exercise 11.9 and, again, we use the same logarithmic transformation. The scatterplot of the first two PCs of the transformed data set was already given in the same exercise in Figure 11.9. The transformed variables are centered and standardized so that the scales of measurement of all variables become comparable; the standardization is not crucial for SIR, but it simplifies the interpretation of the resulting coefficients.

Two companies, IBM and General Electric, have extremely large market value and we removed them from the next analysis as outliers. Without these two observations, the rest of the data set is more "spread out" in the scatterplots in Figure 18.5.

After the removal of the two outliers, there are 77 observations left in the data set. For the SIR, we have created 7 slices with 11 observations each. The eigenvalues are $\widehat{\lambda} = (0.70, 0.19, 0.07, 0.03, 0.01)$ and it seems that only one factor explains "larger than average" amount $(1/5 = 0.20)$ of the conditional variance of the IR curve. In Figure 18.5 we plot the corresponding screeplot and scatterplots of the response, market value, against the first three resulting projections of the five explanatory variables in Figure 18.5.

The scatterplot of the market value against the first factor, $\mathcal{X}\widehat{\beta}_1$, shows strong nonlinear relationship. The coefficients of the first factor are given by $\widehat{\beta}_1 = (0.35, 0.14, 0.19, 0.03, 0.91)^\top$. Clearly, most of the factor is given by the 5th explanatory variable, number of employees. Important role is played also by the first variable, assets. Less important are the second and the third variable, i.e., sales and profits.

The second factor, explaining 19% of the variance, is given by coefficients $\widehat{\beta}_2 = (0.25, 0.55, 0.38, -0.05, -0.70)^\top$. It could be described as "large sales, profits, and assets with small number of employees" factor. The scatterplot in Figure 18.5 does not seem to show any clear relationship between the market

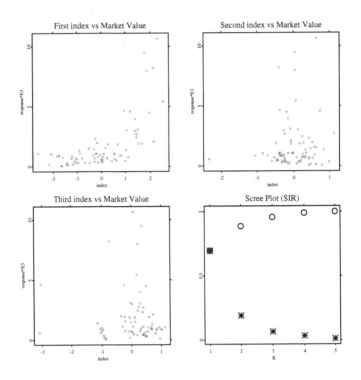

Fig. 18.5. SIR applied on the U.S. companies data set (without IBM and General Electric). Screeplot and scatterplots of the first three indices against the response. Q SMSsiruscomp

value and the second factor. However, 3-dimensional plot would reveal rather complicated nonlinear dependency of the market value on the first two factors. Unfortunately, for obvious reasons, a 3-dimensional plot cannot be printed in a book.

We can conclude that the market value of a company is a nonlinear function of a factor given mainly by number of employees and assets of the company.

EXERCISE 18.5. *Simulate a data set with $X \sim N_4(0, I_4), Y = (X_1 + 3X_2)^2 + (X_3 - X_4)^4 + \varepsilon$ and $\varepsilon \sim N(0,1)$ and use SIR and SIR II technique to find the EDR directions.*

We have simulated altogether 200 observations from the nonlinear regression model. The true response variable depends on the explanatory variables nonlinearly through the linear combinations $X\beta_1 = X_1 + 3X_2$ and $X\beta_2 = X_3 - X_4$, where $\beta_1 = (1, 3, 0, 0)^\top$ and $\beta_2 = (0, 0, 3, -4)^\top$.

The screeplot and the scatterplots of the response Y against the estimated projections obtained by SIR algorithm are plotted in Figure 18.6. The screeplot,

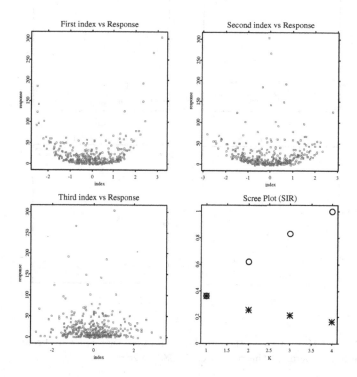

Fig. 18.6. SIR applied on the simulated data set. Screeplot and scatterplots of first three indices against the response. ◌ SMSsirsimu

corresponding to eigenvalues $\widehat{\lambda} = (0.36, 0.26, 0.21, 0.17)^{\top}$ does not really show any truly dominant direction. However, choosing the first two factors would lead $\widehat{\beta}_1 = (-0.16, -0.17, 0.76, -0.60)^{\top}$ and $\widehat{\beta}_2 = (0.49, 0.81, 0.31, 0.07)^{\top}$ which are not too far from the original β_2 and β_1.

In Härdle & Simar (2003, example 18.2) it is demonstrated that the SIR algorithm does not work very well if the response variable is symmetric as in this exercise. In such situations, the SIR II algorithm should be able to provide more reliable results. The results of the SIR II algorithm, based on conditional variance rather than on conditional expectations, are graphically displayed in Figure 18.7.

Clearly, the factors of the SIR II algorithm plotted in Figure 18.7 are very similar to the factors obtained by the SIR algorithm in Figure 18.6. The main

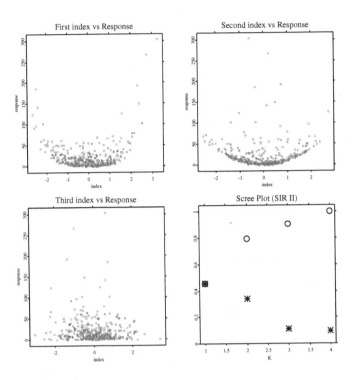

Fig. 18.7. SIR II applied on the simulated data set. Screeplot and scatterplots of first three indices against the response. ◻ SMSsir2simu

difference is in the screeplot which now more strongly suggests that two factors are appropriate. The eigenvalues are $\widehat{\lambda} = (0.45, 0.34, 0.11, 0.09)^{\top}$ and the first two factors here explain 79% of the variance.

The coefficients of the first two factors, $\widehat{\beta}_1 = (-0.12, -0.16, 0.81, -0.55)$ and $\widehat{\beta}_2 = (0.31, 0.93, 0.12, -0.15)$, are also very close to the true values of β_2 and β_1.

The SIR II algorithm basically provides the same directions, but the result can be seen more clearly. Better results from the SIR algorithms might be expected for monotone relationships.

The dependency of the response on the first two factors is actually stronger than it appears from the two-dimensional scatterplots in Figures 18.6 and 18.7. Plotting the dependency of the response on the first two factors in three-dimensional graphics shows very clear three-dimensional surface. In Figure 18.7, we can see only two side views of the "bowl". However, with some effort, it is not impossible to imagine how this surface actually looks like.

EXERCISE 18.6. *Apply the SIR and SIR II technique on the car data set in Table A.4 with Y = "price".*

The 9 explanatory variables in the cars data set are: mileage, headroom, rear seat clearance, trunk space, weight, length, turning diameter, displacement and gear ratio. We have dropped the variables measuring repair record since they were containing missing values. The variable "company headquarters" is used to define the plotting symbols in the resulting graphics.

Before running the SIR algorithm, the explanatory variables were centered and standardized.

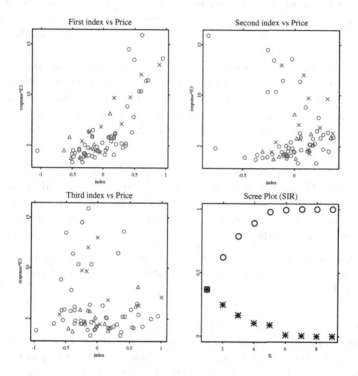

Fig. 18.8. SIR applied on the cars data set. Screeplot and scatterplots of first three indices against the response. The plotting symbol denotes company headquarters (circle=USA, cross=Europe, triangle=Japan). ▢ SMSsircars

The screeplot and the scatterplots of the response versus the first three indices are plotted in Figure 18.8. Considering the number of explanatory variables and the obtained eigenvalues,

$$\widehat{\lambda} = (0.37, 0.25, 0.16, 0.11, 0.09, 0.01, 0.01, 0.00, 0.00)^\top,$$

we should keep three or four factors. The corresponding coefficients are:

$$\widehat{\beta}_1 = (-0.10, -0.20, 0.12, 0.09, 0.83, -0.26, -0.37, 0.05, 0.18)^\top,$$
$$\widehat{\beta}_2 = (-0.08, -0.06, -0.08, 0.17, -0.67, 0.67, -0.16, 0.14, 0.09)^\top,$$
$$\widehat{\beta}_3 = (-0.16, -0.23, 0.16, -0.24, -0.61, 0.20, 0.56, 0.18, 0.28)^\top,$$
$$\widehat{\beta}_4 = (-0.03, 0.20, 0.08, -0.25, 0.08, -0.10, -0.35, 0.64, 0.59)^\top.$$

The first factor seems to assign most of the weight to the variable "weight".The second factor is contrast between the length and the weight of the car.

In Figure 18.6, the increasing price as a function of weight is clearly visible in the first scatterplot. The most expensive and heaviest cars come from USA (circles) and Europe (crosses). The dependency of the price on the second factor seems to be more complicated. Again, as in the previous exercises, the first two factors have to be considered jointly and the best visualization would be achieved by interactive (rotating) three-dimensional plot: in such graphical device it can be clearly seen that the graph of the price plotted against the first two factors can be described as three-dimensional "twisted tube". In the upper two scatterplots in Figure 18.6, this "twisted tube" can be seen only from the front and the side view.

In Figure 18.9, we can see the results of the SIR II algorithm applied on the same data set. The screeplot immediately suggests that the SIR II algorithm is not appropriate for this data and that it does not find any interesting directions. In the scatterplots, we do not see any clear relationship between the response and the indices.

In this situation, better results are provided by the SIR algorithm which discovers an interesting nonlinear relationship of the price of the car on its weight and length.

EXERCISE 18.7. *Generate four regions on the two-dimensional unit square by sequentially cutting parallel to the coordinate axes. Generate 100 two-dimensional Uniform random variables and label them according to their presence in the above regions. Apply the CART algorithm to find the regions bound and to classify the observations.*

The example has been generated by cutting first the unit square at $x_2 = 0.5$ and then dividing each half at $x_1 = 0.75$. The class assignment is displayed graphically in the left plot in Figure 18.10, where the classes 1, 2, and 3 are respectively denoted by triangles, squares, and diamonds.

The CART procedure finds the tree displayed in Figure 18.10. One sees that CART almost perfectly reproduces the split points.

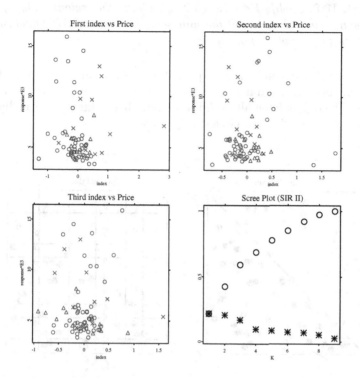

Fig. 18.9. SIR II applied on the cars data set. Screeplot and scatterplots of first three indices against the response. The plotting symbol denotes company headquarters (circle=USA, cross=Europe, triangle=Japan). ◙ SMSsir2cars

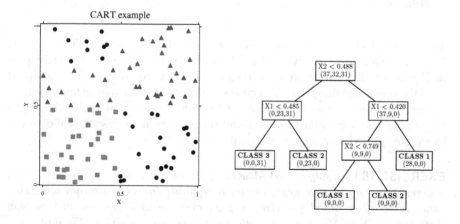

Fig. 18.10. Classification tree applied on the example data set. ◙ SMScartsq

EXERCISE 18.8. *Modify Exercise 18.7 by defining the regions as lying above and below the main diagonal of the unit square. Make a CART analysis and comment on the complexity of the tree.*

The design of this example is not optimal for CART since the optimal split does not lie along a coordinate axis. A simulated data set is plotted in Figure 18.11. The points lying above (group 1) and below (group 2) the diagonal are denoted by triangles and circles, respectively.

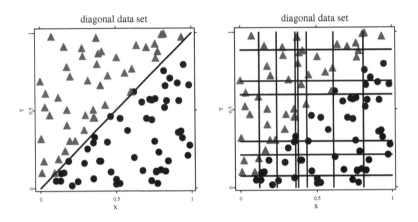

Fig. 18.11. The diagonal data set. The points from groups 1 (triangles) and 2 (circles) are separated by a single line, the diagonal, in the left plot. The horizontal and vertical lines in the right plot are the thresholds obtained by CART algorithm.
Q SMScartdiag

The diagonal of the unit square is plotted in the left plot in Figure 18.11. The thresholds obtained by the CART algorithm are plotted in the right plot in Figure 18.11. Clearly, many thresholds are needed to split the simulated data set across the diagonal. As a consequence, the CART algorithm produces an oversized tree, plotted in Figure 18.12. One can see that the CART algorithm tries to approximate the optimal diagonal split by a sequence of small rectangles placed around the diagonal.

EXERCISE 18.9. *Apply the SVM with different radial basis parameter r and different capacity parameter c in order to separate two circular data sets. This example is often called the Orange Peel exercise and involves two normal distributions $N(\mu, \Sigma_i)$, $i = 1, 2$, with covariance matrices $\Sigma_1 = 2\mathcal{I}_2$ and $\Sigma_2 = 0.5\mathcal{I}_2$.*

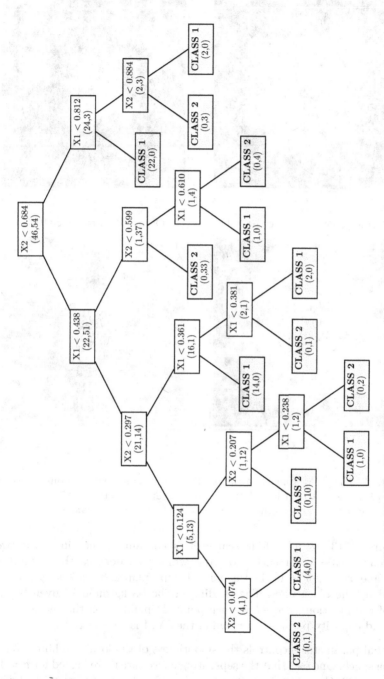

Fig. 18.12. Classification tree applied on the diagonal data set. ◯ SMScartdiag

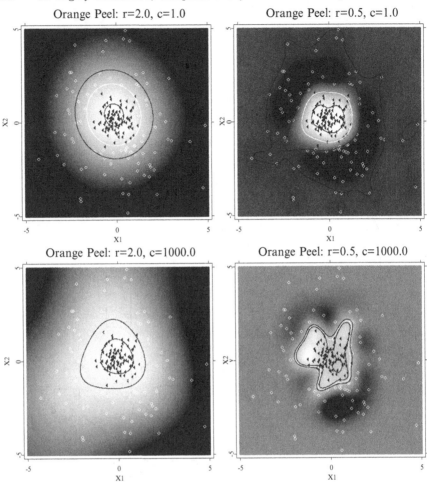

Fig. 18.13. The SVM applied to the orange peel data set with various choices of parameters r and c. Upper left plot: $r = 2$, $c = 1$, upper right: $r = 0.5$, $c = 1$, lower left: $r = 2$, $c = 1000$, lower right: $r = 0.5$, $c = 1000$. ☯ SMSsvmorange

In Figure 18.13, we plot four scatterplots containing the simulated two-dimensional dataset. In each plot, the white line denotes the separating hyperplane $x^\top w + b = 0$. The canonical hyperplanes, $x^\top w + b = \pm 1$, are denoted by the black lines. The shading of the background is given by the value of the function $x^\top w + b$ in each point. Depending on the choice of the radial and capacity parameters r and c, the SVM is very flexible.

The radial parameter r controls the smoothness of the local neighborhood in the data space. One sees that the separating curves are more jagged for $r = 0.5$ than for $r = 2$. Compare the pictures in the left column of Figure 18.13 with those in the right column.

The capacity parameter c controls the amount of nonseparable observations Letting c grow makes the SVM more sensitive to the classification error as can be seen from 18.3. The SVM therefore yields smaller margins.

For the orange peel data involving two circular covariance structures, the parameter constellation $r = 2$, $c = 1$ gives the best separability results.

EXERCISE 18.10. *The noisy spiral data set consists of two intertwining spirals that need to be separated by a nonlinear classification method. Apply the SVM with different radial basis parameter r and capacity parameter c in order to separate the two spiral datasets.*

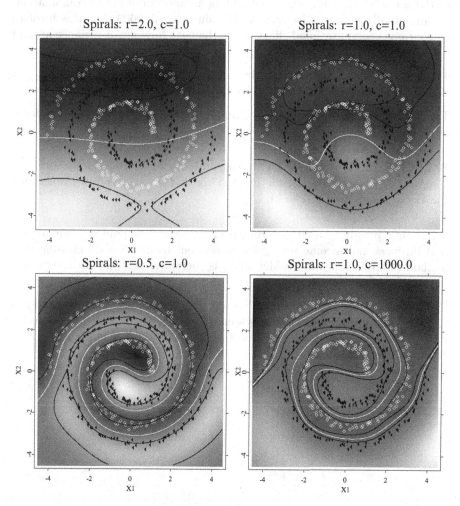

Fig. 18.14. The SVM applied on the spiral data set with various choices of parameters r and c. Upper left plot: $r = 2$, $c = 1$, upper right: $r = 1$, $c = 1$, lower left: $r = 0.5$, $c = 1$, lower right: $r = 1$, $c = 1000$. ⓠ SMSsvmspiral

The simulated data set, plotted in Figure 18.14, was generated by adding a random noise, $N_2(0_2, \mathcal{I}/100)$, to regularly spaced points lying on the spirals E_1 and E_2:

$$E_1 = \left\{ \begin{pmatrix} (1+x)\sin(x) \\ (1+x)\cos(x) \end{pmatrix}, x \in (0, 3\pi) \right\},$$

$$E_2 = \left\{ \begin{pmatrix} (1+x)\sin(x+\pi) \\ (1+x)\cos(x+\pi) \end{pmatrix}, x \in (0, 3\pi) \right\}.$$

The noisy spiral data is certainly hard to separate for only linear classification method. For SVM, it is a matter of finding an appropriate (r, c) combination. It can actually be found by cross validation (Vapnik 2000) but this involves an enormous computational effort. Since the data have small variance around the spiral, we can work with big capacity c entailing small margins.

The local sensitivity is controlled by r as can be seen from the upper row of Figure 18.14. Lowering the sensitivity increases the correct classifications.

The best result is obtained for $r = 1$ and $c = 1000$, see the lower right corner of Figure 18.14.

EXERCISE 18.11. *Apply the SVM to separate the bankrupt from the surviving (profitable) companies using the profitability and leverage ratios given in the Bankruptcy data set in Table A.3.*

Separating possibly bankrupt from profit making companies is an important business and income source for investment banks. A good classification method (Härdle, Moro & Schäfer 2005) is therefore vital also for the performance of a bank.

Figure 18.15 shows the variation of r and c over the range $r = 2, 0.5, 5$ and $c = 1, 1000$. The capacity parameter c is seen to produce the best classifications for $c = 1$. The radial parameter $r = 2$—see the upper left corner in Figure 18.15—gives the best classification result.

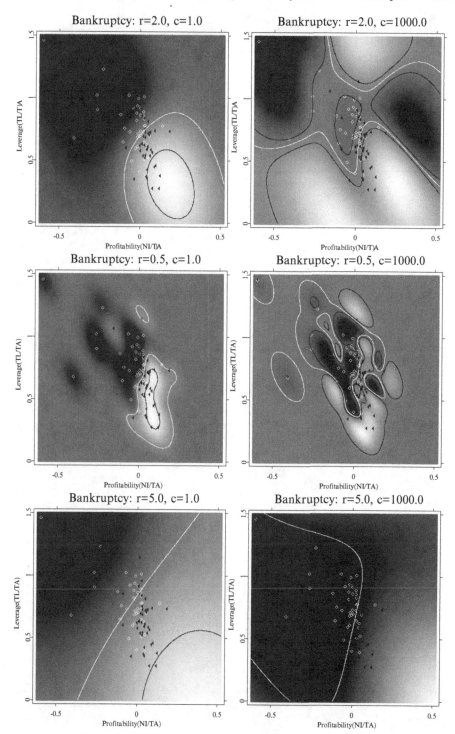

Fig. 18.15. The SVM technique applied on the Bankruptcy data set with various choices of parameters r and c. ◖ SMSsvmbankrupt

A

Data Sets

All data sets are available on the authors' Web page.

A.1 Athletic Records Data

This data set provides data on athletic records for 55 countries.

Country	100m (s)	200m (s)	400m (s)	800m (s)	1500m (min)	5000m (min)	10000m (min)	Marathon (min)
Argentina	10.39	20.81	46.84	1.81	3.70	14.04	29.36	137.71
Australia	10.31	20.06	44.84	1.74	3.57	13.28	27.66	128.30
Austria	10.44	20.81	46.82	1.79	3.60	13.26	27.72	135.90
Belgium	10.34	20.68	45.04	1.73	3.60	13.22	27.45	129.95
Bermuda	10.28	20.58	45.91	1.80	3.75	14.68	30.55	146.61
Brazil	10.22	20.43	45.21	1.73	3.66	13.62	28.62	133.13
Burma	10.64	21.52	48.30	1.80	3.85	14.45	30.28	139.95
Canada	10.17	20.22	45.68	1.76	3.63	13.55	28.09	130.15
Chile	10.34	20.80	46.20	1.79	3.71	13.61	29.30	134.03
China	10.51	21.04	47.30	1.81	3.73	13.90	29.13	133.53
Colombia	10.43	21.05	46.10	1.82	3.74	13.49	27.88	131.35
Cook Island	12.18	23.20	52.94	2.02	4.24	16.70	35.38	164.70
Costa Rica	10.94	21.90	48.66	1.87	3.84	14.03	28.81	136.58
Czech Rep	10.35	20.65	45.64	1.76	3.58	13.42	28.19	134.32
Denmark	10.56	20.52	45.89	1.78	3.61	13.50	28.11	130.78
Dom Rep	10.14	20.65	46.80	1.82	3.82	14.91	31.45	154.12
Finland	10.43	20.69	45.49	1.74	3.61	13.27	27.52	130.87
France	10.11	20.38	45.28	1.73	3.57	13.34	27.97	132.30
GDR	10.12	20.33	44.87	1.73	3.56	13.17	27.42	129.92
FRG	10.16	20.37	44.50	1.73	3.53	13.21	27.61	132.23
GB	10.11	20.21	44.93	1.70	3.51	13.01	27.51	129.13

continues on next page ⟶

Country	100m	200m	400m	800m	1500m	5000m	10000m	Marathon
Greece	10.22	20.71	46.56	1.78	3.64	14.59	28.45	134.60
Guatemala	10.98	21.82	48.40	1.89	3.80	14.16	30.11	139.33
Hungary	10.26	20.62	46.02	1.77	3.62	13.49	28.44	132.58
India	10.60	21.42	45.73	1.76	3.73	13.77	28.81	131.98
Indonesia	10.59	21.49	47.80	1.84	3.92	14.73	30.79	148.83
Ireland	10.61	20.96	46.30	1.79	3.56	13.32	27.81	132.35
Israel	10.71	21.00	47.80	1.77	3.72	13.66	28.93	137.55
Italy	10.01	19.72	45.26	1.73	3.60	13.23	27.52	131.08
Japan	10.34	20.81	45.86	1.79	3.64	13.41	27.72	128.63
Kenya	10.46	20.66	44.92	1.73	3.55	13.10	27.80	129.75
Korea	10.34	20.89	46.90	1.79	3.77	13.96	29.23	136.25
P Korea	10.91	21.94	47.30	1.85	3.77	14.13	29.67	130.87
Luxemburg	10.35	20.77	47.40	1.82	3.67	13.64	29.08	141.27
Malaysia	10.40	20.92	46.30	1.82	3.80	14.64	31.01	154.10
Mauritius	11.19	33.45	47.70	1.88	3.83	15.06	31.77	152.23
Mexico	10.42	21.30	46.10	1.80	3.65	13.46	27.95	129.20
Netherlands	10.52	29.95	45.10	1.74	3.62	13.36	27.61	129.02
NZ	10.51	20.88	46.10	1.74	3.54	13.21	27.70	128.98
Norway	10.55	21.16	46.71	1.76	3.62	13.34	27.69	131.48
PNG	10.96	21.78	47.90	1.90	4.01	14.72	31.36	148.22
Philippines	10.78	21.64	46.24	1.81	3.83	14.74	30.64	145.27
Poland	10.16	20.24	45.36	1.76	3.60	13.29	27.89	131.58
Portugal	10.53	21.17	46.70	1.79	3.62	13.13	27.38	128.65
Romania	10.41	20.98	45.87	1.76	3.64	13.25	27.67	132.50
Singapore	10.38	21.28	47.40	1.88	3.89	15.11	31.32	157.77
Spain	10.42	20.77	45.98	1.76	3.55	13.31	27.73	131.57
Sweden	10.25	20.61	45.63	1.77	3.61	13.29	27.94	130.63
Switzerland	10.37	20.45	45.78	1.78	3.55	13.22	27.91	131.20
Tapei	10.59	21.29	46.80	1.79	3.77	14.07	30.07	139.27
Thailand	10.39	21.09	47.91	1.83	3.84	15.23	32.56	149.90
Turkey	10.71	21.43	47.60	1.79	3.67	13.56	28.58	131.50
USA	9.93	19.75	43.86	1.73	3.53	13.20	27.43	128.22
USSR	10.07	20.00	44.60	1.75	3.59	13.20	27.53	130.55
W Samoa	10.82	21.86	49.00	2.02	4.24	16.28	34.71	161.83

A.2 Bank Notes Data

Six variables measured on 100 genuine and 100 counterfeit old Swiss 1000-franc bank notes. The data stem from Flury & Riedwyl (1988). The columns correspond to the following 6 variables.

X_1: length of the bank note
X_2: height of the bank note, measured on the left
X_3: height of the bank note, measured on the right
X_4: distance of the inner frame to the lower border
X_5: distance of the inner frame to the upper border
X_6: length of the diagonal

Observations 1–100 are the genuine bank notes and the other 100 observations are the counterfeit bank notes.

Length	Height (left)	Height (right)	Inner Frame (lower)	Inner Frame (upper)	Diagonal
214.8	131.0	131.1	9.0	9.7	141.0
214.6	129.7	129.7	8.1	9.5	141.7
214.8	129.7	129.7	8.7	9.6	142.2
214.8	129.7	129.6	7.5	10.4	142.0
215.0	129.6	129.7	10.4	7.7	141.8
215.7	130.8	130.5	9.0	10.1	141.4
215.5	129.5	129.7	7.9	9.6	141.6
214.5	129.6	129.2	7.2	10.7	141.7
214.9	129.4	129.7	8.2	11.0	141.9
215.2	130.4	130.3	9.2	10.0	140.7
215.3	130.4	130.3	7.9	11.7	141.8
215.1	129.5	129.6	7.7	10.5	142.2
215.2	130.8	129.6	7.9	10.8	141.4
214.7	129.7	129.7	7.7	10.9	141.7
215.1	129.9	129.7	7.7	10.8	141.8
214.5	129.8	129.8	9.3	8.5	141.6
214.6	129.9	130.1	8.2	9.8	141.7
215.0	129.9	129.7	9.0	9.0	141.9
215.2	129.6	129.6	7.4	11.5	141.5
214.7	130.2	129.9	8.6	10.0	141.9
215.0	129.9	129.3	8.4	10.0	141.4
215.6	130.5	130.0	8.1	10.3	141.6
215.3	130.6	130.0	8.4	10.8	141.5
215.7	130.2	130.0	8.7	10.0	141.6
215.1	129.7	129.9	7.4	10.8	141.1
215.3	130.4	130.4	8.0	11.0	142.3
215.5	130.2	130.1	8.9	9.8	142.4
215.1	130.3	130.3	9.8	9.5	141.9
215.1	130.0	130.0	7.4	10.5	141.8
214.8	129.7	129.3	8.3	9.0	142.0
215.2	130.1	129.8	7.9	10.7	141.8
214.8	129.7	129.7	8.6	9.1	142.3
215.0	130.0	129.6	7.7	10.5	140.7
215.6	130.4	130.1	8.4	10.3	141.0
215.9	130.4	130.0	8.9	10.6	141.4
214.6	130.2	130.2	9.4	9.7	141.8
215.5	130.3	130.0	8.4	9.7	141.8
215.3	129.9	129.4	7.9	10.0	142.0
215.3	130.3	130.1	8.5	9.3	142.1
213.9	130.3	129.0	8.1	9.7	141.3
214.4	129.8	129.2	8.9	9.4	142.3

continues on next page ⟶

Length	Height (left)	Height (right)	Inner Frame (lower)	Inner Frame (upper)	Diagonal
214.8	130.1	129.6	8.8	9.9	140.9
214.9	129.6	129.4	9.3	9.0	141.7
214.9	130.4	129.7	9.0	9.8	140.9
214.8	129.4	129.1	8.2	10.2	141.0
214.3	129.5	129.4	8.3	10.2	141.8
214.8	129.9	129.7	8.3	10.2	141.5
214.8	129.9	129.7	7.3	10.9	142.0
214.6	129.7	129.8	7.9	10.3	141.1
214.5	129.0	129.6	7.8	9.8	142.0
214.6	129.8	129.4	7.2	10.0	141.3
215.3	130.6	130.0	9.5	9.7	141.1
214.5	130.1	130.0	7.8	10.9	140.9
215.4	130.2	130.2	7.6	10.9	141.6
214.5	129.4	129.5	7.9	10.0	141.4
215.2	129.7	129.4	9.2	9.4	142.0
215.7	130.0	129.4	9.2	10.4	141.2
215.0	129.6	129.4	8.8	9.0	141.1
215.1	130.1	129.9	7.9	11.0	141.3
215.1	130.0	129.8	8.2	10.3	141.4
215.1	129.6	129.3	8.3	9.9	141.6
215.3	129.7	129.4	7.5	10.5	141.5
215.4	129.8	129.4	8.0	10.6	141.5
214.5	130.0	129.5	8.0	10.8	141.4
215.0	130.0	129.8	8.6	10.6	141.5
215.2	130.6	130.0	8.8	10.6	140.8
214.6	129.5	129.2	7.7	10.3	141.3
214.8	129.7	129.3	9.1	9.5	141.5
215.1	129.6	129.8	8.6	9.8	141.8
214.9	130.2	130.2	8.0	11.2	139.6
213.8	129.8	129.5	8.4	11.1	140.9
215.2	129.9	129.5	8.2	10.3	141.4
215.0	129.6	130.2	8.7	10.0	141.2
214.4	129.9	129.6	7.5	10.5	141.8
215.2	129.9	129.7	7.2	10.6	142.1
214.1	129.6	129.3	7.6	10.7	141.7
214.9	129.9	130.1	8.8	10.0	141.2
214.6	129.8	129.4	7.4	10.6	141.0
215.2	130.5	129.8	7.9	10.9	140.9
214.6	129.9	129.4	7.9	10.0	141.8
215.1	129.7	129.7	8.6	10.3	140.6
214.9	129.8	129.6	7.5	10.3	141.0
215.2	129.7	129.1	9.0	9.7	141.9
215.2	130.1	129.9	7.9	10.8	141.3
215.4	130.7	130.2	9.0	11.1	141.2
215.1	129.9	129.6	8.9	10.2	141.5
215.2	129.9	129.7	8.7	9.5	141.6
215.0	129.6	129.2	8.4	10.2	142.1
214.9	130.3	129.9	7.4	11.2	141.5
215.0	129.9	129.7	8.0	10.5	142.0
214.7	129.7	129.3	8.6	9.6	141.6
215.4	130.0	129.9	8.5	9.7	141.4
214.9	129.4	129.5	8.2	9.9	141.5
214.5	129.5	129.3	7.4	10.7	141.5
214.7	129.6	129.5	8.3	10.0	142.0
215.6	129.9	129.9	9.0	9.5	141.7
215.0	130.4	130.3	9.1	10.2	141.1
214.4	129.7	129.5	8.0	10.3	141.2
215.1	130.0	129.8	9.1	10.2	141.5
214.7	130.0	129.4	7.8	10.0	141.2
214.4	130.1	130.3	9.7	11.7	139.8
214.9	130.5	130.2	11.0	11.5	139.5
214.9	130.3	130.1	8.7	11.7	140.2
215.0	130.4	130.6	9.9	10.9	140.3
214.7	130.2	130.3	11.8	10.9	139.7

continues on next page $\longrightarrow$

Length	Height (left)	Height (right)	Inner Frame (lower)	Inner Frame (upper)	Diagonal
215.0	130.2	130.2	10.6	10.7	139.9
215.3	130.3	130.1	9.3	12.1	140.2
214.8	130.1	130.4	9.8	11.5	139.9
215.0	130.2	129.9	10.0	11.9	139.4
215.2	130.6	130.8	10.4	11.2	140.3
215.2	130.4	130.3	8.0	11.5	139.2
215.1	130.5	130.3	10.6	11.5	140.1
215.4	130.7	131.1	9.7	11.8	140.6
214.9	130.4	129.9	11.4	11.0	139.9
215.1	130.3	130.0	10.6	10.8	139.7
215.5	130.4	130.0	8.2	11.2	139.2
214.7	130.6	130.1	11.8	10.5	139.8
214.7	130.4	130.1	12.1	10.4	139.9
214.8	130.5	130.2	11.0	11.0	140.0
214.4	130.2	129.9	10.1	12.0	139.2
214.8	130.3	130.4	10.1	12.1	139.6
215.1	130.6	130.3	12.3	10.2	139.6
215.3	130.8	131.1	11.6	10.6	140.2
215.1	130.7	130.4	10.5	11.2	139.7
214.7	130.5	130.5	9.9	10.3	140.1
214.9	130.0	130.3	10.2	11.4	139.6
215.0	130.4	130.4	9.4	11.6	140.2
215.5	130.7	130.3	10.2	11.8	140.0
215.1	130.2	130.2	10.1	11.3	140.3
214.5	130.2	130.6	9.8	12.1	139.9
214.3	130.2	130.0	10.7	10.5	139.8
214.5	130.2	129.8	12.3	11.2	139.2
214.9	130.5	130.2	10.6	11.5	139.9
214.6	130.2	130.4	10.5	11.8	139.7
214.2	130.0	130.2	11.0	11.2	139.5
214.8	130.1	130.1	11.9	11.1	139.5
214.6	129.8	130.2	10.7	11.1	139.4
214.9	130.7	130.3	9.3	11.2	138.3
214.6	130.4	130.4	11.3	10.8	139.8
214.5	130.5	130.2	11.8	10.2	139.6
214.8	130.2	130.3	10.0	11.9	139.3
214.7	130.0	129.4	10.2	11.0	139.2
214.6	130.2	130.4	11.2	10.7	139.9
215.0	130.5	130.4	10.6	11.1	139.9
214.5	129.8	129.8	11.4	10.0	139.3
214.9	130.6	130.4	11.9	10.5	139.8
215.0	130.5	130.4	11.4	10.7	139.9
215.3	130.6	130.3	9.3	11.3	138.1
214.7	130.2	130.1	10.7	11.0	139.4
214.9	129.9	130.0	9.9	12.3	139.4
214.9	130.3	129.9	11.9	10.6	139.8
214.6	129.9	129.7	11.9	10.1	139.0
214.6	129.7	129.3	10.4	11.0	139.3
214.5	130.1	130.1	12.1	10.3	139.4
214.5	130.3	130.0	11.0	11.5	139.5
215.1	130.0	130.3	11.6	10.5	139.7
214.2	129.7	129.6	10.3	11.4	139.5
214.4	130.1	130.0	11.3	10.7	139.2
214.8	130.4	130.6	12.5	10.0	139.3
214.6	130.6	130.1	8.1	12.1	137.9
215.6	130.1	129.7	7.4	12.2	138.4
214.9	130.5	130.1	9.9	10.2	138.1
214.6	130.1	130.0	11.5	10.6	139.5
214.7	130.1	130.2	11.6	10.9	139.1
214.3	130.3	130.0	11.4	10.5	139.8
215.1	130.3	130.6	10.3	12.0	139.7
216.3	130.7	130.4	10.0	10.1	138.8
215.6	130.4	130.1	9.6	11.2	138.6
214.8	129.9	129.8	9.6	12.0	139.6

continues on next page $\longrightarrow$

Length	Height (left)	Height (right)	Inner Frame (lower)	Inner Frame (upper)	Diagonal
214.9	130.0	129.9	11.4	10.9	139.7
213.9	130.7	130.5	8.7	11.5	137.8
214.2	130.6	130.4	12.0	10.2	139.6
214.8	130.5	130.3	11.8	10.5	139.4
214.8	129.6	130.0	10.4	11.6	139.2
214.8	130.1	130.0	11.4	10.5	139.6
214.9	130.4	130.2	11.9	10.7	139.0
214.3	130.1	130.1	11.6	10.5	139.7
214.5	130.4	130.0	9.9	12.0	139.6
214.8	130.5	130.3	10.2	12.1	139.1
214.5	130.2	130.4	8.2	11.8	137.8
215.0	130.4	130.1	11.4	10.7	139.1
214.8	130.6	130.6	8.0	11.4	138.7
215.0	130.5	130.1	11.0	11.4	139.3
214.6	130.5	130.4	10.1	11.4	139.3
214.7	130.2	130.1	10.7	11.1	139.5
214.7	130.4	130.0	11.5	10.7	139.4
214.5	130.4	130.0	8.0	12.2	138.5
214.8	130.0	129.7	11.4	10.6	139.2
214.8	129.9	130.2	9.6	11.9	139.4
214.6	130.3	130.2	12.7	9.1	139.2
215.1	130.2	129.8	10.2	12.0	139.4
215.4	130.5	130.6	8.8	11.0	138.6
214.7	130.3	130.2	10.8	11.1	139.2
215.0	130.5	130.3	9.6	11.0	138.5
214.9	130.3	130.5	11.6	10.6	139.8
215.0	130.4	130.3	9.9	12.1	139.6
215.1	130.3	129.9	10.3	11.5	139.7
214.8	130.3	130.4	10.6	11.1	140.0
214.7	130.7	130.8	11.2	11.2	139.4
214.3	129.9	129.9	10.2	11.5	139.6

A.3 Bankruptcy Data

The data are the profitability, leverage, and bankruptcy indicators for 84 companies.

The data set contains information on 42 of the largest companies that filed for protection against creditors under Chapter 11 of the U.S. Bankruptcy Code in 2001–2002 after the stock market crash of 2000. The bankrupt companies were matched with 42 surviving companies with the closest capitalizations and the same US industry classification codes available through the Division of Corporate Finance of the Securities and Exchange Commission (SEC 2004).

The information for each company was collected from the annual reports for 1998–1999 (SEC 2004), i.e., three years prior to the defaults of the bankrupt companies. The following data set contains profitability and leverage ratios calculated, respectively, as the ratio of net income (NI) and total assets (TA) and the ratio of total liabilities (TL) and total assets (TA).

Profitability (NI/TA)	Leverage (TL/TA)	Bankruptcy
0.022806	0.7816	1
−0.063584	0.7325	1
−0.227860	1.2361	1
0.021364	0.7350	1
0.042058	0.6339	1
0.021662	0.8614	1
0.023952	0.6527	1
0.000005	0.7385	1
0.020702	0.8954	1
−0.006640	0.7009	1
0.021634	0.7338	1
−0.023206	0.8226	1
−0.263667	0.9085	1
−0.047161	0.9275	1
−0.098931	0.7617	1
0.140857	0.7802	1
−0.018031	0.6187	1
0.047647	0.5294	1
−0.267393	1.0289	1
−0.105816	0.6542	1
−0.013929	0.7181	1
0.001215	0.8653	1
−0.012698	1.0000	1
−0.021990	0.8811	1
0.039147	0.5987	1
−0.000300	0.9213	1
−0.071546	1.0254	1
0.004256	0.7058	1
−0.003599	0.4086	1
−0.029814	0.6017	1
0.030197	0.7866	1
−0.016912	0.9428	1
0.026682	0.5173	1
−0.052413	0.4983	1
−0.408583	0.6821	1
−0.015960	0.6915	1
0.022807	0.8348	1
0.055888	0.6986	1

continues on next page ⟶

Profitability (NI/TA)	Leverage (TL/TA)	Bankruptcy
0.025634	1.0152	1
−0.599016	1.4633	1
0.044064	0.4001	1
−0.121531	0.8697	1
−0.000172	0.6131	−1
−0.156216	1.0584	−1
0.012473	0.6254	−1
0.081731	0.2701	−1
0.080826	0.5593	−1
0.033538	0.7468	−1
0.036645	0.5338	−1
0.052686	0.7101	−1
0.122404	0.2700	−1
0.068682	0.7182	−1
0.030576	0.6175	−1
0.094346	0.7293	−1
0.091535	0.4425	−1
0.058916	0.6997	−1
0.186226	0.7254	−1
0.072777	0.5797	−1
0.101209	0.4526	−1
0.015374	0.8504	−1
0.047247	0.5918	−1
−0.085583	0.5945	−1
0.033137	0.5160	−1
0.016055	1.1353	−1
0.008357	0.9068	−1
0.034960	0.7169	−1
0.046514	0.3473	−1
−0.084510	0.8422	−1
0.029492	0.8319	−1
0.045271	0.5813	−1
0.041463	0.3619	−1
0.030059	0.9479	−1
0.023445	0.6856	−1
0.046705	0.4164	−1
0.127897	0.3694	−1
0.050956	0.6073	−1
0.020425	0.5295	−1
0.035311	0.6796	−1
0.066434	0.5303	−1
0.066550	0.7194	−1
0.055333	0.8063	−1
0.015738	0.6294	−1
−0.034455	0.3446	−1
0.004824	0.5705	−1

A.4 Car Data

The car data set (Chambers, Cleveland, Kleiner & Tukey 1983) consists of 13 variables measured for 74 car types. The abbreviations in the table are as follows:

X_1: P price
X_2: M mileage (in miles per gallon)
X_3: R78 repair record 1978 (rated on a 5-point scale: 5 best, 1 worst)
X_4: R77 repair record 1977 (scale as before)
X_5: H headroom (in inches)
X_6: R rear seat clearance (in inches)
X_7: Tr trunk space (in cubic feet)
X_8: W weight (in pound)
X_9: L length (in inches)
X_{10}: T turning diameter (clearance required to make a U-turn, in feet)
X_{11}: D displacement (in cubic inches)
X_{12}: G gear ratio for high gear
X_{13}: C company headquarters (1 United States, 2 Japan, 3 Europe)

Model	P	M	R78	R77	H	R	Tr	W	L	T	D	G	C
AMC Concord	4099	22	3	2	2.5	27.5	11	2930	186	40	121	3.58	1
AMC Pacer	4749	17	3	1	3.0	25.5	11	3350	173	40	258	2.53	1
AMC Spirit	3799	22	–		3.0	18.5	12	2640	168	35	121	3.08	1
Audi 5000	9690	17	5	2	3.0	27.0	15	2830	189	37	131	3.20	1
Audi Fox	6295	23	3	3	2.5	28.0	11	2070	174	36	97	3.70	3
BMW 320i	9735	25	4	4	2.5	26.0	12	2650	177	34	121	3.64	3
Buick Century	4816	20	3	3	4.5	29.0	16	3250	196	40	196	2.93	1
Buick Electra	7827	15	4	4	4.0	31.5	20	4080	222	43	350	2.41	1
Buick Le Sabre	5788	18	3	4	4.0	30.5	21	3670	218	43	231	2.73	1
Buick Opel	4453	26	–		3.0	24.0	10	2230	170	34	304	2.87	1
Buick Regal	5189	20	3	3	2.0	28.5	16	3280	200	42	196	2.93	1
Buick Riviera	10372	16	3	4	3.5	30.0	17	3880	207	43	231	2.93	1
Buick Skylark	4082	19	3	3	3.5	27.0	13	3400	200	42	231	3.08	1
Cadillac Deville	11385	14	3	3	4.0	31.5	20	4330	221	44	425	2.28	1
Cadillac El Dorado	14500	14	2	2	3.5	30.0	16	3900	204	43	350	2.19	1
Cadillac Seville	15906	21	3	3	3.0	30.0	13	4290	204	45	350	2.24	1
Chevrolet Chevette	3299	29	3	3	2.5	26.0	9	2110	163	34	231	2.93	1
Chevrolet Impala	5705	16	4	4	4.0	29.5	20	3690	212	43	250	2.56	1
Chevrolet Malibu	4504	22	3	3	3.5	28.5	17	3180	193	41	200	2.73	1
Chevrolet Monte Carlo	5104	22	2	3	2.0	28.5	16	3220	200	41	200	2.73	1
Chevrolet Monza	3667	24	2	2	2.0	25.0	7	2750	179	40	151	2.73	1
Chevrolet Nova	3955	19	3	3	3.5	27.0	13	3430	197	43	250	2.56	1
Datsun 200-SX	6229	23	4	3	1.5	21.0	6	2370	170	35	119	3.89	2
Datsun 210	4589	35	5	5	2.0	23.5	8	2020	165	32	85	3.70	2
Datsun 510	5079	24	4	4	2.5	22.0	8	2280	170	34	119	3.54	2
Datsun 810	8129	21	4	4	2.5	27.0	8	2750	184	38	146	3.55	2
Dodge Colt	3984	30	5	4	2.0	24.0	8	2120	163	35	98	3.54	2
Dodge Diplomat	5010	18	2	2	4.0	29.0	17	3600	206	46	318	2.47	1
Dodge Magnum XE	5886	16	2	2	3.5	26.0	16	3870	216	48	318	2.71	1
Dodge St. Regis	6342	17	2	2	4.5	28.0	21	3740	220	46	225	2.94	1
Fiat Strada	4296	21	3	1	2.5	26.5	16	2130	161	36	105	3.37	3
Ford Fiesta	4389	28	4	–	1.5	26.0	9	1800	147	33	98	3.15	1
Ford Mustang	4187	21	3	3	2.0	23.0	10	2650	179	42	140	3.08	1
Honda Accord	5799	25	5	5	3.0	25.5	10	2240	172	36	107	3.05	2
Honda Civic	4499	28	4	4	2.5	23.5	5	1760	149	34	91	3.30	2

continues on next page ⟶

Model	P	M	R78	R77	H	R	Tr	W	L	T	D	G	C
Lincoln Continental	11497	12	3	4	3.5	30.5	22	4840	233	51	400	2.47	1
Lincoln Cont Mark V	13594	12	3	4	2.5	28.5	18	4720	230	48	400	2.47	1
Lincoln Versailles	13466	14	3	3	3.5	27.0	15	3830	201	41	302	2.47	1
Mazda GLC	3995	30	4	4	3.5	25.5	11	1980	154	33	86	3.73	1
Mercury Bobcat	3829	22	4	3	3.0	25.5	9	2580	169	39	140	2.73	1
Mercury Cougar	5379	14	4	3	3.5	29.5	16	4060	221	48	302	2.75	1
Mercury Cougar XR-7	6303	14	4	3	3.0	25.0	16	4130	217	45	302	2.75	1
Mercury Marquis	6165	15	3	2	3.5	30.5	23	3720	212	44	302	2.26	1
Mercury Monarch	4516	18	3	–	3.0	27.0	15	3370	198	41	250	2.43	1
Mercury Zephyr	3291	20	3	3	3.5	29.0	17	2830	195	43	140	3.08	1
Oldsmobile 98	8814	21	4	4	4.0	31.5	20	4060	220	43	350	2.41	1
Oldsmobile Cutlass	4733	19	3	4	4.5	28.0	16	3300	198	42	231	2.93	1
Oldsmobile Cutlass Supreme	5172	19	3	4	2.0	28.0	16	3310	198	42	231	2.93	1
Oldsmobile Delta 88	5890	18	4	4	4.0	29.0	20	3690	218	42	231	2.73	1
Oldsmobile Omega	4181	19	3	3	4.5	27.0	14	3370	200	43	231	3.08	1
Oldsmobile Starfire	4195	24	1	1	2.0	25.5	10	2720	180	40	151	2.73	1
Oldsmobile Tornado	10371	16	3	3	3.5	30.0	17	4030	206	43	350	2.41	1
Peugeot 604 SL	12990	14	–	–	3.5	30.5	14	3420	192	38	163	3.58	3
Plymouth Arrow	4647	28	3	2	2.0	21.5	11	2360	170	37	156	3.05	1
Plymouth Champ	4425	34	5	4	2.5	23.0	11	1800	157	37	86	2.97	1
Plymouth Horizon	4482	25	3	–	4.0	25.0	17	2200	165	36	105	3.37	1
Plymouth Sapporo	6486	26	–	–	1.5	22.0	8	2520	182	38	119	3.54	1
Plymouth Volare	4060	18	2	2	5.0	31.0	16	3330	201	44	225	3.23	1
Pontiac Catalina	5798	18	4	4	4.0	29.0	20	3700	214	42	231	2.73	1
Pontiac Firebird	4934	18	1	2	1.5	23.5	7	3470	198	42	231	3.08	1
Pontiac Grand Prix	5222	19	3	3	2.0	28.5	16	3210	201	45	231	2.93	1
Pontiac Le Mans	4723	19	3	3	3.5	28.0	17	3200	199	40	231	2.93	1
Pontiac Phoenix	4424	19	–	–	3.5	27.0	13	3420	203	43	231	3.08	1
Pontiac Sunbird	4172	24	2	2	2.0	25.0	7	2690	179	41	151	2.73	1
Renault Le Car	3895	26	3	3	3.0	23.0	10	1830	142	34	79	3.72	3
Subaru	3798	35	5	4	2.5	25.5	11	2050	164	36	97	3.81	2
Toyota Celica	5899	18	5	5	2.5	22.0	14	2410	174	36	134	3.06	2
Toyota Corolla	3748	31	5	5	3.0	24.5	9	2200	165	35	97	3.21	2
Toyota Corona	5719	18	5	5	2.0	23.0	11	2670	175	36	134	3.05	2
VW Rabbit	4697	25	4	3	3.0	25.5	15	1930	155	35	89	3.78	3
VW Rabbit Diesel	5397	41	5	4	3.0	25.5	15	2040	155	35	90	3.78	3
VW Scirocco	6850	25	4	3	2.0	23.5	16	1990	156	36	97	3.78	3
VW Dasher	7140	23	4	3	2.5	37.5	12	2160	172	36	97	3.74	3
Volvo 260	11995	17	5	3	2.5	29.5	14	3170	193	37	163	2.98	3

A.5 Car Marks

The data are averaged marks for 24 car types from a sample of 40 persons. The marks range from 1 (very good) to 6 (very bad) like German school marks. The variables are:

X_1: A cconomy
X_2: B service
X_3: C nondepreciation of value
X_4: D price, mark 1 for very cheap cars
X_5: E design
X_6: F sporty car
X_7: G safety
X_8: H easy handling

Type	Model	Economy	Service	Value	Price	Design	Sport	Safety	Easy Handling
Audi	100	3.9	2.8	2.2	4.2	3.0	3.1	2.4	2.8
BMW	5 series	4.8	1.6	1.9	5.0	2.0	2.5	1.6	2.8
Citroen	AX	3.0	3.8	3.8	2.7	4.0	4.4	4.0	2.6
Ferrari		5.3	2.9	2.2	5.9	1.7	1.1	3.3	4.3
Fiat	Uno	2.1	3.9	4.0	2.6	4.5	4.4	4.4	2.2
Ford	Fiesta	2.3	3.1	3.4	2.6	3.2	3.3	3.6	2.8
Hyundai		2.5	3.4	3.2	2.2	3.3	3.3	3.3	2.4
Jaguar		4.6	2.4	1.6	5.5	1.3	1.6	2.8	3.6
Lada	Samara	3.2	3.9	4.3	2.0	4.3	4.5	4.7	2.9
Mazda	323	2.6	3.3	3.7	2.8	3.7	3.0	3.7	3.1
Mercedes	200	4.1	1.7	1.8	4.6	2.4	3.2	1.4	2.4
Mitsubishi	Galant	3.2	2.9	3.2	3.5	3.1	3.1	2.9	2.6
Nissan	Sunny	2.6	3.3	3.9	2.1	3.5	3.9	3.8	2.4
Opel	Corsa	2.2	2.4	3.0	2.6	3.2	4.0	2.9	2.4
Opel	Vectra	3.1	2.6	2.3	3.6	2.8	2.9	2.4	2.4
Peugeot	306	2.9	3.5	3.6	2.8	3.2	3.8	3.2	2.6
Renault	19	2.7	3.3	3.4	3.0	3.1	3.4	3.0	2.7
Rover		3.9	2.8	2.6	4.0	2.6	3.0	3.2	3.0
Toyota	Corolla	2.5	2.9	3.4	3.0	3.2	3.1	3.2	2.8
Volvo		3.8	2.3	1.9	4.2	3.1	3.6	1.6	2.4
Trabant	601	3.6	4.7	5.5	1.5	4.1	5.8	5.9	3.1
VW	Golf	2.4	2.1	2.0	2.6	3.2	3.1	3.1	1.6
VW	Passat	3.1	2.2	2.1	3.2	3.5	3.5	2.8	1.8
Wartburg	1.3	3.7	4.7	5.5	1.7	4.8	5.2	5.5	4.0

A.6 Classic Blue Pullover Data

This is a data set consisting of 10 measurements of 4 variables. A textile shop manager is studying the sales of "classic blue" pullovers over 10 periods. He uses three different marketing methods and hopes to understand his sales as a fit of these variables using statistics. The variables measured are

X_1: number of sold pullovers
X_2: price (in EUR)
X_3: advertisement costs in local newspapers (in EUR)
X_4: presence of a sales assistant (in hours per period)

	Sales	Price	Advert.	Asst Hours
1	230	125	200	109
2	181	99	55	107
3	165	97	105	98
4	150	115	85	71
5	97	120	0	82
6	192	100	150	103
7	181	80	85	111
8	189	90	120	93
9	172	95	110	86
10	170	125	130	78

A.7 Fertilizer Data

The yields of wheat have been measured in 30 parcels, which have been randomly attributed to 3 lots prepared by one of 3 different fertilizers A, B, and C.

X_1: fertilizer A
X_2: fertilizer B
X_3: fertilizer C

Yield A	Yield B	Yield C
4	6	2
3	7	1
2	7	1
5	5	1
4	5	3
4	5	4
3	8	3
3	9	3
3	9	2
1	6	2

A.8 French Baccalauréat Frequencies

The data consist of observations of 202100 French baccalauréats in 1976 and give the frequencies for different sets of modalities classified into regions. For a reference, see Bouroche & Saporta (1980). The variables (modalities) are:

X_1: A philosophy letters
X_2: B economics and social sciences
X_3: C mathematics and physics
X_4: D mathematics and natural sciences
X_5: E mathematics and techniques
X_6: F industrial techniques
X_7: G economic techniques
X_8: H computer techniques

Abbrev.	Region	A	B	C	D	E	F	G	H	total
ILDF	Ile de France	9724	5650	8679	9432	839	3353	5355	83	43115
CHAM	Champagne Ardennes	924	464	567	984	132	423	736	12	4242
PICA	Picardie	1081	490	830	1222	118	410	743	13	4907
HNOR	Haute Normandie	1135	587	686	904	83	629	813	13	4850
CENT	Centre	1482	667	1020	1535	173	629	989	26	6521
BNOR	Basse Normandie	1033	509	553	1063	100	433	742	13	4446
BOUR	Bourgogne	1272	527	861	1116	219	769	1232	13	6009
NOPC	Nord Pas de Calais	2549	1141	2164	2752	587	1660	1951	41	12845
LORR	Lorraine	1828	681	1364	1741	302	1289	1683	15	8903
ALSA	Alsace	1076	443	880	1121	145	917	1091	15	5688
FRAC	Franche Comté	827	333	481	892	137	451	618	18	3757
PAYL	Pays de la Loire	2213	809	1439	2623	269	990	1783	14	10140
BRET	Bretagne	2158	1271	1633	2352	350	950	1509	22	10245
PCHA	Poitou Charentes	1358	503	639	1377	164	495	959	10	5505
AQUI	Aquitaine	2757	873	1466	2296	215	789	1459	17	9872
MIDI	Midi Pyrénées	2493	1120	1494	2329	254	855	1565	28	10138
LIMO	Limousin	551	297	386	663	67	334	378	12	2688
RHOA	Rhônes Alpes	3951	2127	3218	4743	545	2072	3018	36	19710
AUVE	Auvergne	1066	579	724	1239	126	476	649	12	4871
LARO	Languedoc Roussillon	1844	816	1154	1839	156	469	993	16	7287
PROV	Provence Alpes Côte d'Azur	3944	1645	2415	3616	343	1236	2404	22	15625
CORS	Corse	327	31	85	178	9	27	79	0	736
	total	45593	21563	32738	46017	5333	19656	30749	451	202100

A.9 French Food Data

The data set consists of the average expenditures on food for several different types of families in France (manual workers = MA, employees = EM, managers = CA) with different numbers of children (2, 3, 4, or 5 family members). The data are taken from Lebart, Morineau & Fénelon (1982).

		Bread	Vegetables	Fruit	Meat	Poultry	Milk	Wine
1	MA2	332	428	354	1437	526	247	427
2	EM2	293	559	388	1527	567	239	258
3	CA2	372	767	562	1948	927	235	433
4	MA3	406	563	341	1507	544	324	407
5	EM3	386	608	396	1501	558	319	363
6	CA3	438	843	689	2345	1148	243	341
7	MA4	534	660	367	1620	638	414	407
8	EM4	460	699	484	1856	762	400	416
9	CA4	385	789	621	2366	1149	304	282
10	MA5	655	776	423	1848	759	495	486
11	EM5	584	995	548	2056	893	518	319
12	CA5	515	1097	887	2630	1167	561	284
$\bar{x}$		446.7	737.8	505.0	1886.7	803.2	358.2	368.6
$s_{X_i X_i}$		102.6	172.2	158.1	378.9	238.9	112.1	68.7

A.10 Geopol Data

This data set contains a comparison of 41 countries according to 10 different political and economic parameters:

X_1: popu population
X_2: giph gross internal product per habitant
X_3: ripo rate of increase of the population
X_4: rupo rate of urban population
X_5: rlpo rate of illiteracy in the population
X_6: rspo rate of students in the population
X_7: eltp expected lifetime of people
X_8: rnnr rate of nutritional needs realized
X_9: nunh number of newspapers and magazines per 1000 habitants
X_{10}: nuth number of television per 1000 habitants

AFS	South Africa	DAN	Denmark	MAR	Morocco
ALG	Algeria	EGY	Egypt	MEX	Mexico
BRD	Germany	ESP	Spain	NOR	Norway
GBR	Great Britain	FRA	France	PER	Peru
ARS	Saudi Arabia	GAB	Gabon	POL	Poland
ARG	Argentina	GRE	Greece	POR	Portugal
AUS	Australia	HOK	Hong Kong	SUE	Sweden
AUT	Austria	HON	Hungary	SUI	Switzerland
BEL	Belgium	IND	India	THA	Thailand
CAM	Cameroon	IDO	Indonesia	URS	USSR
CAN	Canada	ISR	Israel	USA	USA
CHL	Chile	ITA	Italy	VEN	Venezuela
CHN	China	JAP	Japan	YOU	Yugoslavia
CUB	Cuba	KEN	Kenya		

	popu	giph	ripo	rupo	rlpo	rspo	eltp	rnnr	nunh	nuth
AFS	37	2492	2	58.9	44	1.08	60	120	48	98
ALG	24.6	1960	3	44.7	50.4	0.73	64	112	21	71
BRD	62	19610	0.4	86.4	2	2.72	72	145	585	759
GBR	57.02	14575	0.04	92.5	2.2	1.9	75	128	421	435
ARS	14.4	5980	2.7	77.3	48.9	0.91	63	125	34	269
ARG	32.4	2130	1.6	86.2	6.1	2.96	71	136	82	217
AUS	16.81	16830	1.4	85.5	5	2.5	76	125	252	484
AUT	7.61	16693	0	57.7	1.5	2.52	74	130	362	487
BEL	9.93	15243	0.2	96.9	3	2.56	74	150	219	320

continues on next page ⟶

	popu	giph	ripo	rupo	rlpo	rspo	eltp	rnnr	nunh	nuth
CAM	11	1120	2.7	49.4	58.8	0.17	53	88	6	12
CAN	26.25	20780	0.9	76.4	1	6.89	77	129	321	586
CHL	12.95	1794	1.6	85.6	8.9	1.73	71	106	67	183
CHN	1119	426	1.1	21.4	34.5	0.16	69	111	36	24
CUB	10.5	1050	0.8	74.9	3.8	2.38	75	135	129	203
DAN	5.13	20570	0.4	86.4	1.5	2.38	75	131	359	526
EGY	52.52	665	2.5	48.8	61.8	1.67	59	132	39	84
ESP	39.24	9650	0.4	78.4	4.2	2.55	77	137	75	380
FRA	56.1	16905	0.4	74.1	2	2.63	76	130	193	399
GAB	1.1	3000	4	45.7	60	0.36	52	107	14	23
GRE	10	5370	0.3	62.6	9.5	1.89	76	147	102	175
HOK	5.75	10900	0	100	22.7	1.34	77	121	521	247
HON	10.6	2330	−0.1	60.3	1.1	0.93	70	135	273	404
IND	810	317	1.9	28	59.2	0.55	57	100	28	7
IDO	179	454	2	28.8	32.7	0.55	60	116	21	41
ISR	4.47	9800	1.4	91.6	8.2	2.62	75	118	253	276
ITA	57.55	15025	0.1	68.6	3.5	2.25	75	139	105	419
JAP	123.2	22825	0.6	77	3	2.1	78	122	566	589
KEN	23.88	400	3.8	23.6	69	0.11	58	92	13	6
MAR	24.51	800	2.2	48.5	78.6	0.86	61	118	12	55
MEX	84.3	2096	2.5	72.6	17	1.55	68	120	124	124
NOR	4.2	22060	0.3	74.4	2	2.74	77	124	551	350
PER	21.75	1899	2.1	70.2	18.1	2.04	61	93	31	85
POL	38	1740	0.9	63.2	1.2	1.3	71	134	184	263
POR	10.5	4304	0.6	33.3	20.6	1.99	74	128	70	160
SUE	8.47	22455	0.1	84	1.5	2.21	77	113	526	395
SUI	6.7	26025	0.5	59.6	1	1.87	77	128	504	408
THA	55.45	1130	1.9	22.6	12	1.59	65	105	46	104
URS	289	6020	0.8	67.5	2	1.76	69	133	474	319
USA	247.5	20765	1	74	0.5	5.01	75	138	259	812
VEN	19.2	3220	2.5	90	15.3	2.6	69	102	164	147
YOU	23.67	2599	0.7	50.2	10.4	1.44	72	139	100	179

A.11 German Annual Population Data

The data set shows yearly average population rates for the old federal states in Germany (given in 1000 inhabitants).

Year	Inhabitants	Unemployed
1960	55433	271
1961	56158	181
1962	56837	155
1963	57389	186
1964	57971	169
1965	58619	147
1966	59148	161
1967	59268	459
1968	59500	323
1969	60067	179
1970	60651	149
1971	61302	185
1972	61672	246
1973	61976	273
1974	62054	582
1975	61829	1074
1976	61531	1060
1977	61400	1030
1978	61327	993
1979	61359	876
1980	61566	889
1981	61682	1272
1982	61638	1833
1983	61423	2258
1984	61175	2266
1985	61024	2304
1986	61066	2228
1987	61077	2229
1988	61449	2242
1989	62063	2038
1990	63254	1883
1991	64074	1689
1992	64865	1808
1993	65535	2270
1994	65858	2556
1995	66156	2565
1996	66444	2796
1997	66648	3021

A.12 Journals Data

This is a data set that was created from a survey completed in the 1980's in Belgium questioning people's reading habits. They were asked where they live (10 regions comprising 7 provinces and 3 regions around Brussels) and what kind of newspaper they read on a regular basis. The 15 possible answers belong to 3 classes: Flemish newspapers (first letter v), French newspapers (first letter f) and both languages (first letter b).y

X_1: WaBr Walloon Brabant
X_2: Brar Brussels area
X_3: Antw Antwerp
X_4: FlBr Flemish Brabant
X_5: OcFl Occidental Flanders
X_6: OrFl Oriental Flanders
X_7: Hain Hainaut
X_8: Lièg Liège
X_9: Limb Limburg
X_{10}: Luxe Luxembourg

	WaBr	Brar	Antw	FlBr	OcFl	OrFl	Hain	Lièg	Limb	Luxe
v_a	1.8	7.8	9.1	3.0	4.3	3.9	0.1	0.3	3.3	0.0
v_b	0.1	3.4	17.8	1.0	0.7	4.1	0.0	0.0	0.2	0.0
v_c	0.1	9.4	4.6	7.7	4.4	5.8	1.6	0.1	1.4	0.0
v_d	0.5	15.6	6.1	12.0	10.5	10.2	0.7	0.3	5.4	0.0
v_e	0.1	5.2	3.3	4.8	1.6	1.4	0.1	0.0	3.5	0.0
f_f	5.6	13.7	3.1	2.4	0.5	1.7	1.9	2.3	0.2	0.2
f_g	4.1	16.5	1.9	1.0	1.0	0.9	2.4	3.2	0.1	0.3
f_h	8.3	29.5	1.8	7.3	0.8	0.4	5.1	3.2	0.2	0.3
f_i	0.9	7.8	0.2	2.6	0.1	0.1	5.6	3.8	0.1	0.8
b_j	6.1	18.2	10.8	4.1	4.5	5.3	2.0	2.6	3.4	0.2
b_k	8.3	35.4	6.2	11.0	5.0	6.1	5.5	3.3	1.5	0.3
b_l	4.4	9.9	6.7	3.4	1.1	3.9	2.1	1.5	2.1	0.0
v_m	0.3	11.6	14.2	4.7	5.1	7.9	0.3	0.5	3.0	0.0
f_n	5.1	21.0	1.3	3.4	0.2	0.2	2.3	4.4	0.0	0.4
f_o	2.2	9.8	0.1	0.3	0.0	0.7	2.3	3.0	0.3	1.0

A.13 NYSE Returns Data

This data set consists of returns of seven stocks traded on the New York Stock Exchange (Berndt 1990). The monthly returns of IBM, PanAm, Delta Airlines, Consolidated Edison, Gerber, Texaco, and Digital Equipment Company are stated from January 1978 to December 1987.

IBM	PanAm	Delta	Edison	Gerber	Texaco	DEC
−0.029	0.025	−0.028	−0.079	−0.048	−0.054	−0.100
−0.043	−0.073	−0.033	−0.003	0.160	−0.010	−0.063
−0.063	0.184	0.070	0.022	−0.036	0.015	0.010
0.130	0.089	0.150	−0.005	0.004	0.000	0.165
−0.018	0.082	−0.031	−0.014	0.046	−0.029	0.038
−0.004	0.019	0.023	0.034	0.028	−0.025	−0.021
0.092	0.204	0.185	0.011	−0.012	0.042	0.107
0.049	0.031	−0.021	0.024	−0.079	0.000	−0.017
−0.051	0.075	−0.081	0.048	0.104	0.010	−0.037
−0.046	−0.250	−0.153	−0.067	−0.138	−0.066	−0.077
0.031	0.000	0.055	0.035	0.078	0.055	0.064
0.108	−0.019	−0.023	0.005	−0.086	0.000	0.117
0.034	0.019	−0.054	0.076	0.042	0.037	−0.012
−0.017	−0.130	−0.060	−0.011	−0.023	−0.010	−0.066
0.052	0.043	0.098	0.000	0.065	0.068	0.088
−0.004	0.020	−0.056	−0.057	−0.088	0.059	0.005
−0.022	−0.060	0.063	0.032	−0.023	−0.040	−0.028
−0.035	0.000	−0.006	0.066	0.095	0.083	0.059
−0.049	0.319	0.075	0.015	−0.096	0.032	0.009
0.016	−0.065	0.021	−0.021	0.148	0.041	0.140
−0.032	−0.034	−0.026	0.000	−0.009	0.030	−0.027
−0.079	−0.089	−0.147	−0.049	−0.090	−0.053	−0.010
0.060	−0.098	0.063	0.109	−0.014	0.067	0.095
−0.013	0.043	0.020	0.005	−0.036	−0.029	0.018
0.066	−0.042	0.022	−0.039	0.048	0.229	0.058
−0.062	−0.109	−0.093	−0.061	−0.004	0.161	0.034
−0.122	−0.195	−0.031	0.006	−0.237	−0.179	−0.182
−0.016	0.000	−0.018	0.140	0.027	0.082	0.047
0.025	0.121	0.144	0.043	0.233	0.007	0.016
0.061	−0.027	0.010	0.040	0.011	0.032	0.021
0.111	0.278	0.283	−0.027	0.005	0.003	0.183
0.017	−0.043	−0.056	−0.005	−0.008	0.031	0.081
−0.021	−0.091	−0.053	−0.010	0.066	−0.037	0.045
0.039	−0.025	0.046	−0.021	0.026	0.087	−0.028
0.035	0.026	0.220	−0.035	0.023	0.399	0.056
−0.004	−0.150	0.040	0.131	0.070	−0.109	0.035
−0.052	0.118	0.112	−0.015	0.056	−0.145	−0.089
0.011	−0.079	0.031	−0.021	−0.020	−0.012	0.006

continues on next page ⟶

IBM	PanAm	Delta	Edison	Gerber	Texaco	DEC
−0.029	0.143	0.024	0.151	0.023	−0.063	0.075
−0.060	0.025	0.062	0.061	0.031	−0.003	0.075
0.017	0.122	0.105	0.017	0.008	−0.055	0.107
−0.015	−0.196	−0.114	0.022	0.066	0.025	−0.112
−0.030	−0.216	−0.094	0.026	0.021	0.045	−0.014
−0.002	−0.069	−0.072	0.021	0.031	0.003	−0.065
−0.018	−0.111	−0.013	−0.013	0.000	−0.093	−0.019
−0.048	0.000	−0.072	0.112	−0.012	0.008	0.102
0.075	0.167	−0.032	0.038	0.011	0.065	−0.065
0.044	−0.214	−0.062	−0.008	−0.077	−0.047	−0.060
0.119	0.091	0.056	0.042	−0.004	−0.045	0.027
−0.014	−0.042	0.145	0.036	−0.111	−0.004	−0.049
−0.034	0.087	0.038	0.022	0.136	−0.029	−0.104
0.075	0.160	−0.025	0.050	0.044	−0.008	0.054
−0.029	0.000	0.042	0.016	0.043	0.034	−0.056
−0.014	−0.103	0.106	−0.024	−0.033	−0.017	−0.073
0.082	0.077	−0.118	−0.032	0.019	−0.060	−0.055
0.087	−0.036	0.055	0.133	0.130	0.056	0.273
0.041	−0.259	−0.139	0.039	0.209	0.027	−0.061
0.089	0.350	0.171	−0.050	−0.009	0.056	0.133
0.094	0.074	0.289	−0.011	−0.072	0.012	0.175
0.113	0.000	0.093	0.123	0.015	0.029	−0.052
0.027	0.103	0.040	−0.012	0.015	0.036	0.225
0.010	0.406	0.027	0.060	0.024	0.008	−0.010
0.028	−0.067	−0.016	0.048	0.084	0.039	0.034
0.150	0.024	−0.043	0.045	0.119	0.098	−0.060
−0.041	0.186	−0.045	−0.012	0.016	−0.038	−0.052
0.081	0.275	0.012	0.000	0.114	0.018	0.075
0.001	−0.015	−0.259	0.017	−0.007	0.036	−0.142
0.001	−0.047	0.080	−0.023	0.062	0.059	0.007
0.062	−0.066	0.041	0.087	0.049	−0.037	−0.005
−0.001	0.035	0.039	0.101	0.000	−0.014	−0.364
−0.066	0.119	0.120	−0.025	0.077	0.011	0.065
0.039	−0.015	−0.028	0.005	0.063	0.021	0.034
−0.065	0.031	−0.013	0.005	0.065	0.108	0.208
−0.026	−0.179	−0.117	−0.069	−0.091	0.151	−0.024
0.034	−0.018	0.065	0.055	−0.003	−0.122	0.057
−0.002	−0.185	−0.085	0.031	−0.025	0.022	0.053
−0.044	−0.045	−0.070	0.021	−0.087	−0.105	−0.071
−0.019	−0.024	−0.012	0.020	0.105	−0.046	−0.043
0.047	−0.024	0.045	0.054	−0.112	−0.044	−0.009
0.127	0.000	0.040	0.029	0.018	0.140	0.159
0.004	0.000	0.008	0.051	0.165	0.045	−0.025
0.012	−0.050	0.161	0.019	−0.160	−0.080	0.093
−0.023	−0.026	−0.026	0.004	0.094	0.007	0.006
0.011	0.000	0.156	0.084	−0.005	0.000	0.070
0.108	−0.027	−0.010	−0.021	0.091	0.044	0.084

continues on next page ⟶

IBM	PanAm	Delta	Edison	Gerber	Texaco	DEC
−0.009	−0.028	0.087	0.034	0.006	0.022	−0.067
−0.052	0.086	−0.003	0.057	0.130	0.014	−0.071
−0.004	0.053	−0.123	0.019	−0.037	0.111	−0.050
0.025	0.350	0.179	0.098	0.234	−0.065	0.057
−0.038	0.056	0.021	0.046	−0.031	0.031	−0.101
0.062	0.000	0.008	−0.084	−0.036	−0.030	0.080
−0.028	0.088	−0.066	0.043	0.025	0.021	0.032
−0.022	−0.065	−0.112	−0.032	−0.048	−0.007	0.036
0.048	0.069	−0.083	0.066	0.097	0.099	0.040
0.085	0.016	0.020	0.032	0.137	−0.175	0.073
0.113	−0.016	0.030	0.082	0.063	−0.077	0.095
−0.026	0.129	0.122	0.022	−0.088	−0.038	0.162
0.003	−0.029	−0.055	0.048	0.034	0.071	0.093
0.004	−0.074	0.076	0.021	0.174	−0.004	−0.063
0.031	−0.206	0.059	−0.006	0.113	0.050	0.119
−0.018	0.060	−0.043	0.042	−0.040	0.069	0.037
−0.039	−0.094	−0.070	0.017	−0.038	−0.042	−0.063
−0.096	−0.063	0.018	0.125	−0.105	−0.036	0.066
0.055	0.022	0.018	0.061	0.111	0.135	0.105
−0.031	−0.065	0.026	−0.139	0.037	0.026	−0.110
−0.081	0.023	0.134	0.045	−0.069	0.043	0.103
0.037	0.023	−0.018	0.070	−0.020	−0.028	0.048
−0.056	−0.244	−0.010	−0.046	−0.060	0.047	0.008
0.073	0.353	0.161	0.040	0.057	0.049	0.385
0.092	−0.152	0.133	−0.067	0.019	−0.080	0.056
0.076	−0.103	−0.129	−0.050	0.040	0.103	0.061
0.067	0.114	−0.121	0.020	−0.063	−0.094	0.055
0.006	0.000	0.151	−0.012	0.138	0.114	−0.082
0.016	0.103	0.014	0.059	0.005	0.073	0.041
−0.009	0.000	0.043	−0.039	0.232	0.142	0.000
0.053	−0.093	−0.037	0.043	−0.113	−0.076	0.157
−0.105	−0.051	−0.067	−0.006	−0.061	−0.053	0.001
−0.187	−0.270	−0.260	−0.017	−0.288	−0.194	−0.281
−0.087	0.185	−0.137	−0.012	−0.085	−0.031	−0.127
0.043	−0.313	0.121	−0.006	0.070	0.178	0.134

A.14 Plasma Data

In Olkin & Veath (1980), the evolution of citrate concentration in the plasma is observed at 3 different times of day for two groups of patients. Each group follows a different diet.

X_1: 8 AM
X_2: 11 AM
X_3: 3 PM

Group	(8 AM)	(11 AM)	(3 PM)
	125	137	121
	144	173	147
I	105	119	125
	151	149	128
	137	139	109
	93	121	107
	116	135	106
II	109	83	100
	89	95	83
	116	128	100

A.15 Time Budget Data

In Volle (1985), we can find data on 28 individuals identified according to gender, country where they live, professional activity, and matrimonial status, which indicates the amount of time each person spent on 10 categories of activities over 100 days (100·24h = 2400 hours total in each row) in 1976.

X_1:   prof : professional activity
X_2:   tran : transportation linked to professional activity
X_3:   hous : household occupation
X_4:   kids : occupation linked to children
X_5:   shop : shopping
X_6:   pers : time spent for personal care
X_7:   eat :   eating
X_8:   slee : sleeping
X_9:   tele :   watching television
X_{10}: leis :   other leisure activities

maus: active men in the United States
waus: active women in the United States
wnus: nonactive women in the United States
mmus: married men in United States
wmus: married women in United States
msus: single men in United States
wsus: single women in United States
mawe: active men from Western countries
wawe: active women from Western countries
wnwe: nonactive women from Western countries
mmwe: married men from Western countries
wmwe: married women from Western countries
mswe: single men from Western countries
wswe: single women from Western countries
mayo: active men from Yugoslavia
wayo: active women from Yugoslavia
wnyo: nonactive women from Yugoslavia
mmyo: married men from Yugoslavia
wmyo: married women from Yugoslavia
msyo: single men from Yugoslavia
wsyo: single women from Yugoslavia
maes: active men from Eastern countries
waes: active women from Eastern countries
wnes: nonactive women from Eastern countries
mmes: married men from Eastern countries
wmes: married women from Eastern countries
mses: single men from Eastern countries
wses: single women from Eastern countries

	prof	tran	hous	kids	shop	pers	eat	slee	tele	leis
maus	610	140	60	10	120	95	115	760	175	315
waus	475	90	250	30	140	120	100	775	115	305
wnus	10	0	495	110	170	110	130	785	160	430
mmus	615	140	65	10	115	90	115	765	180	305
wmus	179	29	421	87	161	112	119	776	143	373
msus	585	115	50	0	150	105	100	760	150	385
wsus	482	94	196	18	141	130	96	775	132	336
mawe	653	100	95	7	57	85	150	808	115	330
wawe	511	70	307	30	80	95	142	816	87	262
wnwe	20	7	568	87	112	90	180	843	125	368
mmwe	656	97	97	10	52	85	152	808	122	321
wmwe	168	22	528	69	102	83	174	824	119	311
mswe	643	105	72	0	62	77	140	813	100	388
wswe	429	34	262	14	92	97	147	849	84	392
mayo	650	140	120	15	85	90	105	760	70	365
wayo	560	105	375	45	90	90	95	745	60	235
wnyo	10	10	710	55	145	85	130	815	60	380
mmyo	650	145	112	15	85	90	105	760	80	358
wmyo	260	52	576	59	116	85	117	775	65	295
msyo	615	125	95	0	115	90	85	760	40	475
wsyo	433	89	318	23	112	96	102	774	45	408
maea	650	142	122	22	76	94	100	764	96	334
waea	578	106	338	42	106	94	92	752	64	228
wnea	24	8	594	72	158	92	128	840	86	398
mmea	652	133	134	22	68	94	102	763	122	310
wmea	436	79	433	60	119	90	107	772	73	231
msea	627	148	68	0	88	92	86	770	58	463
wsea	434	86	297	21	129	102	94	799	58	380

A.16 Unemployment Data

This data set provides unemployment rates in all federal states of Germany in September 1999.

No. Federal State	Unemployment Rate
1 Schleswig-Holstein	8.7
2 Hamburg	9.8
3 Mecklenburg-Vorpommern	17.3
4 Niedersachsen	9.8
5 Bremen	13.9
6 Nordrhein-Westfalen	9.8
7 Hessen	7.9
8 Rheinland-Pfalz	7.7
9 Saarland	10.4
10 Baden-Württemberg	6.2
11 Bayern	5.8
12 Berlin	15.8
13 Brandenburg	17.1
14 Sachsen-Anhalt	19.9
15 Thüringen	15.1
16 Sachsen	16.8

A.17 U.S. Companies Data

The data set consists of measurements for 79 U.S. companies. The abbreviations are as follows:

X_1: A   assets (USD)
X_2: S   sales (USD)
X_3: MV   market value (USD)
X_4: P   profits (USD)
X_5: CF   cash flow (USD)
X_6: E   employees

Company	A	S	MV	P	CF	E	Sector
Bell Atlantic	19788	9084	10636	1092.9	2576.8	79.4	Communication
Continental Telecom	5074	2557	1892	239.9	578.3	21.9	Communication
American Electric Power	13621	4848	4572	485.0	898.9	23.4	Energy
Brooklyn Union Gas	1117	1038	478	59.7	91.7	3.8	Energy
Central Illinois Publ. Serv.	1633	701	679	74.3	135.9	2.8	Energy
Cleveland Electric Illum.	5651	1254	2002	310.7	407.9	6.2	Energy
Columbia Gas System	5835	4053	1601	−93.8	173.8	10.8	Energy
Florida Progress	3494	1653	1442	160.9	320.3	6.4	Energy
Idaho Power	1654	451	779	84.8	130.4	1.6	Energy
Kansas Power & Light	1679	1354	687	93.8	154.6	4.6	Energy
Mesa Petroleum	1257	355	181	167.5	304.0	0.6	Energy
Montana Power	1743	597	717	121.6	172.4	3.5	Energy
People's Energy	1440	1617	639	81.7	126.4	3.5	Energy
Phillips Petroleum	14045	15636	2754	418.0	1462.0	27.3	Energy
Publ. Serv. Coop New Mexico	3010	749	1120	146.3	209.2	3.4	Energy
San Diego Gas & Electric	3086	1739	1507	202.7	335.2	4.9	Energy
Valero Energy	1995	2662	341	34.7	100.7	2.3	Energy
American Savings Bank FSB	3614	367	90	14.1	24.6	1.1	Finance
Bank South	2788	271	304	23.5	28.9	2.1	Finance
H&R Block	327	542	959	54.1	72.5	2.8	Finance
California First Bank	5401	550	376	25.6	37.5	4.1	Finance
Cigna	44736	16197	4653	−732.5	−651.9	48.5	Finance
Dreyfus	401	176	1084	55.6	57.0	0.7	Finance
First American	4789	453	367	40.2	51.4	3.0	Finance
First Empire State	2548	264	181	22.2	26.2	2.1	Finance
First Tennessee National Bank	5249	527	346	37.8	56.2	4.1	Finance
Marine Corp	3720	356	211	26.6	34.8	2.4	Finance
Mellon Bank	33406	3222	1413	201.7	246.7	15.8	Finance
National City Bank	12505	1302	702	108.4	131.4	9.0	Finance
Norstar Bancorp	8998	882	988	93.0	119.0	7.4	Finance
Norwest Bank	21419	2516	930	107.6	164.7	15.6	Finance
Southeast Banking	11052	1097	606	64.9	97.6	7.0	Finance
Sovran Financial	9672	1037	829	92.6	118.2	8.2	Finance
United Financial Group	4989	518	53	−3.1	−0.3	0.8	Finance
Apple Computer	1022	1754	1370	72.0	119.5	4.8	Hi-Tech
Digital Equipment	6914	7029	7957	400.6	754.7	87.3	Hi-Tech
EG&G	430	1155	1045	55.7	70.8	22.5	Hi-Tech
General Electric	26432	28285	33172	2336.0	3562.0	304.0	Hi-Tech
Hewlett-Packard	5769	6571	9462	482.0	792.0	83.0	Hi-Tech
IBM	52634	50056	95697	6555.0	9874.0	400.2	Hi-Tech
NCR	3940	4317	3940	315.2	566.3	62.0	Hi-Tech
Telex	478	672	866	67.1	101.6	5.4	Hi-Tech
Armstrong World Industries	1093	1679	1070	100.9	164.5	20.8	Manufacturing
CBI Industries	1128	1516	430	−47.0	26.7	13.2	Manufacturing
Fruehauf	1804	2564	483	70.5	164.9	26.6	Manufacturing
Halliburton	4662	4781	2988	28.7	371.5	66.2	Manufacturing
LTV	6307	8199	598	−771.5	−524.3	57.5	Manufacturing

continues on next page →

Company	A	S	MV	P	CF	E	Sector
Owens-Corning Fiberglas	2366	3305	1117	131.2	256.5	25.2	Manufacturing
PPG Industries	4084	4346	3023	302.7	521.7	37.5	Manufacturing
Textron	10348	5721	1915	223.6	322.5	49.5	Manufacturing
Turner	752	2149	101	11.1	15.2	2.6	Manufacturing
United Technologies	10528	14992	5377	312.7	710.7	184.8	Manufacturing
Commun. Psychiatric Centers	278	205	853	44.8	50.5	3.8	Medical
Hospital Corp of America	6259	4152	3090	283.7	524.5	62.0	Medical
AH Robins	707	706	275	61.4	77.8	6.1	Medical
Shared Medical Systems	252	312	883	41.7	60.6	3.3	Medical
Air Products	2687	1870	1890	145.7	352.2	18.2	Other
Allied Signal	13271	9115	8190	−279.0	83.0	143.8	Other
Bally Manufacturing	1529	1295	444	25.6	137.0	19.4	Other
Crown Cork & Seal	866	1487	944	71.7	115.4	12.6	Other
Ex-Cell-0	799	1140	633	57.6	89.2	15.4	Other
Liz Claiborne	223	557	1040	60.6	63.7	1.9	Other
Warner Communications	2286	2235	2306	195.3	219.0	8.0	Other
Dayton-Hudson	4418	8793	4459	283.6	456.5	128.0	Retail
Dillard Department Stores	862	1601	1093	66.9	106.8	16.0	Retail
Giant Food	623	2247	797	57.0	93.8	18.6	Retail
Great A&P Tea	1608	6615	829	56.1	134.0	65.0	Retail
Kroger	4178	17124	2091	180.8	390.4	164.6	Retail
May Department Stores	3442	5080	2673	235.4	361.5	77.3	Retail
Stop & Shop Cos	1112	3689	542	30.3	96.9	43.5	Retail
Supermarkets General	1104	5123	910	63.7	133.3	48.5	Retail
Wickes Cos	2957	2806	457	40.6	93.5	50.0	Retail
FW Woolworth	2535	5958	1921	177.0	288.0	118.1	Retail
AMR	6425	6131	2448	345.8	682.5	49.5	Transportation
IU International	999	1878	393	−173.5	−108.1	23.3	Transportation
PanAm	2448	3484	1036	48.8	257.1	25.4	Transportation
Republic Airlines	1286	1734	361	69.2	145.7	14.3	Transportation
TWA	2769	3725	663	−208.4	12.4	29.1	Transportation
Western AirLines	952	1307	309	35.4	92.8	10.3	Transportation

A.18 U.S. Crime Data

This is a data set consisting of 50 measurements of 7 variables. It states for one year (1985) the reported number of crimes in the 50 states of the United States classified according to 7 categories (X_3–X_9):

X_1: land area (land)
X_2: population 1985 (popu 1985)
X_3: murder (murd)
X_4: rape
X_5: robbery (robb)
X_6: assault (assa)
X_7: burglary (burg)
X_8: larceny (larc)
X_9: auto theft (auto)
X_{10}: U.S. states region number (reg)
X_{11}: U.S. states division number (div)

Division Numbers Region Numbers

New England	1	Northeast	1
Mid-Atlantic	2	Midwest	2
E N Central	3	South	3
W N Central	4	West	4
S Atlantic	5		
E S Central	6		
W S Central	7		
Mountain	8		
Pacific	9		

State	land	popu 1985	murd	rape	robb	assa	burg	larc	auto	reg	div
ME	33265	1164	1.5	7.0	12.6	62	562	1055	146	1	1
NH	9279	998	2.0	6	12.1	36	566	929	172	1	1
VT	9614	535	1.3	10.3	7.6	55	731	969	124	1	1
MA	8284	5822	3.5	12.0	99.5	88	1134	1531	878	1	1
RI	1212	968	3.2	3.6	78.3	120	1019	2186	859	1	1
CT	5018	3174	3.5	9.1	70.4	87	1084	1751	484	1	1
NY	49108	17783	7.9	15.5	443.3	209	1414	2025	682	1	2
NJ	7787	7562	5.7	12.9	169.4	90	1041	1689	557	1	2
PA	45308	11853	5.3	11.3	106.0	90	594	11	340	1	2
OH	41330	10744	6.6	16.0	145.9	116	854	1944	493	2	3
IN	36185	5499	4.8	17.9	107.5	95	860	1791	429	2	3
IL	56345	11535	9.6	20.4	251.1	187	765	2028	518	2	3
MI	58527	9088	9.4	27.1	346.6	193	1571	2897	464	2	3
WI	56153	4775	2.0	6.7	33.1	44	539	1860	218	2	3
MN	84402	4193	2.0	9.7	89.1	51	802	1902	346	2	4

continues on next page ⟶

State	land	popu 1985	murd	rape	robb	assa	burg	larc	auto	reg	div
IA	56275	2884	1.9	6.2	28.6	48	507	1743	175	2	4
MO	69697	5029	10.7	27.4	2.8	167	1187	2074	538	2	4
ND	70703	685	0.5	6.2	6.5	21	286	1295	91	2	4
SD	77116	708	3.8	11.1	17.1	60	471	1396	94	2	4
NE	77355	1606	3.0	9.3	57.3	115	505	1572	292	2	4
KS	82277	2450	4.8	14.5	75.1	108	882	2302	257	2	4
DE	2044	622	7.7	18.6	105.5	196	1056	2320	559	3	5
MD	10460	4392	9.2	23.9	338.6	253	1051	2417	548	3	5
VA	40767	5706	8.4	15.4	92.0	143	806	1980	297	3	5
WV	24231	1936	6.2	6.7	27.3	84	389	774	92	3	5
NC	52669	6255	11.8	12.9	53.0	293	766	1338	169	3	5
SC	31113	3347	14.6	18.1	60.1	193	1025	1509	256	3	5
GA	58910	5976	15.3	10.1	95.8	177	9	1869	309	3	5
FL	58664	11366	12.7	22.2	186.1	277	1562	2861	397	3	5
KY	40409	3726	11.1	13.7	72.8	123	704	1212	346	3	6
TN	42144	4762	8.8	15.5	82.0	169	807	1025	289	3	6
AL	51705	4021	11.7	18.5	50.3	215	763	1125	223	3	6
MS	47689	2613	11.5	8.9	19.0	140	351	694	78	3	6
AR	53187	2359	10.1	17.1	45.6	150	885	1211	109	3	7
LA	47751	4481	11.7	23.1	140.8	238	890	1628	385	3	7
OK	69956	3301	5.9	15.6	54.9	127	841	1661	280	3	7
TX	266807	16370	11.6	21.0	134.1	195	1151	2183	394	3	7
MT	147046	826	3.2	10.5	22.3	75	594	1956	222	4	8
ID	83564	15	4.6	12.3	20.5	86	674	2214	144	4	8
WY	97809	509	5.7	12.3	22.0	73	646	2049	165	4	8
CO	104091	3231	6.2	36.0	129.1	185	1381	2992	588	4	8
NM	121593	1450	9.4	21.7	66.1	196	1142	2408	392	4	8
AZ	1140	3187	9.5	27.0	120.2	214	1493	3550	501	4	8
UT	84899	1645	3.4	10.9	53.1	70	915	2833	316	4	8
NV	110561	936	8.8	19.6	188.4	182	1661	3044	661	4	8
WA	68138	4409	3.5	18.0	93.5	106	1441	2853	362	4	9
OR	97073	2687	4.6	18.0	102.5	132	1273	2825	333	4	9
CA	158706	26365	6.9	35.1	206.9	226	1753	3422	689	4	9
AK	5914	521	12.2	26.1	71.8	168	790	2183	551	4	9
HI	6471	1054	3.6	11.8	63.3	43	1456	3106	581	4	9

A.19 U.S. Health Data

This is a data set consisting of 50 measurements of 13 variables. It states for one year (1985) the reported number of deaths in the 50 states of the U.S. classified according to 7 categories:

X_1: land area (land)
X_2: population 1985 (popu)
X_3: accident (acc)
X_4: cardiovascular (card)
X_5: cancer (canc)
X_6: pulmonary (pul)
X_7: pneumonia flu (pneu)
X_8: diabetes (diab)
X_9: liver (liv)
X_{10}: doctors (doc)
X_{11}: hospitals (hosp)
X_{12}: U.S. states region number (reg)
X_{13}: U.S. states division number (div)

Division Numbers		Region Numbers	
New England	1	Northeast	1
Mid-Atlantic	2	Midwest	2
E N Central	3	South	3
W N Central	4	West	4
S Atlantic	5		
E S Central	6		
W S Central	7		
Mountain	8		
Pacific	9		

State	land	popu 1985	acc	card	canc	pul	pneu	diab	liv	doc	hosp	reg	div
ME	33265	1164	37.7	466.2	213.8	33.6	21.1	15.6	14.5	1773	47	1	1
NH	9279	998	35.9	395.9	182.2	29.6	20.1	17.6	10.4	1612	34	1	1
VT	9614	535	41.3	433.1	188.1	33.1	24.0	15.6	13.1	1154	19	1	1
MA	8284	5822	31.1	460.6	219.0	24.9	29.7	16.0	13.0	16442	177	1	1
RI	1212	968	28.6	474.1	231.5	27.4	17.7	26.2	13.4	2020	21	1	1
CT	5018	3174	35.3	423.8	205.1	23.2	22.4	15.4	11.7	8076	65	1	1
NY	49108	17783	31.5	499.5	209.9	23.9	26.0	17.1	17.7	49304	338	1	2
NJ	7787	7562	32.2	464.7	216.3	23.3	19.9	17.3	14.2	15120	131	1	2
PA	45308	11853	34.9	508.7	223.6	27.0	20.1	20.4	12.0	23695	307	1	2
OH	41330	10744	33.2	443.1	198.8	27.4	18.0	18.9	10.2	18518	236	2	3
IN	36185	5499	37.7	435.7	184.6	27.2	18.6	17.2	8.4	7339	133	2	3
IL	56345	11535	32.9	449.6	193.2	22.9	21.3	15.3	12.5	22173	279	2	3
MI	58527	9088	34.3	420.9	182.3	24.2	18.7	14.8	13.7	15212	231	2	3

continues on next page ⟶

State	land	popu	1985	acc	card	canc	pul	pneu	diab	liv	doc	hosp	reg	div
WI	56153	4775	33.8	444.3	189.4	22.5	21.2	15.7	8.7	7899	163	2	3
MN	84402	4193	35.7	398.3	174.0	23.4	25.6	13.5	8.1	8098	181	2	4
IA	56275	2884	38.6	490.1	199.1	31.2	28.3	16.6	7.9	3842	140	2	4
MO	69697	5029	42.2	475.9	211.1	29.8	25.7	15.3	9.6	8422	169	2	4
ND	70703	685	48.2	401.0	173.7	18.2	25.9	14.9	7.4	936	58	2	4
SD	77116	708	53.0	495.2	182.1	30.7	32.4	12.8	7.2	833	68	2	4
NE	77355	1606	40.8	479.6	187.4	31.6	28.3	13.5	7.8	2394	110	2	4
KS	82277	2450	42.9	455.9	183.9	32.3	24.9	16.9	7.8	3801	165	2	4
DE	2044	622	38.8	404.5	202.8	25.3	16.0	25.0	10.5	1046	14	3	5
MD	10460	4392	35.2	366.7	195.0	23.4	15.8	16.1	9.6	11961	85	3	5
VA	40767	5706	37.4	365.3	174.4	22.4	20.3	11.4	9.2	9749	135	3	5
MV	24231	1936	46.7	502.7	199.6	35.2	20.1	18.4	10.0	2813	75	3	5
NC	52669	6255	45.4	392.6	169.2	22.6	19.8	13.1	10.2	9355	159	3	5
SC	31113	3347	47.8	374.4	156.9	19.6	19.2	14.8	9.0	4355	89	3	5
GA	58910	5976	48.2	371.4	157.9	22.6	20.5	13.2	10.4	8256	191	3	5
FL	58664	11366	46.0	501.8	244.0	34.0	18.3	16.1	17.2	18836	254	3	5
KY	40409	3726	48.8	442.5	194.7	29.8	22.9	15.9	9.1	5189	120	3	6
TN	42144	4762	45.0	427.2	185.6	27.0	20.8	12.0	8.3	7572	162	3	6
AL	51705	4021	48.9	411.5	185.8	25.5	16.8	16.1	9.1	5157	146	3	6
MS	47689	2613	59.3	422.3	173.9	21.7	19.5	14.0	7.1	2883	118	3	6
AR	53187	2359	51.0	482.0	202.1	29.0	22.7	15.0	8.7	2952	97	3	7
LA	47751	4481	52.3	390.9	168.1	18.6	15.8	17.8	8.3	7061	158	3	7
OK	69956	3301	62.5	441.4	182.4	27.6	24.5	15.3	9.6	4128	143	3	7
TX	266807	16370	48.9	327.9	146.5	20.7	17.4	12.1	8.7	23481	562	3	7
MT	147046	826	59.0	372.2	170.7	33.4	25.1	14.4	11.1	1058	67	4	8
ID	83564	15.0	51.5	324.8	140.4	29.9	22.3	12.4	9.2	1079	52	4	8
WY	97809	509	67.6	264.2	112.2	27.7	18.5	9.2	9.2	606	31	4	8
CO	104091	3231	44.7	280.2	125.1	29.9	22.8	9.6	9.5	5899	98	4	8
NM	121593	1450	62.3	235.6	137.2	28.7	17.8	17.5	13.1	2127	56	4	8
AZ	1140	3187	48.3	331.5	165.6	36.3	21.2	12.6	13.1	5137	79	4	8
UT	84899	1645	39.3	242.0	93.7	17.6	14.5	11.1	7.3	2563	44	4	8
NV	110561	936	57.3	299.5	162.3	32.3	13.7	11.1	15.4	1272	26	4	8
WA	68138	4409	41.4	358.1	171.0	31.1	21.2	13.0	10.9	7768	122	4	9
OR	97073	2687	41.6	387.8	179.4	33.8	23.1	11.2	10.4	4904	83	4	9
CA	158706	26365	40.3	357.8	173.0	26.9	22.2	10.7	16.7	57225	581	4	9
AK	5914	521	85.8	114.6	76.1	8.3	12.4	3.4	11.0	545	26	4	9
HI	6471	1054	32.5	216.9	125.8	16.0	16.8	12.7	6.2	1953	26	4	9

A.20 Vocabulary Data

This example of the evolution of the vocabulary of children can be found in Bock (1975). Data are drawn from test results on file in the Records Office of the Laboratory School of the University of Chicago. They consist of scores, obtained from a cohort of pupils from the 8th through 11th grade levels, on alternative forms of the vocabulary section of the Coorperative Reading Test. It provides the following scaled scores shown for the sample of 64 subjects (the origin and units are fixed arbitrarily).

Subjects	Grade 8	Grade 9	Grade 10	Grade 11	Mean
1	1.75	2.60	3.76	3.68	2.95
2	0.90	2.47	2.44	3.43	2.31
3	0.80	0.93	0.40	2.27	1.10
4	2.42	4.15	4.56	4.21	3.83
5	−1.31	−1.31	−0.66	−2.22	−1.38
6	−1.56	1.67	0.18	2.33	0.66
7	1.09	1.50	0.52	2.33	1.36
8	−1.92	1.03	0.50	3.04	0.66
9	−1.61	0.29	0.73	3.24	0.66
10	2.47	3.64	2.87	5.38	3.59
11	−0.95	0.41	0.21	1.82	0.37
12	1.66	2.74	2.40	2.17	2.24
13	2.07	4.92	4.46	4.71	4.04
14	3.30	6.10	7.19	7.46	6.02
15	2.75	2.53	4.28	5.93	3.87
16	2.25	3.38	5.79	4.40	3.96
17	2.08	1.74	4.12	3.62	2.89
18	0.14	0.01	1.48	2.78	1.10
19	0.13	3.19	0.60	3.14	1.77
20	2.19	2.65	3.27	2.73	2.71
21	−0.64	−1.31	−0.37	4.09	0.44
22	2.02	3.45	5.32	6.01	4.20
23	2.05	1.80	3.91	2.49	2.56
24	1.48	0.47	3.63	3.88	2.37
25	1.97	2.54	3.26	5.62	3.35
26	1.35	4.63	3.54	5.24	3.69
27	−0.56	−0.36	1.14	1.34	0.39
28	0.26	0.08	1.17	2.15	0.92
29	1.22	1.41	4.66	2.62	2.47
30	−1.43	0.80	−0.03	1.04	0.09
31	−1.17	1.66	2.11	1.42	1.00
32	1.68	1.71	4.07	3.30	2.69
33	−0.47	0.93	1.30	0.76	0.63
34	2.18	6.42	4.64	4.82	4.51
35	4.21	7.08	6.00	5.65	5.73

continues on next page →

Subjects	Grade 8	Grade 9	Grade 10	Grade 11	Mean
36	8.26	9.55	10.24	10.58	9.66
37	1.24	4.90	2.42	2.54	2.78
38	5.94	6.56	9.36	7.72	7.40
39	0.87	3.36	2.58	1.73	2.14
40	−0.09	2.29	3.08	3.35	2.15
41	3.24	4.78	3.52	4.84	4.10
42	1.03	2.10	3.88	2.81	2.45
43	3.58	4.67	3.83	5.19	4.32
44	1.41	1.75	3.70	3.77	2.66
45	−0.65	−0.11	2.40	3.53	1.29
46	1.52	3.04	2.74	2.63	2.48
47	0.57	2.71	1.90	2.41	1.90
48	2.18	2.96	4.78	3.34	3.32
49	1.10	2.65	1.72	2.96	2.11
50	0.15	2.69	2.69	3.50	2.26
51	−1.27	1.26	0.71	2.68	0.85
52	2.81	5.19	6.33	5.93	5.06
53	2.62	3.54	4.86	5.80	4.21
54	0.11	2.25	1.56	3.92	1.96
55	0.61	1.14	1.35	0.53	0.91
56	−2.19	−0.42	1.54	1.16	0.02
57	1.55	2.42	1.11	2.18	1.82
58	0.04	0.50	2.60	2.61	1.42
59	3.10	2.00	3.92	3.91	3.24
60	−0.29	2.62	1.60	1.86	1.45
61	2.28	3.39	4.91	3.89	3.62
62	2.57	5.78	5.12	4.98	4.61
63	−2.19	0.71	1.56	2.31	0.60
64	−0.04	2.44	1.79	2.64	1.71
Mean	1.14	2.54	2.99	3.47	2.53

A.21 WAIS Data

Morrison (1990) compares the results of 4 subtests of the Wechsler Adult Intelligence Scale (WAIS) for 2 categories of people. In group 1 are $n_1 = 37$ people who do not present a senile factor; in group 2 are those ($n_2 = 12$) presenting a senile factor.

WAIS subtests:

X_1:	information
X_2:	similarities
X_3:	arithmetic
X_4:	picture completion

	Group I			
Subject	Information	Similarities	Arithmetic	Picture Completion
1	7	5	9	8
2	8	8	5	6
3	16	18	11	9
4	8	3	7	9
5	6	3	13	9
6	11	8	10	10
7	12	7	9	8
8	8	11	9	3
9	14	12	11	4
10	13	13	13	6
11	13	9	9	9
12	13	10	15	7
13	14	11	12	8
14	15	11	11	10
15	13	10	15	9
16	10	5	8	6
17	10	3	7	7
18	17	13	13	7
19	10	6	10	7
20	10	10	15	8
21	14	7	11	5
22	16	11	12	11
23	10	7	14	6
24	10	10	9	6
25	10	7	10	10
26	7	6	5	9
27	15	12	10	6
28	17	15	15	8
29	16	13	16	9
30	13	10	17	8
31	13	10	17	10
32	19	12	16	10
33	19	15	17	11
34	13	10	7	8
35	15	11	12	8
36	16	9	11	11
37	14	13	14	9
Mean	12.57	9.57	11.49	7.97

		Group II		
Subject	Information	Similarities	Arithmetic	Picture Completion
1	9	5	10	8
2	10	0	6	2
3	8	9	11	1
4	13	7	14	9
5	4	0	4	0
6	4	0	6	0
7	11	9	9	8
8	5	3	3	6
9	9	7	8	6
10	7	2	6	4
11	12	10	14	3
12	13	12	11	10
Mean	8.75	5.33	8.50	4.75

References

Andrews, D. (1972), 'Plots of high-dimensional data', *Biometrics* **28**, 125–136.

Bartlett, M. S. (1954), 'A note on multiplying factors for various chi-squared approximations', *Journal of the Royal Statistical Society, Series B* **16**, 296–298.

Berndt, E. R. (1990), *The Practice of Econometrics: Classic and Contemporary*, Addison-Wesley, Reading.

Bock, R. D. (1975), *Multivariate Statistical Methods in Behavioral Research*, McGraw-Hill, New York.

Bouroche, J.-M. & Saporta, G. (1980), *L'analyse des données*, Presses Universitaires de France, Paris.

Breiman, L. (1973), *Statistics: With a View Towards Application*, Houghton Mifflin Company, Boston.

Breiman, L., Friedman, J. H., Olshen, R. & Stone, C. J. (1984), *Classification and Regression Trees*, Wadsworth, Belmont.

Chambers, J. M., Cleveland, W. S., Kleiner, B. & Tukey, P. A. (1983), *Graphical Methods for Data Analysis*, Duxbury Press, Boston.

Chernoff, H. (1973), 'Using faces to represent points in k-dimensional space graphically', *Journal of the American Statistical Association* **68**, 361–368.

Cook, R. D. & Weisberg, S. (1991), 'Comment on sliced inverse regression for dimension reduction', *Journal of the American Statistical Association* **86**(414), 328–332.

Duan, N. & Li, K.-C. (1991), 'Slicing regression: A link-free regression method', *Annals of Statistics* **19**(2), 505–530.

Feller, W. (1966), *An Introduction to Probability Theory and Its Application*, Vol. 2, Wiley & Sons, New York.

Flury, B. & Riedwyl, H. (1981), 'Graphical representation of multivariate data by means of asymmetrical faces', *Journal of the American Statistical Association* **76**, 757–765.

Flury, B. & Riedwyl, H. (1988), *Multivariate Statistics, A practical Approach*, Chapman and Hall, London.

Franke, J., Härdle, W. & Hafner, C. (2004), *Statistics of Financial Markets: An Introduction*, Springer, Berlin.

Friedman, J. H. & Tukey, J. W. (1974), 'A projection pursuit algorithm for exploratory data analysis', *IEEE Transactions on Computers* **C 23**, 881–890.

Hall, P. & Li, K.-C. (1993), 'On almost linearity of low dimensional projections from high dimensional data', *Annals of Statistics* **21**(2), 867–889.

Härdle, W., Moro, R. A. & Schäfer, D. (2005), Predicting bankruptcy with support vector machines, Discussion Paper 2005-009, SFB 649, Humboldt-Universität zu Berlin.

Härdle, W., Müller, M., Sperlich, S. & Werwatz, A. (2004), *Nonparametric and Semiparametric Models*, Springer, Berlin.

Härdle, W. & Simar, L. (2003), *Applied Multivariate Statistical Analysis*, Springer, Berlin.

Harville, D. A. (1997), *Matrix Algebra from a Statistician's Perspective*, Springer, New York.

Harville, D. A. (2001), *Matrix Algebra: Exercises and Solutions*, Springer, New York.

Johnson, R. A. & Wichern, D. W. (1998), *Applied Multivariate Analysis, 4th ed.*, Prentice Hall, Englewood Cliffs.

Jones, M. C. & Sibson, R. (1987), 'What is projection pursuit? (with discussion)', *Journal of the Royal Statistical Society, Series A* **150**(1), 1–36.

Kaiser, H. F. (1985), 'The varimax criterion for analytic rotation in factor analysis', *Psychometrika* **23**, 187–200.

Kötter, T. (1996), Entwicklung statistischer Software, PhD thesis, Institut für Statistik und Ökonometrie, Humboldt-Universität zu Berlin.

Kruskal, J. B. (1965), 'Analysis of factorial experiments by estimating a monotone transformation of data', *Journal of the Royal Statistical Society, Series B* **27**, 251–263.

Lebart, L., Morineau, A. & Fénelon, J. P. (1982), *Traitement des Donnés Statistiques: méthodes et programmes*, Dunod, Paris.

Li, K.-C. (1991), 'Sliced inverse regression for dimension reduction (with discussion)', *Journal of the American Statistical Association* **86**(414), 316–342.

Lütkepohl, H. (1996), *Handbook of Matrices*, John Wiley & Sons, Chichester.

Mardia, K. V., Kent, J. T. & Bibby, J. M. (1979), *Multivariate Analysis*, Academic Press, Duluth, London.

Morrison, D. F. (1990), *Multivariate Statistical Methods*, McGraw-Hill, NewYork.

Olkin, I. & Veath, M. (1980), 'Maximum likelihood estimation in a two-way analysis with correlated errors in one classification', *Biometrika* **68**, 653–660.

Schott, J. R. (1994), 'Determining the dimensionality in sliced inverse regression', *Journal of the American Statistical Association* **89**(425), 141–148.

SEC (2004), Archive of historical documents, Securities and Exchange Commission, www.sec.gof/info/cgi-bin/srch-edgar.

Serfling, R. J. (2002), *Approximation Theorems of Mathematical Statistics.*, John Wiley & Sons, New York.

Sobel, R. (1988), *Panic on Wall Street: A Classic History of America's Financial Disasters with a New Exploration of the Crash of 1987*, Truman Talley Books/Dutton, New York.

Vapnik, V. N. (2000), *The Nature of Statistical Learning Theory. 2nd ed.*, Springer, New York.

Volle, V. (1985), *Analyse des Données*, Economica, Paris.

Ward, J. H. (1963), 'Hierarchical grouping methods to optimize an objective function', *Journal of the American Statistical Association* **58**, 236–244.

Index